Institutions and the Environment

Institutions and the Environment

Arild Vatn

Professor of Environmental Sciences, Norwegian University of Life Sciences, Ås, Norway

Edward Elgar

Cheltenham, UK • Northampton, MA, USA

Published by
Edward Elgar Publishing Limited
The Lypiatts
15 Lansdown Road
Cheltenham
Glos GL50 2JA
UK

Edward Elgar Publishing, Inc.
William Pratt House
9 Dewey Court
Northampton
Massachusetts 01060
USA

This book has been printed on demand to keep the title in print.

A catalogue record for this book
is available from the British Library

ISBN 978 1 84376 100 6 (cased)
978 1 84720 121 8 (paperback)

MIX
Paper from
responsible sources
FSC® C013604

Contents

PART II FROM INSTITUTIONS TO ACTION

PART III FROM ACTION TO INSTITUTIONS

PART IV INSTITUTIONS, ENVIRONMENT AND POLICY

Preface

The immediate background for this book is experience from teaching a course in institutional economics at the Norwegian University of Life Sciences for almost ten years. While the variation in positions and perspectives in this field is a tremendous basis for creative thinking, the students have been faced with great demands that this heterogeneity puts on their ability to synthesize and structure. Not only are there several institutionalist traditions: as important are the distinctions between these positions and standard neoclassical economics. So the students, while very engaged, gave me some negative feedback on the texts I used. I compensated by writing a substantial number of lecture notes. This led me to the idea of putting this material into a book. Others, I thought, might have the same problem that I had experienced.

The final product differs substantially from the original notes. The writing has in itself been a great learning process about a complex, but exciting literature asking for some 'bold syntheses'. Thus, I hope the book may offer support both to students of institutional and environmental economics, and to researchers in the field of public and environmental policy who have recognized the importance of institutional issues, but who are perhaps rather daunted by a voluminous and heterogeneous literature. I hope also that practitioners in these fields may find the text to be useful.

The aim of the book is threefold. First, it is devoted to categorizing and comparing different positions in the field of institutional economics. Second, it engages in developing one specific position – that of classical institutionalism. Finally, it applies the insights developed to public policy making – specifically to the area of environmental policy. My contention is that institutional economics has a lot to offer to this field.

The book can be read in two different ways. Those interested mainly in institutional issues may think of the environmental material included as exemplifications of the theoretical perspectives offered. Those interested mainly in the applied issues – in environmental or public policy making – may similarly view institutional theory as a good basis for understanding the more fundamental challenges these policy areas raise for humanity.

Arild Vatn

Acknowledgements

This book grew out of ideas I had already grappled with as a student of economics in the late 1970s. While I was intrigued by the sophistication of neoclassical microeconomics, I was also concerned by the fact that textbooks tended to pass over the assumptions so quickly. What were they really? How relevant and realistic were they, and what would the consequences be if some of them were changed?

Professors Sigmund Borgan, Fritz Holte, Per Ove Røkholt and Stein Tveite supported me early on in following up such questions and looking into alternatives. A visit to Daniel Bromley at the University of Wisconsin in 1991/92 gave me the opportunity to concentrate more consistently on these issues. His insights, patience and great willingness to engage were extremely valuable.

The writing has also given me the opportunity to engage more systematically in discussions with colleagues and friends about the various issues. Their inputs have been very valuable, even though I know that I have not been able to follow up adequately on all the responses. I would like to thank Olvar Bergland, Kjell Arne Brekke, Erling Krogh, Terje Kvilhaug, Ståle Navrud, John O'Neill, Karen Refsgaard, Eirik Romstad, Per Kristian Rørstad and Pål Vedeld for engaging in the process and delivering valuable and thoughtful comments to earlier drafts of various chapters.

The list of other people whose competence I have utilized is vast, and I regret any omissions. I would like to thank Federico Aguilera Klink, Claudia Carter, Per Ove Eikeland, Kate Farrell, John Gowdy, Frode Gundersen, Juha Hiedanpää, Alan Holland, Klaus Mittenzwei, Richard Norgaard, Martin O'Connor, Gisli Pálsson, Jouni Paavola, John Proops, Clive Spash, Sigrid Stagl, Peter Söderbaum and Mark van Vugt for their active engagement in discussions over issues raised in this book. I would also like to thank the members of the Socio-Economic Research Programme at the Macaulay Institute (Aberdeen) for their responses to many of the ideas as presented at a seminar series there in April 2003. Special thanks go to my friend and former colleague Geir Isaksen for accompanying me on long walks during which many of the issues in the present book were discussed, and to the students I have had over the years whose responses have been very valuable. Some have even commented on drafts of chapters as used in class over the last two years.

Finally, I would like to thank Valborg Kvakkestad, Marianne Aasen and Gunnstein Rudjord for all their inputs. Valborg Kvakkestad has read and commented upon all chapters and revealed several inconsistencies throughout the text. Her encouraging attitude in the process has been very helpful. Marianne Aasen has helped in collecting material used to formulate many of the boxes found in the second half of the book. Finally, my thanks to Gunnstein Rudjord for assisting with the formalities, setting up the index, checking the reference list and also making valuable comments.

Since the book owes a great deal to my habit of spending evenings, weekends and some holidays 'hiding' in my home office, my greatest gratitude goes to my family, especially to you Gabriella, not only for tolerating this rather 'irrational' use of time, but also for supporting me strongly in the process.

Further acknowledgements

The publishers wish to thank the following who have kindly given permission for the use of copyright material.

University of Wisconsin Press for: 'Environmental Valuation and Rationality', by Arild Vatn, *Land Economics*, **80**(1): 1–18 © 2004. Reproduced by permission of the University of Wisconsin Press.

Elsevier Ltd for: 'Choices without Prices without Apologies', by Arild Vatn and Daniel Bromley, *Journal of Environmental Economics and Management*, **26**(2): 129–48.

Every effort has been made to trace all the copyright holders but if any have been inadvertently overlooked the publishers will be pleased to make the necessary arrangements at the first opportunity.

Abbreviations

CAFSAC	Canadian Atlantic Fisheries Scientific Advisory Committee
CBA	cost–benefit analysis
CBD	Convention on Biological Diversity
CDM	clean development mechanism
CFC	chlorofluorocarbon
CIC	community indifference curve
CITES	Convention on International Trade in Endangered Species
CO_2	carbon dioxide
COP	Conference of the Parties
CPR	common property regime
CV	contingent valuation
DDT	dichlorodiphenyltrichlorethane
DM	decision maker
EEZ	exclusive economic zone
EIA	Environmental Impact Assessment
EU	European Union
GATT	General Agreement on Tariffs and Trade
GMO	genetically modified organism
GUPF	grand utility possibilities frontier
HP	hedonic pricing
IMF	International Monetary Fund
IMO	International Maritime Organization
IPCC	International Panel on Climate Change
ITQ	individually traded quota
JI	joint implementation
MAC	marginal abatement cost
MAUT	multiattribute utility theory
MCA	multicriteria analysis
MEC	marginal environmental cost
MNOO	marginal net offer from owners
MNOP	marginal net offer for polluters
MNOV	marginal net offer for victims
MNOW	marginal net offer from workers
MRPT	marginal rate of physical transformation
MRS	marginal rate of substitution

MRTS	marginal rate of technical substitution
MSY	maximum sustainable yield
NPV	net present value
OA	open access
PIA	point of instrument application
PPF	production possibilities frontier
PPI	potential Pareto improvement
PPP	polluter pays principle
R_O	Owner's right
R_P	Polluter's right
R_V	Victim's right
R_W	Worker's right
SMS	safe minimum standard
SO_2	sulphur dioxide
TAC	total allowable catch
TC	transaction cost
TCM	travel cost method
TRIPS	Trade-Related Aspects of Intellectual Property Rights
UNCLOS	United Nations Convention on the Law of the Sea
UNEP	United Nations Environment Programme
UPF	utility possibilities frontier
VAI	value articulating institution
WMO	World Meteorological Organization
WTA	willingness to accept
WTO	World Trade Organization
WTO/SPS	WTO agreement on sanitary and phytosanitary measures
WTP	willingness to pay

1. Institutions: the web of human life

Living is to choose. By choosing we make our living, our own future, as we also affect the conditions for others. Due to the form and size of our activities, we increasingly shape the possibilities for people even far away from where we live and for people not yet born. This is visible in the international markets for goods and capital. It is evident in the increasing worries concerning, for example, global warming and biodiversity loss.

The natural environment in which we live is a *common good*. We share its qualities. What one person or firm does influences the opportunities for others. If a firm pollutes a lake, the inhabitants around the lake may not find it possible to swim or fish there. The wider ecosystem functions will be damaged, creating future problems for human and other life forms. If I drive a car, I emit carbon dioxide which is a greenhouse gas. The emissions contribute to changes in the composition of the atmosphere, most probably implying higher future temperatures, greater variability in weather patterns and so on. In this case small individual emissions over time aggregate and change the functioning of even global systems.

Several questions are of importance in relation to these simple examples (see also Box 1.1). First, how should societies make decisions about the common good? How should the various and often conflicting interests be taken into account? Should people's willingness or ability to pay be decisive? What role should arguments and collective reasoning play?

Second, after deciding what to do, we need to establish structures that motivate people to act in accordance with what we have found to be collectively wise. This implies that we need to know what motivates people when they are making a choice. Do they consider only what affects themselves, or do they also take the consequences for others into consideration? Maybe the propensity to act selfishly or in a cooperative way depends on the social context? If so, developing 'good social contexts' becomes crucial. This is the core institutional issue.

In the literature, situations where choices are interlinked as above are called 'collective choice problems'. It is acknowledged that what is individually rational or sensible to do, may in such situations be collectively detrimental. Thus, we observe that local, national and international authorities try to change the conditions for individual choices so that what is collectively reasonable also becomes individually so. The way this is done does, however, vary. Policies shift substantially between sectors and societies. Some policies seem to work rather well while others are a failure. Understanding why and when this is so is important. It is fundamental if we want to be able to formulate better policies and form a better world.

When studying this, we need to understand both what motivates choices and how these motivations are formed. Concerning these issues we observe a rather clear divide within the social sciences. On the one hand we have standard economic theory – neoclassical economics – depicting human beings as self-contained individuals with given preferences, whose choices are driven only by the concern for maximizing individual utility. On the other hand, we have more institutionally orientated social sciences like institutional economics, sociology and social anthropology. Here it is common to view humans more as a product of the social conditions under which they are living. A human being acts as part of social or organized groups. Following from this, choices are understood as influenced also by a concern for the collective – for the other.

Neoclassical economics is a vast endeavour. Despite its impressive models and technical elegancy, however, it is also characterized by some important shortcomings. These become especially evident when studying decisions concerning natural resources and human interaction with and within ecological systems. The institutional perspective, increasingly endorsed by economists, may be developed to give good responses to many of the issues that have been raised over the years.

1.1 THEORIES ABOUT CHOICE

Economics, sociology and anthropology – indeed, any theory or model – represent simplified characterizations of a complex world. As such they emphasize what the theory identifies as the main relationships and dynamics. Such simplifications are necessary both when we act and when we analyse. The crucial point is how well our theory is able to capture the most essential relationships. Formulating good social theories is not easy given the complexity of choice and social relations. Many issues seem to be highly relevant, and a theory about human behaviour and social systems must give answers to a wide set of questions. First, we need to define what characterizes the *process of choosing*. Second, we need to characterize how people and their choices *interact*. Finally, we need a description of *the world* in which people act. Social theories diverge precisely because they give different answers to these questions.

BOX 1.1 INSTITUTIONS AND CHOICES: THE CASE OF MOBILE PHONES

The use of mobile phones has exploded over the last few years, resulting in a series of different effects. It makes it easier to maintain contact with family and friends. It also simplifies job-related communication. Negative effects are, however, visible. We may feel uncomfortable or even embarrassed having to listen to other people discussing private matters over the mobile phone when, for example, travelling by train. There may be a greater risk of accidents on the road as people can now phone while driving.

Ackerman and Heinzerling (2004) indicate that in the United States as many as 1000 people die each year in accidents caused by the use of such phones in cars. They further document that economists have estimated that the value of the right to phone while driving – measured as individual willingness to pay for phoning – is higher than the value of those 1000 lives also measured in economic terms. From this one could conclude that the practice should continue. It is efficient.

We may react differently to the logic of such a calculation. It is interesting to observe that issues of a form that previously would be considered to be about good or bad conduct, increasingly tend to be formulated as a market issue – about what is the most efficient thing to do measured in monetary terms. As the example by Ackerman and Heinzlinger is formulated, most people would still conclude that making a monetary evaluation of what is best – to accept or restrict the use of mobile phones when driving – is bizarre. This is just not the way such an issue should be decided. At the same time, most people support the use of monetary evaluation and of markets in many other instances. One is prompted to ask the following questions. When are markets or market surrogates proper institutional contexts and when are they not? What alternative solutions exist? What are their merits? The aim of this book is to help readers in their search for answers to such questions.

The idea in this book is that by comparing the responses of different theories to the above questions, we get a deeper understanding of the theories – their strengths and shortcomings. It helps us evaluate their relevance in different situations, and it supports us in deciding which theoretical position each of us in the end find reasonable to take on as our core perspective. Such a deepened understanding of social theory should also help to foster communication across the social sciences.

Taking on this job, we are confronted with a large set of questions. With regard to the process of choosing, several issues come to the forefront when characterizing a theory:

- What is the logic of choosing? How is rationality understood and defined?
- What characterizes the motives or preferences of those choosing?
- How are motives and preferences developed – that is, are they purely individually defined or are they also socially contingent?
- What mental capacity do people have when handling complex choice processes?

Concerning the interaction of people, we must ask whether they produce stable (equilibrium) outcomes or involve each other in ongoing changes. People may cooperate or they may fight.

With regard to the description of the external world, a long list of issues is also of importance. In this book we shall focus on the following:

- Is the perspective of the physical world – 'nature' – mechanistic or systems orientated? Is it viewed as a set of items or as a system of, for example, ecological processes? What kinds of complexities are thus allowed for?
- Is information costly – that is, are outcomes uncertain?
- Are communicating and transacting understood to be costly operations?
- What is perceived to be the ideal rights structure and what does it imply for the distribution of resources and power?

The answers to these questions define the character of the social science involved. Gaining insights about this is not just an 'academic issue'. Rather, the theory we use can be viewed as *glasses through which we look*. Since we tend only to see what we are looking for – just the solutions that our theory or model allows us to see – the quality of our theory becomes crucial not least for practice. In the case of environmental problems, how we understand people's motivations and how they will react to various policies becomes especially important. We would think differently about how to solve these problems if we believe that people are pursuing individual gains only – that they think only in 'I' terms – as compared to recognizing that humans may acknowledge the interests of others too, that is, they also reflect in 'We' terms. We would think differently about this issue if we believe that this propensity is furthermore a given fact or if it is something that is in itself dependent on the social context, on the institutions of society.

The theories or 'glasses' we use have taken a long time to develop. In understanding, using and developing a theory further, we need to be aware of its history. Concerning the understanding, we must acknowledge that a theory is not developed in a social vacuum. It is a response to the social circumstances, interests and needs of the societies in which the theory builders lived. It is not accidental that economics became a separate science in the burgeoning period of industrialization and market expansion in the late eighteenth century; or that sociology started to develop 50–100 years later when the vast social dynamics inherent in the same process became visible through the creation of new social classes, new professions, new social relationships and so on. In trying to understand a theory, understanding its history is important. I agree with Habermas when he says:

> [S]ocial-scientific paradigms are internally connected with the social contexts in which they emerge and become influential. In them is reflected the world- and self-understanding of various collectives; mediately [*sic*] they serve the interpretation of social-interest situations, horizons of aspirations and expectation. Thus, for any social theory, linking up with the history of theory is also a kind of test; the more freely it can take up, explain, criticise, and carry on the intentions of earlier theory traditions, the more impervious it is to the danger that particular interests are being brought to bear unnoticed in its own theoretical perspective. (Habermas 1984, p. 140)

This book is fundamentally about understanding choices and the role of institutions concerning the actions we take. While I shall compare various theoretical positions, it should be made clear that this book also represents and develops a specific position – that of *classical institutional economics*. Thus the book not only presents and compares different answers to the list of questions previously offered, but it also develops a set of answers that is seen to be most relevant when studying choices, not least concerning the environment.

The book restricts itself to the economic and environmental spheres or, more precisely, how people organize themselves to utilize resources/the natural environment when sustaining their way of life. The first half – Parts I–III – focuses on this issue in more general terms. In the second half – Part IV – I conduct a more detailed investigation of choices and institutions concerning the utilization and protection of our physical environment. While the general parts can stand on their own, their main motivation is to formulate a basis for the analyses of the environmental issues focused on in Part IV.

The aim of the present chapter is to introduce the reader to the understanding of what institutions are and their importance for economic and environmental activity. The various issues introduced will be examined more comprehensively throughout the following chapters.

1.2 OBSERVING INSTITUTIONS

Choices are made within different types of contexts, both physically and socially. While the physical context defines a set of opportunities and constraints that are basically given by nature, it follows from the term that the social context is constructed by humans and human organizations. One important type of social constructs is institutions. It is not least by developing and changing institutions that we form or change behaviour. The understanding of the role institutions play for individual choices is very different if we compare across the social sciences. This then influences what we understand to be possible, reasonable or efficient policies.

What is an institution? Let me approach the issue by giving some simple examples. When young we learn how to greet others. One pattern may be to give a hug, another to just shake hands or say 'hi'. Children may use different forms of greeting when they meet those of their own age as compared with greeting adults. However, we may not be conscious that these are socially defined rules or 'rituals'. We may not even think about doing it, much less what it implies. It is just something everybody does. Implicit in this example is the fact that institutions are often so 'natural' or fundamental to us, that we actually do not notice that they exist, even less that they are a social construct.

Meeting someone outside our own culture for the first time may be embarrassing. We suddenly realize that these others do 'curious things' when they approach us or each other. At the same time, it is through this kind of comparison that we increase our self-consciousness and become aware of the large body of socially defined rules or norms – regularized behaviour – that structures much of our lives and choices.

Institutions influence choices at all levels of society. They appear as conventions, norms and externally sanctioned rules. *Conventions* have the function of coordinating behaviour through creating regularity – that is, supporting one type of behaviour as opposed to all other possible ways of handling an issue. This just simplifies life. I have already mentioned the conventions concerning how we are supposed to greet each other. Other conventions may concern the various metrics we use such as weight, length, time, money, the directions defined in the sky, dressing codes for various occasions, who does what in a team, how we can behave in traffic, how we can present a bid in the stock market and so on. While some conventions are universal for a whole cultural area, others may be rather local.

The concept of *norms* brings us from just coordinating behaviour to issues where specific values are accentuated or protected. A norm is a response to questions concerning what is considered right or appropriate behaviour. As norms are formed around certain values, they also, when followed, give support to the same values. While greeting with the right hand is a convention, greeting in itself is a norm concerning the importance of showing respect when we meet others. In general, norms concern how we treat our fellows. Norms of good conduct are defined for a variety of circumstances – for example, how we should behave when eating, under what circumstances we are supposed to accept specific offers or perform certain duties, what is considered a proper gift, what is considered the right way to handle various bodily and other emissions (odour, waste, sound), what is good sportsmanship, what are fair business practices, what are environmentally good or acceptable practices and so on.

Finally, *formally sanctioned rules* may cover all levels from the constitution of a society, the civil law, to the laws governing business transactions, rights to resources – property rights – formally defined emission rights and so on. These types of rules play a crucial role in situations where interests are in conflict, which is why formal sanctioning power is necessary. Such rules are backed by the formalized power and sanctions of the collective – of 'third parties' like the state. If an interest is protected by a formally sanctioned rule, the holder of that right expects the sanctioning body to act if the right is not observed by somebody, as when a forest is cut down by someone other than the legal owner.

Most typically we observe the establishment of new institutions around the introduction and use of new technologies. The development of rules concerning the use of mobile phones – see Box 1.1 – is a typical example of how the novel technology establishes new physical relationships between individuals that call for regulation by various norms or by the law. While one would typically expect the use of the mobile phone in public spaces such as a bus or train to be regulated by norms of good conduct, its use while driving is a typical example of regulation where the authority of the law will often be exercised. Simply, more is at stake.

A basic idea underlying this book is that the world is inherently complex. This is the case for both the natural environment and interactions between people. Cooperating with other humans is not easy – the 'simple' act of greeting has already shown this. It could be done in hundreds of ways, and because of this we observe that it is done very differently in different cultures. Nevertheless, for each culture a certain solution is defined. Because of the existing convention, we do not need to think about how the act should be done. The common solution – the institution – is a practical answer to an almost unsolvable coordination problem for the individual.

The fundamental issue emphasized here is that in view of all possible acts that could be done in situations where several people are involved, some kind of regularization is needed both to understand what the situation is about and to coordinate behaviour. Situations may be defined as 'feasts', 'exchanges', 'competitions', 'marriages', 'funerals' and so on, each with its defined meaning and set of expected or accepted behaviour. At the most basic level, this concerns the construction of a language with its common concepts and words. This makes it possible for us to order our experiences within a framework that is common to all. By inventing *the conventions of a language*, a necessary first structure for establishing coordinated behaviour is ascertained – the cognitive. It is on top of this institutional foundation that other conventions, norms and formal rules can then be constructed. Thus, institutions create the regularities necessary to make choices comprehensible and workable. While the creed of modern society is that we are all free to choose, taking such freedom literally would create chaos. Freedom is made possible foremost through the creation of common conventions, norms and rules. This makes the acts of others comprehensible, and it helps us form expectations about what will happen in certain situations. However, it is difficult to understand the role of institutions in creating necessary order since these structures appear to us as given. They are 'the natural order of things'. Looking across societies may help us acknowledge not only the variations in institutions, but also their role and importance.

Institutions vary greatly across social spheres. The rules and norms within the family are generally different from the rules applied in a firm, the marketplace or when we use a common natural resource. To the degree that different disciplines focus on different social spheres, they also focus on different types of institutions, be it anthropology, sociology, political science or economics. In part this has resulted in different theories about the character and role of institutions, a situation which is unsatisfactory.

While sociologists and anthropologists tend to focus mainly on informal institutions and institutions as giving *meaning* to life, economists, when focusing on institutions, tend to look at these more as *formalized rules*, that is, *property rights*. While this book focuses mainly on the economic sphere and the interrelationships between the economy and its environment, it is nevertheless important to ground the theory of institutions in the broader literature. It is the whole institutional setting that defines the characteristics of the economy and whether it can be viewed as a separate sphere or should be understood as being more integrated into society at large. Furthermore, there is no reason to believe that economic institutions are principally different from other types of institutions. Because of this one will gain from utilizing a broader perspective on institutions when building a theory concerning economic activity.

In modern economies, formalized entities like the state, the market and the firm are the main institutional structures. Markets, that is, places where goods are exchanged, have existed for a long time. However, until recently they did not constitute a sphere with institutions separate from the political and social ones. Historically the economic process was 'embedded' in the social structure at large. For example, institutions like kinship and social position played an important role in the process of distributing resources. In modern markets, resources tend to move to those uses that obtain the highest willingness to pay. However, while thus becoming more and more 'disembedded' from social relations, the economy is still best understood as part of the broader social and political framework. In addition, many environmental problems seem to stem from the fact that the 'disembedding' we try to obtain by attempting to turn all natural resources into market goods, presupposes a physical world quite different from the one we actually live in.

In economics we conceptualize goods or resources as *demarcatable items* – that is, as commodities. The natural sciences tell us, however, that the physical world is strongly characterized by *processes* and *interrelated flows* of matter and energy. This establishes important physical interconnections between individuals and groups. This feature has to be incorporated when studying the dynamics between the economy and the natural spheres in which it is embedded. This is a core issue of this book.

1.3 DEFINING INSTITUTIONS

As we have seen, the concept of an institution covers a very diverse set of constructs. Moreover, no common definition is accepted either within or across the various social sciences. This stems from the fact that we are confronted with different interpretations of behaviour. Studying the various positions found in the literature, we realize that there is a necessary relationship between how various theories *understand behaviour* and how they (have to) *define institutions.*

We shall return more fully to this issue (Chapter 2). At this stage I shall just give the reader a first insight into the various perspectives in the literature. Box 1.2 contains a list of definitions of institutions that are representative of various positions.

BOX 1.2 DIFFERENT DEFINITIONS OF AN INSTITUTION

Berger and Luckmann (1967): 'Institutionalization occurs whenever there is a *reciprocal typification* of habitualized actions by types of actors. Put differently, any such typification is an institution' (p. 72).

Scott (1995a): 'Institutions consist of *cognitive, normative, and regulative structures* and activities that provide stability and meaning to social behavior. Institutions are transported by various carriers – cultures, structures, and routines – and they operate at multiple levels of jurisdiction' (p. 33).

Veblen (1919): '[Institutions are] *settled habits of thought* common to the generality of man' (p. 239).

Bromley (1989): '[Institutions are the] *rules and conventions* of society that *facilitate coordination* among people regarding their behavior' (p. 22).

North (1990): 'Institutions are the *rules of the game* in a society or, more formally, are the humanly devised constraints that shape human interaction' (p. 3).

Italics added.

At this stage, we can only scrape the surface of what these differences mean and imply. Concerning the various authors it is important to recognize that Peter Berger and Thomas Luckmann are sociologists. Richard Scott has a basis in sociology, too, but with his main interest in the theory of organizations. Thorstein Veblen is known as the founding father of American or 'classical' institutional economics. Daniel Bromley is a modern representative of the 'Wisconsin school' of the same tradition.[1] Finally, Douglass North is a representative of the school of 'new institutional economics', which in contrast to the 'classical' institutionalist tradition is a rather recent development largely based on a neoclassical economics foundation.

The list in Box 1.2 is far from complete. It is merely intended to form the basis from which some central features of various positions can be discerned. Contrasting the definition from Berger and Luckmann with that of

North may help us see better how differently institutions are understood and consequently how different the underlying model of behaviour is.

According to Berger and Luckmann (1967), people are products of the social conditions under which they grow up and live. They asked the following question: how can people communicate and cooperate in a complex world, one that cannot explain itself directly to us? As they saw it, neither our understanding of the physical world nor our concrete social skills are given to us at birth. However, we have the capacities necessary to learn about the world and develop social abilities. According to these authors, shared concepts in the form of language, action types and mental 'maps' of the world are developed over time and constitute the basis for creating necessary meaning and order so that both understanding and cooperation becomes possible. They call these shared concepts 'reciprocal typifications'. Institutions are such typifications.

According to this perspective, both the *social capabilities* of individuals and the ways they *see* the world are socially constructed. Individuals – as social beings – are constituted through learning the typifications of both the material world and social relations as established by the society. They learn the meanings already created by the society into which they are socialized. They are formed by the institutions of the society in which they are raised. Society itself is likewise perceived through the concepts that are collectively produced. This position is called 'cognitivist' or 'social constructivist'.[2] Institutions enable *people to act* by defining which acts should or could be done in specific situations. Thus, they may even *do the choosing for them* via learned behaviour and so on. In accordance with this, *the role* is a core concept. The role, be it of teacher, policeman, banker or mother, defines the issues for us, what should be preferred and which acts are expected or respected. Our preferences, as they appear, are in a fundamental way influenced by the roles we perform. We do what is expected. The institutional context defines what is rational or, more precisely, reasonable to do.

The perspective of institutions observed among the neoclassically inspired – the 'new institutional economists' – is very different from this. North (1990) is typical when he defines institutions as 'the rules of the game'. Society consists of given individuals. Institutions have no role in forming them. They are just external rules establishing the stage at which these given individuals (inter)act. Individuals have, furthermore, only one kind of goal: they maximize their own utility. Rational action is equated with such maximizing. Preferences are considered stable, that is, unsocialized and immutable.[3] Each individual has, moreover, a predefined ability to understand not only his/her own needs, but also the performance of others and the working of the natural world. In this case, institutions are seen only as *constraints on human choices*, the rules they must follow when playing games with each other, like a game of soccer or 'the Wall Street game'. The most important rules are those defining the rights each individual holds, for example, the rules concerning access to resources. Given these rules and the existing distribution of endowments, individuals transact to get what in the end is considered best for themselves. But transacting is costly. Institutions are, according to this position, invented not least to reduce transaction costs. They are *instruments* that make exchange become more predictable, simple and efficient. The money institution, the contract and the various measurement scales are all understood as invented to simplify transactions.

Moving from the 'new' (North) to the 'classical' institutional economists (Veblen and Bromley), we again observe positions closer to the sociologists. In particular, Veblen's 'settled habits of thought common to the generality of man' is very similar to Berger and Luckmann's 'reciprocal typification'. Bromley is somewhat closer to North in that he also views institutions as mainly external to the individual. He defines institutions as 'choice sets from which individuals, firms, households, and other decision making units choose courses of action' (Bromley 1989, p. 39). Nevertheless, there is a clear difference from North and the 'new' institutional economists in two important ways. First, Bromley also focuses on the role of institutions in facilitating choice. They *enable*, not just constrain, choices. More specifically, they simplify and regularize situations. Thus some of the perspective underlying Berger and Luckmann is taken up by acknowledging that institutions are important in creating common, simplifying frameworks for action. Second, Bromley accentuates the *normative aspect* of institutions. What becomes optimal or efficient depends, according to the tradition in which he stands, on the chosen institutions and the *interests* these are set to defend. In line with this, he emphasizes the importance of the power that various interest groups have – that is, their ability to obtain institutionalized protection of their interests in the form of rights not least to physical resources. This takes us beyond both the cognitive (Berger and Luckmann) and the purely instrumental (North) perspectives on institutions.

There are relevant elements in all the above definitions. Turning finally to Scott, we observe that he acknowledges this and integrates elements from all the others. He still does so within a framework that is very much the same as that of Berger and Luckmann's. I find that Scott offers a good basis for our analyses.[4] From Box 1.2 we also see that he focuses more directly on the various forms that institutions may take. They appear both as internalized conventions and norms, and as external rules. Institutions consist of cognitive,[5] normative and, in his language, 'regulative' structures.[6] The cognitive part concerns our mental structures, how we classify objects, give them meaning and act under their defined domains like that of being a teacher, a daughter, or a judge and so on. As mere typifications they are still not sufficient to guide and assure a certain behaviour. The normative element focuses on the implicit or explicit *values* involved. Formulating the role of the teacher implies choosing among the values that this role should support. Creating the teacher is then not just to define a set of (expected) behaviours – to do the typification. It is also to choose the values s/he is meant to support or sustain. However, in the end, the teacher may not live up to the standards. It may not be enough to define the type or role with its implied value base. It may also be necessary to reward or punish. This is the heart of the regulative element, as Scott calls it. He suggests that while institutions form individuals, it may still be necessary to establish external punishment and reward structures to obtain desired outcomes.

In our context, that of environmental policy, it is important to understand what motivates choices and what are reasonable policies. These policies will have to be different in cases where choices are determined by conventions and norms – that is, *internalized motivations* – as compared to situations where choices can be influenced only by changes in *external reward structures*. Moreover, the incentives used may even influence the logic or rationality that people assign to a certain situation. Using individual incentives may have the capacity to transform an issue that was previously considered a common problem into a question where only individual consequences are thought to be relevant. What previously was perceived to be a normative issue, a 'We' issue, can be turned into an 'I' issue, implying that normative, self-regulating structures are destroyed. This kind of dynamics is invisible if we base our policy recommendations on a model where the motives of individuals are considered to be independent of the institutional context and of the policy itself. However, they become core issues if we accept that such relationships exist.

The analysis undertaken in this book will be based on a social constructivist perspective as briefly outlined above.[7] The insight that *institutions influence individuals and their motivations* will form the basis for our analyses. We may denote this relationship the 'fundamental institutional level'. However, given that basis, we shall integrate two ideas which were central in the positions represented by Bromley and North, respectively. Hence the perspective that institutions protect interests will be given ample consideration. If resource uses are competing, as is dominantly the case with environmental issues, the *distribution of rights* becomes a core issue. Whose interests should get protection by the collective and how the subsequent rights structure influences resource use, are core issues. They form the second level of our institutional analysis. Finally, while the position of 'new institutional economics' is too narrow in my mind, the idea that it is costly to coordinate activities between individual decision makers and that these costs influence resource use, is important. This insight is especially significant in situations where choices are interlinked, as in the case of environmental protection. The way various institutional structures influence these costs – that is, *transaction costs* – defines the third level of our institutional analysis. While the understanding of institutions and choices is very different if we compare new institutional economics with a social constructivist perspective, integrating transaction cost issues into our model of behaviour and institutions is rather straightforward.

1.4 INSTITUTIONS, THE ECONOMY AND THE ENVIRONMENT

Institutions structure the relationships between humans as they utilize their common natural resource base. Today many of these relations are governed by an institutional structure called 'markets'. To understand what this means, what possibilities and restrictions are involved, we need to make comparisons across institutional systems.

Markets are in many ways great creations. They simplify life not least by simplifying many transactions to that of pure exchange. If what is at stake is about exchange and the exchange value a good has, there are many strong reasons for letting markets govern resource allocation. There are also, however, important tensions and problems involved. Institutions like markets favour certain types of motivations and interests and they influence which relations will dominate between humans and between humankind and nature. In his study *The Great Transformation: The Political and Economic Origins of Our Time*, Karl Polanyi ([1944] 1957) shows that establishing the necessary institutional structures of markets has not been a simple task. Creating markets is foremost a transformation of the complex qualities of the involved objects into commodities with a specified price. Markets represent a vast simplification of human interaction and humankind's relation to nature. While there are great gains involved in this, there are also immense problems, not least when environmental issues are implicated. Polanyi draws attention to the problems involved when modern industrialized societies also have to reduce labour and land (nature) to commodities. According to him, it is 'against their nature' and therefore creates several tensions:

> It is with the help of the commodity concept that the mechanism of the market is geared to the various elements of industrial life. Commodities are here empirically defined as objects produced for sale on the market; markets, again, are empirically defined as actual contracts between buyers and sellers. Accordingly, every element of industry is regarded as having been produced for sale. . . . The

> crucial point is this: labor, land, and money are essential elements of industry; they also must be organized in markets; in fact, these markets form an absolutely vital part of the economic system. But labor, land, and money are obviously not commodities; the postulate that anything that is bought and sold must have been produced for sale is emphatically untrue in regard to them. . . . Labor is only another name for a human activity which goes with life itself, which in its turn is not produced for sale but for entirely different reasons, nor can the activity be detached from the rest of life, be stored or mobilized; land is only another name for nature, which is not produced by man; actually money, finally, is merely a token of purchasing power which as a rule, is not produced at all, but comes into being through the mechanism of banking or state finance. None of them is produced for sale. The commodity description of labor, land and money is entirely fictitious. (Ibid.: 72)

As is implicit in this quotation, over time the human species has created vast systems of institutions – order, norms and rights. To be able to evaluate in which directions future development of institutions should take, it is important to understand not only the dynamics of various institutions, but also which of them serves us best in different spheres of life. Let us look at this in an introductory way by comparing two stylized and very contrasting examples of 'stages' in the development of human societies:

1. The 'man of the forest' – the man of the virgin forests who fulfilled his basic needs while utilizing resources in his immediate vicinity – in a way not much different from other species. He took what came before him. Nevertheless, his ability to plan and communicate was the best that natural selection had so far accomplished. He was a representative of the first species who could consciously *construct concepts and rules concerning behaviour* – that is, institutions. Inherent in this was a great potential for the development of the species itself. However, man did not in the beginning deviate from other species in any significant way concerning resource use.
2. The 'man of Manhattan' – the man of the complex urbanized society, who utilizes thousands of times more natural resources than did the man of the forest, but is almost totally detached from the processes delivering these resources. Between him and nature is a very complex set of institutions and organizations, the complex society and its subsystem, the modern economy.

First, there has been a long development – a huge investment in social structures – between these two situations. Second, while this process has created opportunities for the man of Manhattan that the man of the forest could not conceptualize, the development also has some ambiguous aspects attached to it. While the development has created vast opportunities, it has also resulted in great distance or alienation. This may have supported the very pervasive perspective at the root of western culture that nature and society, the ecology and the economy, are two disparate spheres and that human beings are capable of fully controlling nature in the same sense as they have been able to build and manage the Empire State building.

There are important tensions here that go right to the heart of the problems of sustaining modern societies. The most basic tension is between (a) the creation of institutions concerning resource rights and the facilitation of labour division, and (b) the need for keeping interactive, natural processes intact to a degree sufficient to secure the flourishing of all life-supporting natural dynamics. Institutional reform is a constant prerequisite for maintaining the necessary balances here, while at the same time the institutions being built establish a problematic distance between humankind and nature.

I do not intend to go deeply into the developments lying between the man of the forest and the man of Manhattan. I shall merely highlight a few issues in the same metaphoric way in which the vision of these types has already been cast. Despite the immense structures created by humankind, it is one of the youngest species. Studying the life of other primates, one recognizes that it must have been the ability to communicate via symbols that gave the human species its ability to survive and later expand in a rather tough environment. It had few other defensive abilities.

For most of humankind's existence, life has been sustained by hunting and gathering (Goudie 1993). This may primarily have been a continuous fight with other species. Nevertheless, at some point, competition between different groups of humans must also have become significant. Holding on to specific areas for one's own group or tribe may have been based on a balance of threats and direct coercion. Over time, some conventions may have arisen concerning 'ours' and 'yours'. Such agreements may have reduced the need for fighting and more resources could be diverted towards one's own sustenance. However, since societies for a long time lacked third-party institutions to settle resource conflicts – that is, some type of common authority structure like the law – this type of institutionalization was rather weak.

In this phase, humankind predominantly lived in small and socially very cohesive groups. This may have fostered social consciousness, the ability and capacity to cooperate, construct and internalize norms (Ostrom 2000). It seems as if this capacity was a prerequisite for survival in a harsh environment of various threats. Social cohesion and ability and willingness to cooperate gave groups a competitive edge in the fight for survival.

The size of the human species was very small throughout the whole of this period. According to Ehrlich et al. (1977), the number of people living on the earth was in the order of 5 million until the start of the agricultural – the Neolithic – 'revolution' some 10 000 years ago. Agriculture made it possible for the numbers to increase, and changed the way people organized themselves. Some, like North and Thomas (1977), understand this change in resource use to be an effect of an institutional innovation – more specifically, the establishment of private property in land. Thus, the person who invested in clearing, fertilizing and seeding the land, was the one who could harvest the fruits of this investment. By such an institutional change, the incentives necessary for an agricultural type of sustenance, as distinct from hunting and gathering, were created.

While the focus on institutions and incentives is important when understanding this history, private property was not the only option. Indeed, that is mainly an invention of the western world. Furthermore, the first agriculturally sustained societies of any size were the river dynasties of the Near East, India and China (Vidal-Naquet and Bertin 1987). These were complex command systems where the property concept, as understood today, had no real meaning. Much of the power relations were founded on perceptions or 'myths' about the relationships between man and nature, and man and God.

It is no accident that these societies grew up on the banks of large rivers, which represented a continuous supply of basic resources such as water and fertile soil and became the natural forces of necessary abundance to support the development and sustenance of complex and very hierarchical societies.

In the case of these dynasties, one can start talking of humanly created production systems. They were heavily dependent upon natural processes, but they also transformed these processes immensely. We observe the development of social structures specializing in production of grain, cloth, pottery and so on. As Polanyi ([1944] 1957) shows, these economies were for a very long time strongly embedded in the institutions of the society as a whole. There was no autonomous economic sector. The rules concerning who could do what regarding the use of resources were deeply integrated into the broader social structure.

In modern markets, economic transactions constitute in many respects a separate sphere. Here you do not, as an example, have to belong to a certain class or social group to be allowed access to a certain good, and so on. In essence, ability to pay is what counts. Thus, the market transaction as such is based on a form of equality between the parties. If you have the necessary purchasing power, you can buy what is for sale. This equality is still just formal in the sense that a dollar from person A is the same as a dollar from person B. It does not say anything about who has access to the dollars in the first place. In this case, belonging to a certain class or group may be of great advantage.

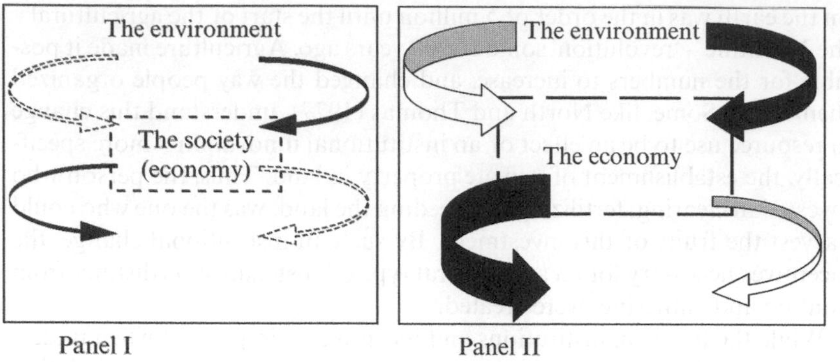

Figure 1.1 The economy and the environment

Figure 1.1 exemplifies the above perspectives of natural resource use and institutional development. It focuses on the increasing influence humans have on the dynamics of natural systems as vast streams of resources are not yet regulated or regulated in a way that captures their values and dynamics.

Panel I may be taken to illustrate the situation of about 10 000 years ago when hunting dominated, but some agriculture was also developed. At this time, humans utilized a small part of the total environmental resources and processes. In addition, the borderline between human societies and their environment – that is, the institutions regulating the use of natural resources – was very rudimentary. The access to and use of some resources (the *white dotted arrow*) is regulated – that is, issues largely concern who is allowed to hunt in certain areas and how the meat should be divided, who should harvest the grain and who should get access to it and so on. A few 'outputs' are also regulated, such as where excrement could be disposed of (*white arrow*). *Black arrows* illustrate unregulated inputs and waste disposal. While still rather small, these are dominating.

In panel II, the society utilizes a much larger fraction of the resources in the biosphere. This can be illustrated by the fact that while the hunter utilized only a few thousand Kcal energy per day, the man of Manhattan utilizes on average 400 000 Kcal (Simmons 1989).[8] In addition, the number of people has increased from a few million to 6 billion. Even if a larger part of the resources are institutionally regulated (*white arrows*) both on the input and to an increasing degree also on the output side, nevertheless a much larger volume goes unregulated in and out of the economy compared to that reflected in panel I. In the situation described by panel II, a vast number of natural resources are owned as private property, state or common property. Air, much of the water and some land is still under open access. Societies are, as we know, trying to get open-sea fisheries under some kind of control. On the 'output' side we observe that emissions of waste – that is, sources of air and water pollution – are increasingly regulated. However, some emissions are disregarded – or perhaps regulation is considered to cost more than it gains. A typical example is the case of greenhouse gas emissions, where the first steps to try to establish regulatory institutions were taken just a few years ago – that is, the Kyoto Protocol.

Integrated with the story of regulating resource use and waste disposal is that of who gets control over resources and waste disposal. Access to resources is both an effect of and a prerequisite for building individual or group influences. While the bands of early hunters and gatherers were probably rather egalitarian in their configuration, the power structures developed in the first 'civilizations', like the river dynastics, seem to have been strong and the execution of power fairly arbitrary. A pronounced stratification existed with slavery forming the 'floor' of these societies. This illustrates that even the definition of what it is to be human, that some people at certain periods have explicitly not been considered human, has been a long and ongoing process. We observe over time a development in the judicial systems so that power itself becomes subject to some restrictions. Roman law represented a substantial development in that respect. However, history had to take a considerable trip through many stages before reaching what we today consider the basic rights of being a human – the UN declaration of human rights – and yet these principles are far from being accepted all over the world.

Access to natural resources is a fundamental issue, and is institutionally defined. The institutions identify who has access, in which form and to what extent the resource may then be utilized. Furthermore, the institutions define whether rights holders can transfer the rights of access, use and so on to others, whether they may decide this themselves or whether it is subjugate to the power of somebody else.

Regulating the relationships between humankind and nature is thus to a very large extent about regulating the relationships between humans. It is a social and political issue. This book is about the various elements involved when defining such relations. The basic task is not a small one. It is to place economics where it belongs – in the interface between social and natural sciences. While this is an enormous task – in all its facets far beyond the capacity of one book – I hope that the steps we are able to take will assist the reader on his or her journey through an important landscape.

1.5 THE STRUCTURE OF THE BOOK

The book is divided into four parts. Part I, 'Understanding Institutions,' represents a more thorough investigation into the issues raised in Section 1.3

above. In Chapter 2 we shall look at the broader social science perspective on institutions and establish the basis for studying economics as a 'special case'. In Chapter 3 we focus more specifically on institutions as a way to handle coordination and conflict. Here we shall also develop a taxonomy for classifying and categorizing institutions. In Chapter 4 we shall move from the broader social science perspective to a presentation and discussion of various positions within institutional economics in more depth.

Institutions influence behaviour, as behaviour also influences institutions. Part II, 'From Institutions to Action', covers the first relationship. Here we shall focus on the role of institutions in both framing and defining choice. Chapter 5 is focused on the concept of rationality and develops the argument that what is rational depends on the institutional structure within which one is acting. Rationality may be individually calculative – that is, imply maximization of individual utility. However, it may also be social, meaning that individuals follow common norms or act reciprocally. Rationality is thus a plural concept and relates to various types of value dimensions. Chapter 6 is about preferences and preference formation. While economists normally tend to look at preferences as a given characteristic of the individual, the alternative perspective is to look also at preferences, values and motivations as social constructs. In particular, reasons for and consequences of the latter position are explored.

In Part III, 'From Action to Institutions', the perspective is turned around. Here we look at how and why people have chosen specific institutions to govern their behaviour. Chapter 7 presents and evaluates different theories why institutions develop and change. Is it a purely spontaneous process, is it guided by efficiency considerations, or is it defined by the power of certain interests? In Chapter 8 the focus is on the normative aspect of this – how can institutional changes be motivated and what is the status of various ideas concerning what are ideal institutions?

Part IV is the largest part of the book. 'Institutions, Environment and Policy' utilizes the insights from the previous chapters in a discussion about environmental issues. We start in Chapter 9 by presenting some core characteristics of the natural environment, its systems character, the role of species and the role that matter and energy flows play in its functioning. Various perceptions of the environment in economic theory are also presented.

Chapter 10 is about resource regimes. Here we link up several issues in a discussion about the way various property regimes such as private, state and common property influence the use and preservation of environmental goods. We shall focus on their capacity to treat both technical and normative aspects. The former concerns their ability to reduce transaction costs. The latter concerns the rationalities, preferences and values that regimes themselves motivate. Given the perspective that people are formed by the institutions they act within, choosing regimes is not only about choosing constraints. It is also about choosing among rationalities and motivations, and which rationalities we more specifically find appropriate for governing the environment. This perspective has profound consequences for how we should think about environmental policy.

Chapter 11 takes this a step further into a more specific discussion about environmental valuation. We shall look at what characterizes the valuation process, how institutional factors influence that process, and finally how well the dominant tradition of economic valuation captures the values involved and the way people think about them. In Chapter 12, the issue of valuation is extended to a systematic comparison of various methods developed to support choices not least in the realm of the environment – that is, cost–benefit analysis, multicriteria analysis and deliberative methods. These tools are built on different assumptions about rationality, values and the characteristics of the goods involved. We draw conclusions concerning their appropriateness for environmental value assessments and decision making.

Chapter 13 focuses on the choice of policy instruments to turn development in the direction of defined environmental goals. The link to the issue of regimes is clarified. The chapter next reviews the capacities of different instruments to solve environmental problems given various assumptions about rights, transaction costs and the physical characteristics of the problems. The relationships between policy instruments and the motivations they evoke are emphasized. If the way policies are formulated influences the type of rationality that people apply when treating an issue, policy measures become more than just instruments. They are also normative structures signalling whether people should look at an issue as, for example, an 'I' or a 'We' problem. This observation has important consequences for what are reasonable policies in short- and long-run perspectives.

Chapter 14 closes the circle by first of all summing up the main findings of the book. On the basis of these we next present a structure that can be used when analysing the choice of institutions governing environmental decision making. Finally, we define a set of challenges for future environmental policy making and our ability to secure a sustainable future. Existing policies deal with environmental issues in a piecemeal way. Even more fundamentally, it is argued that they are caught up in treating symptoms rather than offering a real cure. The normative message of this book is that we need a substantial reorientation of the institutions governing our economy if we are to have any chance of solving increasingly pressing environmental problems. The perspective of humans as fundamentally social offers some hope that this is possible.

NOTES

1. The 'founding father' of this tradition was John R. Commons. We shall return to him and many others in Chapter 4.
2. While some social constructivists – for example, Woolgar (1988) and Tester (1991) – even seems to claim that the external, physical world is a creation of our concepts, I reject such a view. What is constructed is the way we *see* the world. The way social constructivism is defined in this book thus implies that social construction concerns the construction of the social sphere itself and the way we view and understand both the social and physical worlds. The physical world functions independently of our conceptualization of it. I discuss this more in Section 2.4.3.
3. To get the main message as clear as possible, it has been necessary to focus only on the main or dominant position within each of the traditions. There will normally exist some variation across authors of a tradition that will not be covered here. As an example, North (1990) accepts that preferences may change. His tentative explanations for that are, however, economic and not social.
4. Later – in Chapter 3 – I shall present a definition of institutions that will be used for the rest of the book. As the reader will see, it is much influenced by the definition of Scott.
5. Parallel to the concept of conventions as used in Section 1.2.
6. Parallel to the concept of formalized rules as used in Section 1.2.
7. Already at this stage, it should be emphasized that I support a realist as opposed to a relativist interpretation of social constructivism. While some tend to view social constructivism as relativist almost by necessity, I strongly disagree with this. It is one thing to assert that what we see or find to be right is relative to the society in which we are or grow up. It is quite different from claiming that even the external world cannot be seen as independent of our concepts and models, as the relativist position seems to imply.
8. Simmons (1989) gives figures for an average US citizen in the 1980s.

PART I

Understanding Institutions

The idea behind Part I of the book is to give a deeper understanding of the concept of institutions and its applications. While the presentation starts off from the wider scope of the social sciences as a whole, the main focus will still be on institutions as they appear in the economic sphere – broadly understood as the sphere of resource use and management. Part I consists of three chapters, which are intended to give a basis for the issues that are more deeply focused on in the rest of the book.

Chapter 2, 'Institutions: the individual and the society', concentrates on understanding behaviour and thus institutions. It starts out from the problem of understanding how societies and social order can come into being in a complex world with potentially conflicting interests. In doing so it focuses on the main divide in social theory – that is, that between the models of choice as *purely individual* and that of *socially contingent choice*. This split is a fundamental one as it influences the positions taken concerning not only the understanding of institutions, but also the perspectives on social science methodology and the way we explain individual and social phenomena.

Chapter 3, 'Institutions: coordination and conflict', takes these ideas further by focusing more on what institutions look like in practice and *what role they play* in various choice situations. Institutions are categorized as conventions, norms and formally sanctioned rules. Their role in coordinating behaviour, in emphasizing common values and regulating interest conflicts is developed both theoretically and with examples.

In Chapter 4, 'Institutional economics: different positions', the main focus is on how the various perspectives and themes in institutional theory more at large are taken up by economists. While the dominating school of neoclassical economics tends to abstract away from institutional issues altogether, there are several positions that fall under the category *institutional economics*. The aim of the chapter is to clarify the main characteristics of these positions. We shall focus on the way they define institutions and what role they see institutions playing in the economic process.

2. Institutions: the individual and the society

While there are many positions taken in the literature concerning how to understand behaviour and institutions, there are nevertheless basically two main camps. On the one side, we have the position that individuals are self-contained with predefined capabilities. In this case, institutions are seen as constraints within which the given individuals act and choose. Institutions do not influence the characteristics of individuals or the goals they pursue. I shall call this the 'individualist' perspective.

On the other side, we have the position that institutions not only constitute choice sets – or more broadly, the external society – but they also influence individuals with regard to their abilities, ideals and needs. They influence perceptions, values, preferences and capabilities, and thereby the choices that individuals make or can make. Thus society becomes imprinted on individuals. This will here be called the 'social constructivist' position.

The divide in social theory – the individualist and the social constructivist – may be thought of as parallel to a divide between methodological individualism and methodological holism. However, this is not correct. The first suggests that what happens in a society can be explained only by the individuals and their choices. The second suggests that what happens can be explained only by the social structures – the institutions. This book is based on the perspective that 'man is a construct of society' and that 'society is a construct of man'. Hence, humans both influence and are influenced by the institutions. They are both acting agents in the meaning forming the structures of society and influenced by the very same structures. This dialectical perspective (see Figure 2.1) is fundamental to the approach we shall use in understanding the role of institutions.

The aim of this chapter is to describe and evaluate the social constructivist (Section 2.2) and the individualist (Section 2.3) models. Section 2.4 discusses how to explain social phenomena. Before we engage in these issues, however, I shall give a brief introduction to the issue of how social order can come about at all.

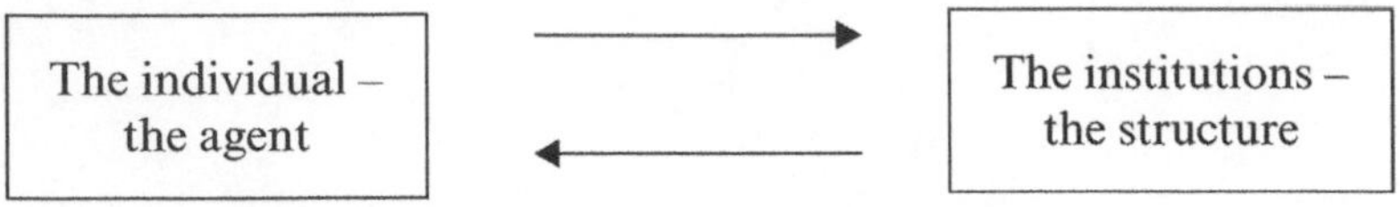

Figure 2.1 The individual and the institutions

2.1 THE PROBLEM OF SOCIAL ORDER

According to Parsons (1937), a prominent American sociologist, the basic issue of the social sciences has been to find a solution to 'the problem of order'.[1] How can it be that we seem to cooperate and not fight all the time? Why don't we just rob each other? This question goes back not least to Thomas Hobbes, whose *Leviathan* ([1651] 1985) played a very important role as the first attempt both to explain order in a situation of conflicting interests and to create the visions of rational individualism as a reaction not least to the dominance of prevailing feudal rule and religious dogma. Hobbes's point of departure was 'the state of nature'. This was a hypothetical situation, a situation that existed if there were no common power that could restrain individuals – that is, no state. This was a situation characterized by the 'war of all against all', following from the supposed nature of human beings to act only according to their self-interest.

According to Hobbes, social order could only appear if people found it in their interest to cooperate. If people recognize conflict – that is, the war of all against all – to be against their interests, they might find it better to accept constraints on their behaviour to secure a cooperative and peaceful future – a future free of continuous terror and fear. Thus, a state or a 'sovereign', with a monopoly on the use of power, is created, and in this way the problem of the war of all against all is avoided. The state is given the power to punish those not cooperating.

The challenge for Hobbes's idea is not to explain how a state can get people to cooperate. The problem is to explain how the establishment of the sovereign comes about, starting off from the 'state of nature'. According to Hobbes, it is created through a process of contracting between the individual citizens, a so-called 'social contract', where the individuals refrain from using force. They give away some of their freedom to the sovereign whose assignment is to secure order. At the same time there are, according to Hobbes, no common norms existing in the state of nature. The only act the atomistic individuals can agree on is to *restrain their use of force in order to obtain that others do likewise*. Hobbes refers to enlightened self-interest to explain this.

BOX 2.1 THE PROBLEM OF SOCIAL ORDER FORMULATED AS A PRISONER'S DILEMMA GAME

Panel I below illustrates the issue of 'the war of all against all' formulated within the modern language of game theory – as a prisoner's dilemma game. While there is a gain for all involved in cooperation – square (I) of panel I – this is not the rational solution for isolated individuals 'playing' against all others. The situation is such that individual A will gain from defecting – 'playing war'/robbing – independently of what others do. If others cooperate, A will gain 15 by defecting as compared to 10 if cooperating her-/himself. If the others also defect, A will reduce her/his losses from −10 to −5 by also defecting. The result of individually rational choices is that we end up in square (IV), which is clearly worse for all than, for example, square (I).

Panel I

		Individual A Cooperate	**Individual A** Defect
All others	Cooperate	(I) 10 10	(II) 15 −10
All others	Defect	(III) −10 15	(IV) −5 −5

Panel II

		Individual A Cooperate	**Individual A** Defect
All others	Cooperate	(I) 10 10	(II) 5 −10
All others	Defect	(III) −10 5	(IV) −15 −15

Transforming Hobbes's solution into this structure, a sovereign (a state) is established with the capacity to change the payoffs of the game. Panel II illustrates this by introducing a 'fine' at a level of 10 on those not cooperating. In this case what is individually and collectively rational to do is harmonized. Square (I) of panel II is a stable solution.

This self-interest could, however, equally well result in free riding in the very process of formulating the social contract – see panel I of Box 2.1. If everybody else relinquishes their right to use force, each individual would observe that s/he could gain from not doing so her-/himself – that is, free ride. This would ruin the whole process of establishing the sovereign power. It has thus been argued that to be able to make binding social contracts in the Hobbesian sense, a pre-existing moral community of norms and social ties must exist. According to Parsons (1937) social order depends on a pre-existing normative order that counteracts free riding. Societies exist because norms and obligations create a community on which also the legitimacy of coercive power like that of the state and the law ultimately rests. This indicates a circularity problem in Hobbes. Contracting away the 'right' to use individual force depends on the existence of a moral order – a state. At the same time, these contracts are, according to Hobbes, the very basis for establishing a state.

In the years following the publication of *Leviathan*, efforts were made to solve the problems inherent in Hobbes's position. John Locke ([1690] 1994) made one important attempt. He too referred to a state of nature implying the absence of state power. However, according to him this situation was already characterized by a set of moral values, a set of rights and duties. The state of nature was characterized by the universal values of freedom and equality. People are, according to Locke, 'born' with a sense of equality and freedom, and the state is foremost an articulation of these human values or human rights.

The ideas developed by Locke played an important role in the democratization process of western societies of his time. He inspired the revolutions of the eighteenth century and more specifically the growing movements against slavery. However, can rights really be given to us by nature itself? No rights are written in the face of newborn children. There must be some other origin. In defending his position, Locke refers partly to reason. His dominant reference is, however, to religious motivations – the punishment of God. It is the potential punishment of God that might deter us from carrying out immoral acts. We see also that the Lockean solution is caught in some circularity, where the creation of the law becomes dependent on the existence of the law itself, even if that law has its basis in an apparently external moral authority in the form of God. Locke's solution was actually problematic in two senses. First, it appealed to pre-modern ideas that he himself, as an advocate for individualism and liberalism, was fighting against. Second, it was contrary to the fundamental liberal idea that values and preferences are purely subjective.

A potential solution to the problem of circularity that we encounter here involves two issues. First, people are not just single, given atoms as in 'the state of nature'. Instead, in the very creation of individuals there also lies a creation of a social consciousness and belonging that forms a potential basis for creating higher-order institutional constructs such as a local community and later a state. Second, this creation is gradual, not a one-shot solution as envisaged by Hobbes. The forming of social order and the creation of (the rights of) individuals become a gradual and interlinked process. In this evolutionary development it is, as in the case of the chicken and the egg, not a question of what came first: the individual rights or the social and political order. It is a question of gradual change where the one defines the other in a continuous process from simpler to more complex forms of societies and personalities. This is the perspective of social construction.

2.2 THE THEORY OF SOCIAL CONSTRUCTION

2.2.1 The Basic Perspective

In sociology and social anthropology, individuals are dominantly seen as socially created, meaning that they carry norms, values and expectations that originate in the institutions of a society. The social aspect is an objective reality, meaning that it can be observed as something independent of the subjects – the specific individuals. The social aspect has distinct and common effects on the shaping of the individual.

Jean-Jacques Rousseau ([1762] 1968) was among the first to present the idea that the individual is a creation of social circumstances. It later became a central – if not *the* central – theme especially in sociology, and has been addressed by authors such as Durkheim ([1893] 1964, [1895] 1938), Parsons (1937, 1951) and Berger and Luckmann (1967), who are all theorists greatly influencing twentieth century understanding and debate. Their positions vary somewhat, not least concerning whether the society and its processes could be described in terms only of social relationships. Durkheim, and Parsons even more, took a position very much abstracting away from the individual and focusing on the social structures and their function for the system as a whole. Berger and Luckmann also put greater emphasis on the individual agent. We shall follow the latter perspective.

To illuminate these issues, let us start by reproducing some passages from Berger and Luckmann (1967). As we remember from Chapter 1, they define institutions as 'reciprocal typifications'. It is obvious that it will take two or more people, normally large groups, to construct an institution, in the sense of a common understanding of what should be done. Nevertheless, two people can be sufficient to create social order, and for expositional reasons such a simple situation has advantages. Berger and Luckmann call the two individuals A and B:

> As A and B interact, in whatever manner, typifications will be produced quite quickly. A watches B perform. He attributes motives to B's actions and seeing

> the actions recur, typifies the motives as recurrent. As B goes on performing, A is soon able to say to himself. 'Aha, there he goes again'. . . . From the beginning, both A and B assume this reciprocity of typification. In the course of their interaction these typifications will be expressed in specific patterns of conduct. That is, A and B will begin to play roles *vis-à-vis* each other. (Ibid.: 74)

B may get the idea that eating should be done while relaxing and not while walking around and so on. He may use a flat stone as a 'table' and use a piece of wood as a 'chair' and sit down to eat. Person A recognizes this when it is repeated: 'there he goes again'. He may even want to join in: 'there he goes again' becomes 'there we go again'. The social construction of the meal is under way. Person A may consider that eating directly from the table is inconvenient and by splitting a coconut he has obtained the first bowls as there is also one for B to use. Through these kinds of process a number of typified actions will emerge. This routinization relieves the actors of effort, as it defines the tasks and the relations around these tasks.

I would suggest that A and B may not only participate by copying each other. It is reasonable that they first communicated about what is the sensible thing to do – in which order is it best to do things. B asks A whether eating together is a sensible idea. A might agree, or he might not. Then B may give up or he may force A to join him if he has such power.

A and B represent a somewhat static situation. A next step would be to bring in children.[2] This changes the situation qualitatively. Berger and Luckmann continue:

> The institutional world, which existed in *statu nascendi* in the original situation of A and B, is now passed on to others. In this process institutionalization perfects itself. The habitualizations and typifications undertaken in the common life of A and B, formations that until this point still had the quality of *ad hoc* conceptions of two individuals, now become historical institutions . . . This means that the institutions that have now been crystallized (for instance, the institution of paternity as it is encountered by children) are experienced as existing over and beyond the individuals who 'happens to' embody them at the moment. (p. 76)

Following our example, the meal becomes an institution for both the parents and the children. However, A and B, who constructed the routines, can change them rather easily. Berger and Luckmann continue:

> What is more, since they themselves have shaped this world in the course of a shared bibliography which they can remember, the world thus shaped appears fully transparent to them . . . All this changes in the process of transmission to the new generation. The objectivity of the institutional world 'thickens' and

> 'hardens', not only for the children, but (by a mirror effect) for the parents as well. The 'There we go again' now becomes 'This is how these things are done'. (pp. 76–7)

While the parents recognize that they created the meal, the children will not observe it as socially created. It is just how these things are done. The mirror effect implies that when something becomes expected by the children, the parents cannot just stop doing it. The children expect this to be the 'nature of things' or objective facts.

Certainly, there never was such a situation with a first couple. Things developed gradually. Nevertheless, the simple anecdote by Berger and Luckmann is helpful. It mirrors the fact that *people both shape institutions and are shaped by them.* This will be a recurring theme of this book. The story also shows that there are actually three phases in the process of institutionalization:

- *Externalization* This is the process whereby subjectively constructed routines take form and are expressed. It is the stage of establishing, for example, the meal. It is the stage of 'there we go again'. The routines are visible, but they still belong only to those creating them – whether they concern language, how to produce, who should perform certain jobs and so on. The actors know the origin of a certain routine and can easily change it if found convenient.
- *Objectivation* This is the situation when others observe the routines as existing 'facts'. They have retained an existence independent of those creating them and stand out as 'things'. The children observe the parents gathering everybody for the meal, understanding that 'this is how these things are done'. The routine becomes a 'reciprocal typification'. The newcomers may like the routines or not, but they are still there. The children do not know how the parents came to do the different things in the defined way. What the parents subjectively chose has become objectively real for the children.
- *Internalization* This is the stage where the children of the anecdote pick up the habits and reproduce them. When they play in the garden, they also have 'meals'. It becomes the 'natural order' of things, and increasingly distant from its origin as a social construct. The process of internalization is often called 'socialization'. The literature distinguishes between primary and secondary socialization. Primary socialization concerns the internalization of the general or basic rules of a society. The individual learns to become a member. Secondary socialization concerns the internalization of specialized rules and routines of the sub-societies with which we choose to affiliate ourselves – for example, educating ourselves to acquire the skills of a

certain profession, becoming members of a specific organization or workplace with its special rules and routines.

While these three phases generally appear in sequence for the individual, they are ongoing if we look at society as a whole. The perspective of Berger and Luckmann implies that society is a subjective product of human beings – of groups of people. People may choose to, for example, greet with the right hand, while they could as well have gone for a hug. Nevertheless, this way of greeting becomes objectively real for those next observing that it is always done this way. It takes on an objective form since it comes to exist independently. Finally, it reproduces itself constantly via the social creation of each individual being born into this society of institutions. Subjective forms become objective 'social facts'. They should, according to Durkheim ([1895] 1938), be studied as 'things'.

BOX 2.2 THE HISTORICAL CONTEXT OF BERGER AND LUCKMANN

The position of Berger and Luckmann (1967) was developed as a reaction to ongoing debates and developments in social theory. This is typically the case of any theoretical development. In their book they explicitly refer to three giants in social theory for various important inputs in their undertaking. Émile Durkheim influenced them strongly on the nature of social reality, that social phenomena are objective and can be studied as 'things'. However, they found Durkheim's theory to be too static, so they borrowed the dialectical perspective from Karl Marx. Finally, the emphasis on the constitution of social reality through individual or subjective meanings is derived from Max Weber.

When Berger and Luckmann wrote their book, sociology was accused of having an 'oversocialized' view of the individual emerging from various system-orientated theories, be they 'functionalist', 'structure-functionalist' or 'structuralist' – for example, Parsons (1937, 1951); Malinowski (1944); Radcliffe-Brown (1952); Althusser (1965); Lévi-Strauss (1968). In most of these models there was little left for human choice. Behaviour was determined by the institutions. Berger and Luckmann's book was partly a reaction to this lack of sensitivity to the role of the individual agent, while still emphasizing the important role of the social element.

The above story of the parents and their children captures the time dimension only partially. History is not that of two generations, but of thousands of them. The process of human development and the creation of societies have been going on for hundreds of thousands of years,[3] first in small bands, later in tribes and settlements such as villages and towns. Finally, national states and even today's international organizations and conventions are social constructions. In this process the complexity of social organization is vastly increased. It is not only about creating the everyday practical institutions of meals and greetings. It is also about individual and collective rights and the complex sets of roles that appear in modern societies. Parallel to this, the acting agents have also changed. In the beginning individuals played core roles. Later organizations of different kinds – for example, guilds, firms, civic organizations, political parties, governments and so on – have become much more important as actors in the institutionalization process.

The creation of any institution may have taken a very long time. Since history is evolving gradually and evidence of the different developments is often rather indirect, it is hard to trace the various changes. Some basic observations of importance to the issue of order and conflict should be mentioned, though.

First, the existence of group organization is pervasive. There is strong evidence that humans could not have survived in early times without their ability to organize and cooperate (Ostrom 2000). They were involved in fights against various predators and certainly also other groups of people. Humans were not specially strong or fast. Organizational talent, the talent of social construction, seems to have been an important element in their capacity to survive and expand. For long periods they operated in rather small groups. This implies that for a long time the development of institutions occurred under conditions of great social cohesion. People depended on each other and external pressures from both natural forces and other human groups most probably had the effect of strengthening the internal solidarity. From this we see that the individualist 'state of nature', as envisaged by Hobbes, is not a good metaphor. The individual was social from the very beginning, and this capability was most likely to have strengthened over time. The creation of rules occurred in a situation where trust, obligation and reciprocity had to be important elements (Barkow, et al. 1992). Whether the situation was such that bands consisting of individuals with a greater propensity to cooperate had a selection advantage, is a hypothesis that can never be directly tested. Evolutionary psychology documents, however, that people have inherited the propensity to learn social norms. This is similar to the inherited ability to learn grammatical rules (Pinker 1994).[4]

This change involving nested structures of individual minds and social institutions, accentuates that the move from 'the state of nature' to a 'civilized world' was a gradual or dialectic one. It was not an overnight switch from 'given man', to 'external constraints' set up by the state. According to this perspective, human beings gradually became social beings with a consciousness of and emerging responsibilities towards others. One may, furthermore, envisage that the group and the individual were created together. In contrast to Locke's ideas about creating order, there was no natural right to hand over from the individual to society. Order was gradually established or expanded to new forms or arenas as societies and individuals developed interactively. Hence the law or formalized rules, as they finally appeared, could connect to existing norms of the society; norms that were internalized at the level of the individual.

2.2.2 Power, Conflict and Individuality

Berger and Luckmann's simple story focuses on the element of social cohesion. Certainly, conflicts are also important factors in the development of societies. To some extent these take the form of 'us' against 'them'. We observe this today in the conflicts between communities over access to natural resources like land and water, the conflicts about control of jurisdictions in the Middle East and so on.

However, conflicts also exist within societies or groups. They may exist between social classes or other forms of internal groupings of a society. There is nothing in the theory of social constructivism which says that societies may not be stratified. Person A may be able to secure the easy jobs, while B is left with the heavy and dirty ones. This situation could be sustained on the basis of a visible use of power – direct coercion. The more visible, though, the more vulnerable it would be to rejection. The typification is simply not accepted as mutual or reciprocal. It cannot be sustained, and there are good reasons to expect that prevailing inequalities internally in a society are supported by institutional structures that make their source of power invisible to newcomers. The power relations are what we call 'systems related'. They have become 'invisible' or 'facts' through being incorporated into the institutional structures of society. They have become 'the natural order' of things. Then the question becomes: who has the power to define which interests should get protection from the internal authority system of a society in order to become such a 'natural' order?

While Berger and Luckmann are rightly criticized for putting little weight on conflict and power relations in their analysis, they are nevertheless right in emphasizing that interests are also largely social constructs, and are a function of the institutional structure of a society. The interests of a factory owner are different from those of the workers. The interests of a property owner are different from those of somebody without property. The interests of a teacher are different from those of a student.

Another aspect that is not well captured by Berger and Luckmann's story is that the development of institutions is not only about constructing and reproducing them, but also about interpreting them. Thus, there will be a certain element of subjectivity involved in determining which institution applies to a certain situation (March and Olsen 1989). Individuals must interpret which kind of situation they face and even in some situations question whether they want to conform. For example, when John is invited to Mary's fiftieth birthday party, he has to interpret whether it demands 'jeans' or 'a suit'.

Next, at least in modern societies, individuals can move readily from one society to another. They may choose to do this for various reasons. Perhaps they find the norms of the culture in which they live to be incompatible with their aspirations (Screpanti 1995). Thus people can choose which institutional setting they want to live within. This accentuates individuality, but as long as these individuals do not turn away from society completely, they will still be involved in social construction and the reproduction of social constructs. It is simply about other social constructs than those in which they were raised. The degree to which breaking out is possible is, furthermore, largely defined by the rules of the existing system – that is, do these individuals satisfy the rules of immigration, does the existing society tolerate the type of new society they may be participating in constructing? Modern societies are characterized by greater mobility and change concerning institutional structures. This does not, however, imply that there is no social construction going on. On the contrary, complexity concerning this construction increases. Parallel to this, the options and the responsibilities for individuals increase, too.

2.2.3 The Two Subtraditions: The Cognitive and the Normative

Social constructivism can be divided in two main subtraditions – the cognitive and the normative – see Boxes 2.3 and 2.4. The cognitivist tradition emphasizes the social construction of concepts and the reciprocal expectations of roles. Berger and Luckmann (1967) are placed within this subtradition. The most important aspect of the cognitive tradition is the focus on how institutional frameworks shape our ends and next how we pursue them. Scott (1987: 508) emphasizes:

BOX 2.3 THE COGNITIVE PERSPECTIVE OF SOCIAL CONSTRUCTION

The cognitive position emphasizes the mental aspect of social construction. Objects surrounding people are not direct facts. They have to be transformed into mental constructs. According to Scott:

> Individuals identify objects in their environment as means for their actions, as consequences of their actions, or as supporting elements in the ongoing framework of their activities. The process of interpretation that is inherent in all interaction involves the actor, first, indicating to her or himself those objects in the environment that are regarded as meaningful, and, second, combining these objectives in an appropriate way. (Scott 1995b: 101)

Berger and Luckmann (1967) represent a major accomplishment within the cognitive tradition of social constructivism. They not only draw on Durkheim, Marx and Weber – see main text. They also explicitly acknowledge the importance of George Mead and his concept of symbolic interactionism. Throughout an ordinary day we have to get up, eat, go to work, do our job tasks, return home, eat dinner, maybe go to a movie, take a friend home, go to bed and so on. All this implies identifying a long series of objects: the bed, the table, the door, the bus or bike, the office, the friend and so on. These objects certainly have an existence independent of the specific individual. Nevertheless,

> their relevance as *meaningful phenomena* [added emphasis] consists in the ways in which they have been constructed symbolically. The *cultural* significance of the bus, for example, consists in the expectation that it will stop in particular places, that it will follow a specific timetable. And that people may travel by handing over that symbolic object that we call money. There is nothing in the physical reality of the bus that requires any of these things. (Scott 1995b: 101–2)

This tradition thus takes a position against the older Cartesian view, as continued in neoclassical economics, that the world comes to us without any preconceptions. According to Mead, first there is the concept, then we can observe the object that it describes. Language plays a crucial role in defining the circumstances. Added to this, the roles and positions in society also influence the interests of the actors.

> Institutional frameworks define the ends and shape the means by which interests are determined and pursued. Institutional factors determine that actors in one type of setting called firms, pursue profits; that actors in another setting called agencies, seek larger budgets; that actors in a third setting, called political parties, seek votes; and that actors in an even stranger setting, research universities, pursue publications.

While it may not be so that the logic of agencies is always to increase budgets and political parties merely to seek votes, the main point is clear. Both the actor and the goals are defined by the institutional structure. Not only persons are subject to social construction. Organizations or collective actors such as firms are even more so. Individuals are also constituted by their physical disposition. The firm is solely an institutional creation. Parallel to this, profit is a social construct dependent on a complex set of rules concerning ownership, responsibilities for costs, technicalities concerning book-keeping and so on.

The normative tradition emphasizes that institutions structure life not only via reciprocal expectations; they also carry messages concerning what is right to do. The institution defines the appropriate action. It takes the form of an expectation and behaviour not following what is expected may be sanctioned. Mother may have said 'you ought to' or 'you may not'. The norm becomes the right way of acting, and through internalization, punishment may become redundant. In the widest sense, this mechanism of norm building reflects the process of creating what it means to be human. Given the normative position, it is rational to do the *appropriate* thing – the behaviour that the role and situation demands. It is not the result of a calculation.

While there is some agreement in the literature concerning which author belongs where, one should not make too much of the divide. My personal position is that we need both elements to construct a viable perspective on human agency and institutions. Here the division is made more for expositional reasons, to explain two sets of mechanisms, than to draw a distinction between different authors or ways of looking at institutions. Actually, most institutions have both cognitive and normative elements.

The distinctions that can be made between the cognitive and the normative positions are moreover mainly those of degree. Both focus on how conventions' respective norms help individuals to sort out complex choice situations. However, while the cognitive perspective focuses on the understanding of what kind of situation we are in and on classifying relevant behaviour, the normative perspective is more prescriptive. It stipulates what is right or appropriate behaviour. While the cognitive view focuses on conventions that are practical solutions to pure coordination

BOX 2.4 THE NORMATIVE PERSPECTIVE

The normative perspective on institutions existed prior to the cognitive one (Box 2.3). However, the cognitive position is more basic than the normative. The latter can be viewed more as a type or subgroup of the former. While all institutions have cognitive elements, a substantial number also have normative significance. This was probably what first attracted attention. It was easier to see the normative role of (some) institutions than to observe the importance of institutions for all kinds of interaction.

The main point of the normative position is that not only do institutions structure life through creating reciprocal expectation, but the value of doing *the right thing* is also emphasized. There are two elements here: the *value* that defines the state and the *norm* that defines what should be done to create this state. The value is to respect people, the norm to greet when meeting someone.

Hence, if we meet someone who seems to be in trouble, may have fallen and so on, there is an expectation that we as fellow citizens should help the person. We not only categorize the situation – the cognitive aspect. Some categories also demand something of us. If we simply pass by and ignore the person, we may feel guilt. An internalized force or constraint is involved.

Scott (1995a) emphasizes that it is this normative conception of institutions that was embraced by most early sociologists from Durkheim to Parsons and Philip Selznick, all of whom focused very much on kinship and religion. March and Olsen (1989) represent a recent version of the normative position. They emphasize:

> It is a commonplace observation in empirical social science that behaviour is contained or dictated by cultural dicta and social norms. Action is often based more on identifying the normatively appropriate behaviour than on calculating the return expected from alternative choices. (Ibid.: 22)

However, March and Olsen emphasize that behaviour is not automatic or unreasoned. Every situation must be interpreted and one must always choose a proper response. The point is that the role and the situation define what is acceptable. Nevertheless, as Searing (1991) emphasizes, the focus on roles and rules does not imply that people are slaves of social norms. They are instead 'reasonable people adapting to the rules of institutions'.

problems, norms define which values should be supported in cases where they are conflicting. Further clarifications on this are given in Chapter 3.

2.3 THE INDIVIDUALIST PERSPECTIVE

I started Chapter 2 by saying that the individualist perspective is also a social construct. It is a special way of understanding behaviour, and it has taken a long time to produce the position, as already indicated by our references to the work of Hobbes and Locke dating back to the seventeenth century. Being a social construct, it is still not seen as such 'from within'. Those supporting this position see it rather as 'the natural order of things' or 'how man is'. The individualist perspective is especially endorsed in economics. In particular, the neoclassical economics programme has attempted to clarify what individual maximizing of utility implies. By following such a programme, many insights have been obtained. Nevertheless, there are problems. Although it is a stringent theoretical structure, applying it to real-world phenomena has created inconsistencies. Moreover, its relevance has often been challenged simply because maximization of individual utility is the logic only of a subset of all institutionally created situations.

Even what we here call the individualist perspective can be divided into several positions, two of which stand out. First, there are those who base their view on a methodological reasoning – that is, it is individuals who choose and choice can only be understood by focusing on who chooses. This is the position dominantly taken by neoclassical economists and has been developed since the mid-nineteenth century. Second, we have the stance that individual freedom is or should be the ultimate goal. While this position also has a long history, it has mainly been developed by Friedrich Hayek and the Austrians from the early twentieth century and onwards. The Hayekian position is that individual freedom is its own justification.

Already we can see that there are many value issues involved in defining the individualist position. Its development can be understood as a reaction to the domination by the monarchic state, the church and its medieval heritage (see Hobbes and Locke). The position was not least carried forward by the wave of enlightenment, and the American and French revolutions of the eighteenth century. Interestingly enough this provoked a conservative counterattack focusing on the 'organic' character of society. Edmund Burke was among the most notable writers accentuating this. According to him, society with its various statuses and positions was divinely ordained. It existed prior to its individual members (see Scott (1995b) on this).

Analysing the individualist position, we shall here restrict ourselves to discussing the neoclassical model in economics. Hayek's position will be visited briefly, and in more detail in later chapters. We shall also give a very brief introduction to the main ideas of 'new' institutional economics which bases much of its thinking on a neoclassical foundation. Neoclassical economics does not itself focus on institutional issues. It is a fully generalized model, which abstracts away from variations in context. There is one disclaimer to be made in relation to this. Most representatives of the model do not claim it to have validity beyond economic issues, even though some like Becker (1976) have argued that the neoclassical model has general applicability – that is, also to the institution of the family, to crime and so on.

2.3.1 The Neoclassical Model in Economics

The neoclassical economic model was developed from the 1870s and onwards. As a discipline, economics was about 100 years old by the time the so-called neoclassical revolution took place and superseded the former classical tradition – see Box 2.5. The main idea advocated by the neoclassicists is that value comes from the utility or the happiness that a thing or a service offers to the individual. Value is rooted in his or her subjective mind. Important initial steps in building this theory were made by W.S. Jevons, F.Y. Edgeworth and Léon Walras. Later came the work of Alfred Marshall, Arthur Pigou and Lionel Robbins.[5] The central concept of the neoclassical model is that of rational choice as maximizing individual utility. Following Lakatos (1974), Becker (1976) and Eggertsson (1990) we may define the following *core* of the neoclassical model:[6]

- rational choice as maximizing individual utility;[7]
- stable preferences; and
- equilibrium outcomes.

Choices are understood as rational if preferences are rational and choices are made in accordance with what is preferred the most by the individual. Preferences are rational if they are *complete*, *transitive* and *continuous*[8] (see also Hausman 1992). This links rationality directly with maximization of utility. Nelson and Winter suggest that rational choice is 'the central tenet of orthodoxy' (1982: 8). We would add, rational choice *as maximizing individual utility* holds this position.[9]

Preferences are assumed to be stable or at least as given. This is the essence of individualism, for the individual to be self-contained. It is a perspective very different from that of social constructivism. There are authors within the neoclassical tradition who accept that preferences change. To keep the economic individualist perspective consistent, this change must, however, not be the result of external circumstances. Becker (1976) and Stigler and Becker (1977) are among those arguing strongly for the view that preferences or 'tastes' should be considered stable. More precisely, they run from a basic utility function which is stable.

Finally, when rational agents act on the basis of such preferences, the only acts they can undertake are exchanges. They can exchange goods – that is, any tangible or intangible items that have the capacity to be demarcated and to enhance utility. Rational agents will furthermore exchange goods until a point is reached where no more gain appears. The marginal gain of exchange is zero. Thus equilibrium states are produced. With given and stable preferences, maximization may seem to straightforwardly imply stationary or equilibrium states. This is also a problematic conclusion, and we shall return to it later.

BOX 2.5 THE HISTORICAL CONTEXT OF NEOCLASSICAL ECONOMICS

Economics was, as a discipline, gradually established from the later part of the eighteenth century. The first hundred years – that is, approximately from 1770 until 1870 – has been called the classical era. The classical economists focused on the study of the existing economy trying to understand growth, the role of markets, and issues concerning industrialization and the specialization of labour. All these processes accelerated throughout the eighteenth century – especially in England. Adam Smith, David Ricardo and Thomas Malthus were important scholars of this tradition.

The classical economists specifically engaged in explaining the distribution of the surplus appearing from production among different input factors such as labour and land. This again was an issue directly related to the distribution between different classes of society, as it also reflected the increasing specialization in the economy. Thus, the political aspect of the economic process was very visible in the writings of these authors. Land, labour and capital were not just abstract input factors, but social phenomena. They were owned by different classes.

Specifically, the classical economists based their analyses on the Lockean labour theory of value. Locke had developed that theory from his idea that all humans are born free and equal. We 'own' our own body. This implies in his mind that we must also 'own' the results of what we produce by using our body. He claimed that when we work, we gain property rights to the products we produce.

Many political conflicts over the division of the surplus produced in industrial societies occurred throughout the nineteenth century. As Joan Robinson formulates it, given the harsh class conflicts, the labour theory of value did not over time 'smell too well' (Robinson 1962). It fuelled social revolt, as Marx used the idea to claim that since labour was the origin of value, it was the legitimate appropriator of the surplus of production.

The neoclassical idea that value can be explained by the marginal utility it offers, shifted the 'battleground'. It was an idea that was inspired by another philosophical tradition originating in the eighteenth century, that of utilitarianism. The utilitarians, notably Jeremy Bentham ([1789] 1970), claimed that what motivates humans is individual utility, the gain of pleasure and the avoidance of pain. This implies that the neoclassical position makes a shift from an objectively measurable construct – number of hours – to a subjectivist standard – that of happiness, of utility. Perhaps more fundamentally, it transformed the issues of resource allocation and surplus distribution, so strongly politicized throughout the nineteenth century, to the realms of 'positive' or neutral economics. It was transformed into an issue about marginal utility and marginal productivity of the various input factors that the market had turned labour, land and capital into. The internal problems related to this solution are strongly illustrated by the 'capital controversy' – that is, the debate between economists of Cambridge, Massachusetts (Paul Samuelson and Robert Solow) on the one side and Cambridge, England (Nicholas Kaldor, Joan Robinson and Piero Sraffa) on the other. For those interested in the latter issue, see Harcourt (1972).

The above concept of the core of a science is borrowed from Lakatos (1974), according to whom, each science is characterized by what he terms both a 'hard core' and a 'protective belt'. Since our needs and perspectives differ somewhat from Lakatos,[10] we shall utilize the concept of 'application area' or 'application theorems' rather than the protective belt to describe the context into which the core assumptions are placed when analyses of real-world phenomena are made.[11] Once more following Eggertsson

(1990), the *standard application area* of neoclassical economics can be defined as follows:

- no information costs;
- no transaction costs; and
- private property rights for all goods which are exchanged in competitive markets.

Transaction costs is probably the least familiar of the above. Arrow (1969: 48) has defined transaction costs as the 'costs of running the economic system'. More specifically, transaction costs can be defined as the costs of information gathering, contracting and controlling contracts (Dahlman 1979). Because of this, information costs could be subsumed under that category. It is kept as a separate point since information gathering is also necessary for activities other than transacting.

The only institutional elements appearing in the neoclassical model are those of rights in resources (property rights) and the market. The last is hardly regarded as an institutional structure, though. Given the kind of rationality involved, the only form of interaction implied by the model is the exchange of goods. It will, moreover, appear as long as utility can be increased by such exchange. Thus, the market is 'the natural order of things' in a model based on pure individualism and zero transaction costs. It is not seen as a creation. What is accepted as a social construction, though, is the establishment of rights in resources. It is the task of the state to form and guard these rights. However, to analyse how this comes about, is taken to be outside of the neoclassical model.

With the elements of the core and the standard application areas, as defined above, the neoclassical model actually equips us with a list of answers to almost all questions we raised earlier in Chapter 1 about which issues should be resolved by a theory concerning human behaviour in social systems. The presentation has been very brief, though. We shall provide much more insight into the core elements of the theory, especially in Chapters 5 and 6.

There are several problems with the model that we shall also return to in later chapters. Some have to do with relevance, others with consistency. First, the model rejects rationalities or reasons for action other than that of maximizing individual utility. Second, changes in preferences, if observed, cannot be explained. Third, in real-world circumstances, information gathering and transacting is costly. If information is costly, even problems with the consistency of the core assumption of rational choice appear. Since one cannot know the value of the next piece of information to be gathered, one cannot rationally distribute resources between decision making and information gathering. If transacting is costly, it may furthermore be that markets are not the best allocation mechanism. Then a 'second-level' optimization problem appears, going beyond that of the individual: which economic structures are best at economizing on transaction costs? This has become a core issue not least for 'new' institutionalists like North when establishing a theory of institutions around the neoclassical core.

2.3.2 Building Institutions around the Neoclassical Model

As pointed out above, the only institutions that are necessary given the neoclassical model are those defining property rights. Rights to resources are fundamental to be able to maximize their rents. The self-contained individual, however, does not need any institutional support to understand or to act. As we have just seen, however, if we change some of the application theorems, distinct institutional questions appear. So the development of institutional thinking growing out of the neoclassical school was a reaction not least to the observation that transaction costs are not zero. If it is costly to run the economic system, then the issue becomes: which system is cheapest to use? Williamson (1985) points out that if transaction costs are zero, it is actually impossible to distinguish between competitive markets, oligopolistic markets, planned economies and so on with regard to resource use and efficiency. If these costs are positive, firms or even the state may be cheaper allocation mechanisms than the market.

This debate all started with Coase's (1937) paper, 'The nature of the firm'. He asked, why are there command systems like firms, if markets are costless to run? His answer was that in some situations it is less costly to use command systems within the firm than to operate with exchange within markets. He never used the term transaction costs, but the idea was certainly that of economizing on this type of costs.

It was some 30–40 years later, and after Coase had written his 'complementary' paper, 'The problem of social cost' (1960), that his insights were utilized to start formulating an institutional economics based on the neoclassical core. The idea was to develop a theory that could describe how various economic structures differed concerning the costs of transacting. The focus on institutions as *constraints* was very much a reflection of this. Institutions defined the 'rules of the game'. They constructed the playground in the form of a single firm, a vertically integrated firm, a modern corporation, different types of contractual arrangements in markets and so on.

This tradition of institutional thinking, 'new' institutional economics or transaction cost economics, is consistently based on the individualist perspective of the neoclassical model. This is just the other side of the coin of defining institutions as constraints. The point is not to change the core, but to focus on the second-level optimization problem, that of optimal institutional constraints.

2.4 HOW TO EXPLAIN SOCIAL PHENOMENA

Underlying the two positions described, social constructivist and individualist, there is a difference in basic methodology. We shall therefore take one step back and look at how we can explain and understand social phenomena and look at which methodological basis underpins the two positions.

2.4.1 Different Types of Explanations

In the social sciences, three categories of explanations are usually emphasized: causal, intentional and functional. The presentation here will be very brief. For those wanting to look more deeply into the matters concerned, see Elster (1983a).

- *Causal explanations* Simply formulated, causation has to do with regularities. An event or phenomenon B is explained by another preceding event or phenomenon A. The causing event can be a physical force (natural sciences) or some type of social influence (social sciences). Thus, if a volume of a gas is reduced (A) *ceteris paribus* the pressure increases (B). Certainly, such regularities demand a mechanism. This points beyond just observing the event chain. Hence the fall of an apple from the tree to the ground is explained as caused by gravity. While causal explanations dominate in natural sciences like physics, they are also important for the social sciences. The expression 'John likes to take a walk because his parents went hiking with him in the forest when he was a child' is a typical example. Here we observe a causal explanation, not of the act, but of the motivation behind it. Walking may be explained by positive experience ('walking is good'), habitualization ('John was trained to walk'), or norms ('the parents had persuaded John that walking was good for him'). However, it is important to note that it is difficult to distinguish between these three types of explanation of the motivation on the basis of just observing the regular act.
- *Intentional explanations* The act or the phenomenon is explained on the basis of the intentions of the acting individual. It is explained on the basis of the preferences or the will of the one acting. This type

of explanation also demands that the individual has a belief that doing A causes B. Intentional explanation is exclusive to the social sciences – it demands consciousness and wilful acting. The following explanations are all intentional: 'John walked to work because he wanted some physical exercise and he expected that walking would offer this'; 'Jane took the car instead of the bus because she was short of time and believed that taking the car would be quicker'.[12] Neoclassical microeconomics is based on intentional explanation. The intentions are given by our preferences. Moreover, taking preferences as stable implies that social factors are not allowed to enter as an explanation of these preferences. Causation, if at all relevant, must only relate to the genetics and/or psyche of the individual. Examples like the ones above under causal explanation are ruled out.

- *Functional explanations* In this case we encounter a type of explanation which dominates in biology, and to some extent is also used in the social sciences. In the latter case it is controversial. A functional explanation has the following structure: a phenomenon – let us say the speed of the antelope – is explained by its positive effect on its survival. The faster it can run, the greater the chance of surviving and growing up in an environment of various predators. There is no intention behind this result. It appears through positive feedback mechanisms. The biological type of selection is thus the example *par excellence* of a functional explanation. The selection is an accidental effect of the interplay between a random gene mutation and the environment of the species where this mutation occurs (also including the other members of its own species). Most mutants will not survive. However, occasionally the random change turns out to offer a competitive edge in the given environment for the specific individual – like speed – and the particular quality is reproduced and gradually magnified. Natural selection operates first by chance, then by a positive feedback. Functional explanations are sometimes also used in the social sciences. Understood as above, it implies that new acts, which are preferable for maintaining the social system involved, appear by pure accident – that is, like gene mutation they are not intended.[13] They are next repeated while still creating an unintended advantage for those repeating the act. Such explanations are sometimes used to account for the existence of social structures like norms. There has been a substantial debate about whether one can talk of functional needs in societies as comparable to, for example, the needs of a body. Furthermore, while a norm may have 'a function' let's say to maintain order in a society, it may not have come about in a functionalistic way.

Elster has been very active in criticizing the use of functional explanations in social sciences (see especially Elster 1979, 1983a). According to him, intentional and causal explanations should be enough to explain any phenomenon in social life. His engagement has been a fruitful endeavour against a tendency (not least in parts of sociology) to remove the intentional agent, the subject, totally from the scene.

I agree with Elster that functional explanations are problematical in the social sciences, even though one cannot rule out the possibility of such explanations in the case of some norms. The incest taboo has been put forward as one possible example. It may have been invented for cultural reasons – for example, the idea that certain things that are 'like' should not be mixed. The positive, but unintended effect on the genetic health of a population with such a belief may then explain its ability to grow and conquer other groups not applying this taboo. Thus it is expanded via its unintended, but positive effect on fertility and the strength of the group adhering to this institution.

However, the taboo can also be understood as intentional, as invented on the basis of observed negative effects of near relatives having children. As a social norm it has been reproduced and over the years the reason behind the rule may have been forgotten. Thus it may seem that a functional mechanism is at play, even though it had an intentional origin. This is a typical characteristic of social constructions. The rule 'lives on' independent of insights behind the original intention. At a certain point in time, when the day of invention is forgotten, it may be 'tempting' to give such a norm a purely functionalistic explanation. It may still have been an intended act to solve problems or to realize certain opportunities. My position is that this is by far the most typical situation.

Actually, institutional mechanisms appear as a combination of intent and cause. First, the norm was established to obtain something – the intention. Next, the norm, when internalized, causes the agent to perform certain acts in the actual situation without necessarily reflecting on its basic motivation. This specific combination of intentional and causal explanations will for the rest of this book be called an *institutional explanation*. It is explicitly thought of as an alternative way to explain seemingly unintentional acts without having to invoke functionalist explanations. To be precise, it is not meant to rule out cases where those reproducing an institution are fully aware of its motivation. It is instead meant also to cover those situations where the motive is unobserved or has been forgotten.

Explanation is the fundamental issue of social sciences. Thus, insight into the ways one can understand social phenomena is crucial. Clarifying types of explanations – as above – is one part of this story. Understanding systems of explanations is another. As we proceed to the latter, we observe that these issues are interrelated.

2.4.2 Systems of Explanation: Methodological Individualism and Holism

In the literature we find two dominant systems of explanation. First we have methodological individualism or agent-based explanations. Second, we have methodological holism or structure-based explanations. We shall first present the core postulates of both. Next we shall discuss their relevance for an institutional type of explanation of social phenomena. As will be made clear, I find both methodological individualism and holism to be inadequate.

Methodological individualism

Methodological individualism implies a position where all social phenomena can be explained on the basis of individual behaviour. Individual purpose is the source of all action. Because of this, intentional explanation is the prominent type of explanation.

Methodological individualism can be differentiated in two positions:

1. According to the most radical version all explanations, both of specific acts and of social phenomena that follow from these acts, are to be explained exclusively on the basis of the individuals involved. Social phenomena are a summary product of individual acts and these phenomena or structures in turn do not have any influence on the behaviour of the individuals.
2. According to the second position, real social phenomena exist. Humans may be influenced by such circumstances in the form of norms and so on. These phenomena should, however, be understood as characteristics of the separate individual when acts are to be explained.

The first position implies that individuals and their preferences cannot be moulded as an effect of social circumstances. This is the position on which neoclassical micro theory is based. We find this most clearly in the core assumption of stable (given) preferences. However, a discussion of this methodological basis has only rarely found its way into economic textbooks. It is also interesting to see that as soon as we turn to macroeconomic textbooks or studies, economists very often turn to 'structural' explanations. We shall return to this.

It is foremost the Austrian school, especially Ludvig von Mises and Friedrich Hayek (see Mises (1949) and Hayek (1948)), which has been engaged in producing a philosophical defence for the first form of methodological individualism. Neoclassical economists have tended not to engage so much in these more fundamental issues, and it is not at all clear if they accept the Austrian defence.

The Austrian defence is also characterized by some fundamental problems, and it has a tendency to end in mere assertions: since action is based on individual purpose, individual purpose must explain action. This type of circular reasoning is especially typical of the work of von Mises. Hayek is somewhat more elaborate. He is open to psychological explanations of purpose, but explicitly denies social ones: 'If conscious action can be "explained", this is a task for psychology but not for economics . . . or any other social science' (Hayek 1948: 67). Dividing up the job and sending the baton to another individually orientated science like psychology keeps the methodological basis for economics intact. It is still not much more satisfying, as is strongly emphasized not least by Hodgson (1988).

The second position carries in my mind much more merit. We find it developed both in work by Karl Popper and Jon Elster. Popper (1945) accepts that institutions as social phenomena are part of the explanation of human action. As Weber before him, he still specifically argues that all social categories, like 'the state' or 'capitalism' should be described with reference to real or 'idealized' individuals. Popper, however, is accused of being confused in that he mixes 'political'[14] and methodological individualism (Hodgson 1988).

Elster (1979, 1983a) develops his position quite systematically. In his understanding, the characteristics of the individual are expanded to cover their place in the social system, for example, in the social hierarchy, the team, the profession and so on. He accepts that preferences are changed due to changes in social position. Thus not only intentional, but also causal explanations are utilized: preferences may have (social) causes. Nevertheless, choices themselves are to be understood as individual and purposive acts running from the existing preferences. Elster identifies two departures from this, though. First, a person may not carry the right understanding of the relationship between the act and the (wanted) effect (false consciousness), or s/he may by accident do the wrong thing – like pressing the accelerator when intending to use the brake. Second, we have the 'weakness of will'. A person may be incapable of (always) doing what s/he prefers. This is against the premise of rational choice, but not against the idea of methodological individualism.

Properly defined, this second variant of methodological individualism can be viewed simply as reducing all social relations to individual characteristics. This implies that individual terms presuppose social relationships that reflect the social constitution and status of the individual. On many occasions this perspective may be acceptable. Nevertheless, it complicates many analyses which, much more easily and with at least the same power, could be undertaken on the basis of studying the social relations, the structures, directly. The danger involved in Elster's strategy is that the social aspects become underestimated. The social aspect is essentially *relational*, and it is problematical to reduce a relation fully to individual terms.

Let us look at some examples to clarify the two methodological individualist positions and their limitations:

- The *contract* is an arrangement whereby two or more individuals define reciprocal rights and duties. A business contract between two parties is one example. A marriage is another. The contract is often used as the core example of something that is purely individual. It is A agreeing with B. Contracts do, however, need a cultural and linguistic framework. A framework of trust and/or a third party, which can guarantee the fulfilment of the contract, is also necessary. The business contract is a relationship not between two, but between three parties. The third party must, furthermore, be institutionalized with the power to mediate, control and punish if the contract is not fulfilled. If we are to choose between the two types of methodological individualism, type 2 seems clearly to be the most relevant even in this case.
- In the special case of the *marriage* contract there are strong individual elements involved. A and B love each other and want to marry. Still, why marry? They could just decide to live together. Is the marriage just something in the heads of the two people? Do they really have a choice if they want to live together? Certainly, in some societies there is, at least today, a choice to be made about the form of relationship. However, this is not the case in most societies of the world. The choice for a couple to live together may be understood in individual terms, they love each other, but the form chosen is not individual. It is a relation instituted by society.
- Another example characterizing the difference between the two types of methodological individualism could be the *power relations* at universities, one between professor A and student B. This could be understood as a relation only between the two. Professor A decides the material for the courses, the type of exam and what is a good paper and so on. However, it is the whole structure of rules at the university that gives A this position. It is a type of 'quality control' as A has also been evaluated for the position. S/he has the necessary competence. It is the position of both A and B in this hierarchy, the social structure, that defines their relation.

  Again, methodological individualism of type 2 is clearly the more appropriate of the two. Nevertheless, one may wonder if a

> perspective focusing on the individual professor as a product of the system is better than studying it as a social relationship, the relationship between roles. If we want to study the effect of various responsibility structures, it does not seem wise to study that issue by seeing it just as characteristics of various individuals. Certainly professor A_1 may be grading differently from professor A_2. A_1 may be more interested in involving the students in group work than A_2, and letting them read Mark Twain instead of Charles Dickens. One must not deny the effect of individual preferences, whichever way they are formed. However, the power to make decisions about who to read comes from the structure of roles.

If people construct institutions and then are influenced by them, there is a continuous loop of causation. One may ask, why then reduce it all to the individual. We could equally well have limited the explanation to the social sphere, to the institutions. This is what methodological holism or collectivism does.

Methodological holism

Methodological holism is the opposite position of methodological individualism of type 1. According to the holist position, social phenomena can only be explained by reference to other social phenomena. While methodological individualism was a term coined by Joseph Schumpeter in 1908 (Hodgson 1988), there is no parallel history of the concept of 'holism'. In the writings of Georg Hegel and Auguste Comte in the first part of the nineteenth century, we can already see that 'societies are treated as totalities with distinct properties of their own' (Scott 1995b: 12). Durkheim gave much weight to the idea that 'social facts' should be explained by other 'social facts'.

Later, holism gained a strong position in American sociology, especially in the 1950s and 1960s. This wave was founded on the work of Talcott Parsons and his structural-functionalism. He saw societies as internally related and self-sustaining systems of roles operating within an environment. Hence, he worked not least on identifying functional prerequisites of a society in order to sustain itself. Society was a 'whole' with different 'parts' performing certain tasks of crucial importance to keep the whole system working. These tasks were the functions. In Parsons's view the system of role relations is the structural core of the social system. He saw the social institution as a complex of institutionalized roles that are of 'strategic structural significance' (1951: 39). From a methodological holist perspective, it is the role that forms the motivations and 'makes the act'.

The problem of methodological holism is due to this two-sidedness. Partly, the individual may tend to be 'oversocialized': it often appears that there is no room for real choices to be made. Partly, we often encounter illegitimate use of functional explanation as previously emphasized. While Parsons represents a more conservative version of structural-functionalism where the focus is on the issue of order and maintenance, of stability, parallel types of reasoning are also found in more radical literature such as various Marxist positions. Here different functions, like those of the state, are understood on the basis of the need to keep a basically exploitative society running (Miliband 1969; Poulantzas et al. 1976).

Despite the various problems encountered, explaining social phenomena by other social phenomena may be a reasonable choice in very many situations. Talking about functions may not be an error, at least as long as it does not imply more than saying that something or somebody performs a specific task in a given system, and as the system is set up, these tasks are necessary to keep it going. Let us look at some examples to clarify the situation:

- 'The relatively short time students stay at university (social phenomenon) makes student organizations (social phenomenon) weak.' This is a sound holist proposition. In this case it is clearly meaningful to focus on structural features only. A strong leadership based on personal capacities of chosen leaders may certainly influence the severity of the problems faced by student organizations. Nevertheless, good leadership is more difficult to establish and it is unable to eliminate the structural difficulties. Following Elster's claim and formulating the problem in terms of individuals may be possible, but it is overly complicated to picture what is really going on.
- 'Reduced unemployment (social phenomenon) will result in increased inflation (social phenomenon).' This standard proposition from macroeconomic theory is holist in the above sense. While observations in the 1970s and 1980s may seem to have violated the 'law' since increased unemployment was observed in parallel with increased inflation, the form of the proposition is still principally sound. Macroeconomic theory is filled with propositions of this kind. Certainly, there is no problem with that. If there is a problem it lies at another level – that a micro theory built on methodological individualist principles exists together with a macro theory, built on the opposite principles. As we know, attempts to construct a macroeconomic theory on microeconomic foundations have failed (Spulber 1989). This supports the position that the whole (the economy), is more than the sum of the parts (the individuals).

- 'In capitalist societies (social phenomenon) the state (social phenomenon) functions as a mediator (social phenomenon) between conflicting classes (social phenomenon).' In some Marxist analyses, 'function as' will be understood as something that is unintended, but necessary to keep the system working. As the proposition is formulated here, 'function as' rather implies 'acting as'. Expressed in this way it may be a sound proposition, reflecting that in class-based societies a continuation of that society depends on some sort of acceptance of the social order across conflicting classes. Creating such acceptance may depend on some kind of redistribution of the surplus from production to reduce the inequalities that appear in capitalist societies. However, there are many difficulties involved. Others than the state may cover the role of mediation. And is it mediation that is taking place? Marxists may rather see the state 'functioning' as a 'representative of the ruling classes' and redistribution merely as a way to avoid social revolt – that is, to secure the continuation of the basically unequal system. On the other hand, the state is rarely a homogeneous entity with one will. Because of this it may be difficult to talk about the state as one agent. Even Marxists accepts some autonomy for the state (Miliband 1969).

Following the second example, reduced unemployment, it is clearly most relevant to describe social phenomena as 'a function' of other social phenomena and not as an effect of specific acts. But this does not imply that we can talk about a functional need of the system. It is more a type of causal relationship where the acts of individuals behaving within a certain system are strongly influenced by the type of system they are acting within. In such situations their acts sum up to aggregates whose internal relationships can be studied without going back to the individual acts themselves. Our example of the weakness of student organizations illustrates this well. Actually, studying such phenomena at the individual level is often impossible. The whole is more than the sum of its parts. The Elster project becomes in many situations not only complicated, but actually impossible, wrong or irrelevant.

Towards a methodological institutionalism

Actually, what we have described so far are two reductionist methodologies. One reduces it to the parts – the individual agents. The other reduces it to the wholes – the social structures or institutions. This agent–structure divide is deeply rooted in social science controversies. The agent and the social structure are, however, two distinct levels that cannot be fully reduced to each other. In this 'chicken and egg' problem it becomes untenable to stick to the one side only. It is both about production of new institutional structures and reproduction of these structures. To explain the former, some kind of agency is needed. To explain the latter, social structures must exist independent of the individual.

The social constructivist perspective, as presented here, is explicitly trying to capture both levels. It focuses on the dialectics between agents and structures. Individuals form institutions (externalization). They become objective facts (objectivation). Finally, they form the individuals (internalization/socialization). We should rather follow Giddens (1984) and view social structures as both a medium for acting and an outcome of acting.[15] This is much more productive than to construct a division between two methodological positions that are both untenable.

I would propose the term 'methodological institutionalism' to describe a methodology which focuses on the dialectic process between agency (individuals) and structure (institutions). Its core element is the combination of intentional and causal explanations as previously suggested. Certainly, at the present stage it is more a framework than a full methodology. Nevertheless, there seems to be an ongoing trend in the literature towards such a synthesis. While it is suggested that both structure and agency must always play a role in social research, it may be relevant to put most weight on structure in some cases and in others on agency – that is, on how institutions affect behaviour as compared to the intentional transformation of institutions. The choice of focus depends on the problem at hand.

There is an asymmetry here, though, which implies that the social structure is hard to avoid in any kind of analysis. Even in cases where the focus is reasonably on agency, on the acting individual or group, it will be necessary to involve some analyses of the institutional structures within which the agents act to fully understand what is going on. This argument of two-sidedness has less force in cases where structurally orientated analyses are the most relevant. They have their strength specifically when individual variation can be exempted. The implicit argument here is that while reproducing institutions is a purely structural phenomenon, a transformation of the same institutions does not occur in an institutional vacuum. So while methodological individualism type 2 leans heavily towards the individual, methodological institutionalism should lean moderately towards the social structure side.

Another way to make the above distinction clear, is to say that methodological institutionalism accepts that social phenomena exist independent of individuals. Here it parallels methodological holism, and the statement just emphasizes that institutions are real, irreducible phenomena. It denies, however, that all social phenomena can be explained only by other social phenomena. In relation to explaining change in social structures, agents must play an important role.

2.4.3 Realism versus Relativism

There is one more methodological issue that warrants a comment at this stage. This concerns the argument that social constructivism implies total relativism (Guba 1990). I shall argue that this conclusion is wrong. There are two issues at stake here. First, we have the question of whether there exists a physical world independent of our conceptualization of it. Second, it is argued that since values are socially constructed, 'anything goes'. There is no way to rationally criticize these constructs. Let us look at this, step by step.[16]

While social constructivism implies that we observe and understand the external world via the concepts that are available to us, this position does not imply that the external world is created by our conception of it. While some social constructivists may be understood as supporting such a view (for example, Woolgar 1988; Tester 1991), it is not at all a necessary consequence of the constructivist view. Rather, it is a very problematic one and a realist position has, in my mind, by far the strongest merits. This position implies that the physical world exists independently of the human cognition of it. It is next possible to evaluate different conceptions of this world in a search for what gives the best description and understanding of it. As an example, the old view that the earth was flat and at the centre of the universe was a cognitive model that seemed to fit well. It was, however, over time challenged by a growing number of observations that questioned this model. In the end this information was combined with already existing analogies like that of a 'spinning ball' to produce a new model. The perception shifted as a consequence of learning.

There is an important distinction to be made between the physical and social spheres in this respect. While the physical world exists independently of our concepts, the social world is directly constructed via our acts and concepts. It exists independently of us in the form of social facts, but it is still created and recreated by us.

This takes us to the second issue, that of whether 'anything goes' in the social sphere. Does the fact that social systems are human creations imply that we can choose whatever solution comes to mind? Is there no objective truth about these issues? Are values and the questions of right and wrong just relative and subjective? These are complex issues that will be discussed more thoroughly later in the book. Here I shall just make some short introductory comments.

First, concerning social relations, one cannot talk of objectivity in the social world in the same sense as in the physical. The social structures and relations we create, the values they are based on, are not given. This still does not imply that they are to be treated in completely relativistic terms. On the contrary and exactly because they are human made, they are open to reasoned critique about what is best to do. Such critique is crucial, as made clear not least by Roy Bhaskar in his defence of what he calls 'critical realism' (Bhaskar 1989, 1991). The point is that values and institutions, since they are collectively created, can both be discussed and evaluated across individuals. In this specific sense they are 'objective'.

Second, the social world, the institutional structure we make, is both common to us, and it exists independently of us as specific individuals. This world is thus also 'objective' in the sense that it can be observed and studied as social facts. This point was, as we saw, made both by Durkheim and by Berger and Luckmann.

Nevertheless, it is observed that cultures develop differently. They embody different perceptions of what is a good life. This may support relativism in the sense that across cultures values cannot be critically examined. But even at this level, one should be careful about claiming that social constructivism necessarily implies relativism. Hence, one observes that there are several human needs that still come through as common across civilizations. The set of basic physical needs comes first to mind. Also social needs like care, acknowledgement and so on seem to be common, human needs. These qualities may take on different forms in different societies, and they may be of different importance. Yet no social construction can do away with them. Furthermore, while societies in some cases support clearly different values like, for example, equality versus inequality between the sexes, it does not mean that these institutional structures cannot be reasoned over when people from different cultures meet and communicate. This does not imply that it is easy to reach common conclusions in these matters. Nevertheless, the norms that exist can be evaluated precisely because they are socially constructed. They are based on certain motivations or goals. They can then be open to both internal and external critiques through a discussion about these motivations and goals.

If we instead shift to the individualist perspective, we observe that in this case such critiques make no sense. The individual is given and the preferences of the individual are not open to any form of critique. Because of this, they are not accessible to reasoned evaluation.

2.5 SUMMARY

In this chapter we have explored the concept of institutions by looking at two principally different ways of understanding the relationship between the individual and society: the theory of social construction and the individualism of neoclassical economics. The theory of social construction – as developed here – looks at the institution as the core concept of social theory. Institutions are products of human acts. They are constructed by people. Parallel to this, individuals are a product of the institutions of the society in which they are raised or live. Institutions influence both their goals and expectations. This position then focuses on the dual idea that institutions are a human creation and that the human being is a product of the same institutions. Finally, it is emphasized that both nature and social relations get their meaning, are understood, through socially constructed concepts. While nature exists independently of humankind, the way we interpret it depends on human constructs.

According to the individualist model, the individual is self-contained and independent of the social context. The world is readily interpretable for this individual. Preferences are stable and action is motivated by maximizing individual utility. Individuals engage only in exchanges with each other. Institutions, if at all taken into account, are seen just as external constraints to the maximizing individuals taking part in such exchanges – compare the school of 'new' institutional economics. The role of institutions is to reduce the cost of exchange – the transaction costs. While some look at the individualist model of exchange as relevant only to formalized market structures, others generalize the idea of exchange to all areas of society.

The divide between the individualist and social constructivist models is partly reflected in the distinction between methodological individualism and methodological holism. The first methodology bases its thinking on the idea that social phenomena are the sum only of individual phenomena. There exists a less radical version of methodological individualism, accepting that real social phenomena – that is, phenomena outside or above the individual – exist. These phenomena may even influence or form the individual. It is, however, claimed that the explanation of behaviour should always be based on the (socially influenced) individual.

According to methodological holism, social phenomena exist independently of the individual. Moreover, social phenomena can only be explained by reference to other social phenomena. According to this position it is the role, a social phenomenon, and not the individual, that defines the act.

The distinction between the two methodologies follows a deep divide in much of social science – that between *the agent* and *the structure*. We have suggested that this divide is artificial and unproductive. A good social science must acknowledge both levels. A sound methodology must recognize both the actor and the institutional structure as irreducible entities. This is the core characteristic of an institutional explanation, combining both intentional and causal loops of explanations.

NOTES

1. Those interested in going deeper into the various issues and positions will find interesting and well-written introductions and evaluations in Scott (1995b).
2. The observant reader would have recognized that both A and B are presented as males. Obviously Berger and Luckmann had no intention of creating a homosexual couple with adopted children as the basis for their example. Nevertheless, it is interesting to comment on this, since in the early stages of the twenty-first century we can observe the battles over institutionalizing such a practice taking place in many western countries.
3. The question about the length of the 'human age' and its civilization is an issue of ongoing research and definitions. Lewin (1988) documents evidence of 'human like' individuals dating back at least 2 million years. 'Modern man' in the form of *homo sapiens* dates back some 200–300 thousand years.
4. I am indebted to Ostrom (2000) for this reference.
5. There are certainly some variations across these authors. The concept of utility itself developed over the years. In the hands of Robbins it developed into something very different from the interpretation of the early neoclassicists – that is, the shift from cardinal to ordinal utility (see Chapter 6). Noteworthy here is the fact that Marshall was occupied by the thought that economic action is not simply based on self-interest. It is also shaped by shared value standards. Thus, elements of socially constructed preferences and rationalities beyond utility maximization can be observed. However, they are not developed into a theory and disappear over the years as part of what becomes 'orthodoxy'.
6. Even though neoclassical economics is by far the most formally developed of the social science positions, it is not possible to define its assumptions in such a way that all economists agree (Hausman 1992). Moreover, we observe differences between textbook expositions and the research agenda among many neoclassically orientated economists (see Nelson and Winter 1982). The Lakatos et al. definition still has broad acceptance.
7. Eggertsson, as an example, does not specify maximizing as an explicit part of the definition of rational choice. I believe that in his view, as in the neoclassical position, rationality is simply implying maximization. I do not agree with that – see also the text.
8. An elaboration on the definition of completeness, transitivity and continuity is given in Chapter 5.
9. It is interesting to observe that maximization is the only way the model perceives a rational act. As we shall investigate later (Chapter 5), rational action can very well be defined without adhering to maximization. With reference to the discussion of Section 2.2, following a norm does not in any way involve maximization, but can still be viewed as a rational act.
10. According to Lakatos the *hard core* of a programme must not be rejected or modified. It is thus protected from falsification by a *protective belt* of auxiliary hypotheses, initial conditions and so on.
11. I am indebted to Daniel Bromley for the concept of application theorems. I make this shift because any model, any hard core, will have to be supplemented by descriptions or assumptions about the situation in which it is to be applied. Certainly, these may be developed to protect the hard core. Lakatos may be right that any science is constructed this way, and that changes in the history of science come about when the evolving protecting belt is thought to be too 'nasty' to be able to protect the core any more. This was the case of the Ptolemaic world conception when it was replaced by the Copernican. Here one core model with the earth at the centre of the universe with its extended protective belt of orbit epicycles was exchanged for a much simpler model placing the sun at the centre of the universe. Today we know that even that model was in a sense wrong – the sun is rather placed on the outskirts of one of a billion galaxies. The point here is that protection of the core is still not the only function of these theorems. One simply cannot do without them.
12. To be precise, an intentional explanation demands three steps. First the motive: 'John wanted physical exercise'; next the belief: 'John was of the opinion that walking to work results in such exercise'; then it follows that 'John walked to work'.

13. In the case of human societies it may also be that they appear for reasons other than those that are important for the survival of the social system or group. While the act is intended, the specific effect is not – compare the example with the incest taboo appearing later in the text.
14. Methodological individualism must be differentiated from 'political individualism'. The latter concerns itself with the normative position that institutions of society should be so structured as to secure the free choice of individuals. While there are many common themes involved and much overlapping treatment of these positions in the literature, methodological individualism is also something else. It is descriptive in its intention, a way to understand action, not a normative position about which society is best. Certainly many others as well as Karl Popper can be accused of mixing these two positions.
15. There is a continuing debate as to whether Giddens is balanced in his own analyses. There may be a tendency in his work to lean towards putting most emphasis on agency.
16. Limited space does not allow me to go more deeply into the debate on these issues. See further: Bhaskar (1989, 1991); Pratt (1995); Gandy (1996); Tacconi (1997); O'Neill (1998).

3. Institutions: coordination and conflict

In Chapter 2, the focus was on how to understand and explain the general process of institutionalization. We shall now move to look more directly at the form and normative content of institutions. In that respect, classifying institutions according to the type of problems they are a response to is one important task. Understanding the relationship between institutions and interests is another.

With regard to the categorization of institutions, I shall first offer a definition of an institution which will be used in the rest of this book. The forms of institutions will be divided into 'conventions', 'norms' and 'formally sanctioned rules', and the core characteristics of each group will be outlined (Section 3.1). The next step will be to present a more universal 'grammar' of institutions, drawing heavily on the work by Crawford and Ostrom (1995) (Section 3.2). Next we shall focus on the relationship between institutions and interests and discuss various mechanisms that may make institutions durable (Section 3.3). This is followed by two short comments, one on the issue of rights protection (Section 3.4) and one on the issue of coercion (Section 3.5). Both topics will be covered more substantially in later chapters.

3.1 CATEGORIZING INSTITUTIONS

On the basis of the discussions in Chapters 1 and 2, I have formulated the following definition of an institution:

> Institutions are the conventions, norms and formally sanctioned rules of a society. They provide expectations, stability and meaning essential to human existence and coordination. Institutions regularize life, support values and produce and protect interests.

The definition explicitly defines institutions according to both their forms and their roles or motivations/rationales. This gives us the opportunity to capture the reasons behind their existence and to understand which situations or type of problems they are a response to. Note that the distinction made between conventions, norms and formally sanctioned rules is similar to the one implicit in Scott's (1995a) definition of an institution – see Chapter 1. A similar grouping is also found in Bromley (1989), although he does not make as clear a division between conventions and norms.

Implicit in the definition is an understanding of how institutions come about, which is similar to that of Berger and Luckmann (1967). However, their undifferentiated concept of an institution, their 'reciprocal typifications', makes it difficult to see the various types of motivations that may lie behind the construction of an institution. Actually their reciprocal typification resembles the concept of a convention. As such, it is in many ways neutral concerning values and interests. Values and interests are, however, a core issue in institutional analysis, and we need to cover this aspect explicitly.

The above definition identifies institutions as more than just creating choice sets or external constraints. Institutions not only define the social environment within which the individual is choosing. They also constitute the individuals themselves and their interests. Thus we follow the perspective of Chapter 2, thereby differing from the narrower position taken by most economists.

Furthermore, we sidestep the distinction made by Scott (1995a) between institutions and their 'carriers' – see Chapter 1. Scott mentions cultures, structures and routines as such carriers. I believe that culture can be viewed as a carrier, but the definition becomes unclear when 'structures' and 'routines' also become carriers. One problem is to distinguish 'regulative structures' in the definition of institutions from 'structures' as part of the carriers. The same problem appears when we try to distinguish between 'routines' as carriers and the convention as an institution itself.

One may ask why we should emphasize that institutions are essential also to the human existence or character and not only to human coordination. Can humans not exist without institutions? Robinson Crusoe lived alone for a period and was able to carry on with his life. On reading Daniel Defoe's work, one observes the strong focus on Crusoe as a social construct, not just a biological being, whose thoughts are organized on the basis of conventions that he brought with him from England. He was engaged in keeping track of time as understood by western culture. He even named 'Friday' on the basis of that system. He continued to live by the habits and norms so typical for the English society at that time.

The point is that a person could certainly live physically without internalizing a single institution. What we consider specifically human about Crusoe, however, is the institutions of which he was a carrier. This, combined with the focus on 'meaning', is the main difference between our definition and the ones normally used by new institutional economists, for example, North. A proper definition should emphasize that constructing institutions is also about constructing what it is to be human.

When categorizing institutions, it is reasonable to relate to the type of problem they are meant to solve. They simplify life and coordinate action. They also produce and protect values and interests. As emphasized in Chapter 1, in a world of scarce and interlinked resources, the action of one influences the possibilities for others. Thus, which individual or position gets access to which resource and the way this access is protected becomes a core issue. We shall start by focusing on the role institutions play in simplifying and coordinating human action – that is, conventions – and then move on to the issue of conflict, value and interest protection.

3.1.1 Conventions

Conventions take a variety of forms, but they have one common feature: they simplify by *combining certain situations with a certain act or solution*. We greet each other under certain circumstances and in specific ways. In some countries we drive on the right and in others on the left side of the road. We use money to simplify transactions. We send Christmas cards, if Christmas is instituted as part of our culture. By defining the situation, the individual knows what is the proper act. We sometimes make errors and misunderstand the situation. Everybody has observed this and experienced the confusion that is created. In traffic, such a misunderstanding may create great danger. In other situations it merely causes a slight fuss or inconvenience, which is recognized and excused: 'Oh, she misunderstood what was going on'.

Following Berger and Luckmann (1967), the basic coordinating instrument in a society is language. Languages differ widely with regard to the sounds used to create a word and also the sentence structure. While the object we sit on is a 'chair' in English, it is a 'stool' in German. Nevertheless, the word typifies the same object or concept. Chairs may vary tremendously in form, but, a small wooden object and a large leather one are classified under the same umbrella if they have a back and provide seating space for only one person. Otherwise it may be a bench, a short bench or a sofa and so on.

Language is a type of 'meta' ordering. It provides the necessary structure within which one can formulate other specific rules or institutions. It is both an institution in itself, and it forms a necessary basis for most of the other institutions. It is these 'other' institutions that will interest us the most.

The number of conventions and the form they may take in a society are prodigious. Some areas are specifically evident:

1. the conventions of the language; syntax and semantics;
2. measurement scales; time, temperature, length, weight, volume, value (money) and so on;
3. directions in the sky; north–south, latitude, longitude and so on; and
4. acts in certain situations; types of greeting, clothing codes, food standards, conventions concerning where to dispose waste, how to do specific construction work, how to behave in traffic and so on.

The typical characteristic of a convention is that it solves a *coordination* problem. It simplifies the various complexities of life by structuring and classifying. There are basically no conflicts involved. Passing on the left- or the right-hand side functions equally well, if we all follow the same rule. This understanding does not preclude that deviating from the convention may create dangerous situations, though.

While conventions generally simplify life and make coordination in a complex world possible, we observe a slight difference between coordination instruments such as the metric system and clothing codes. While being able to measure the length of a piece of cloth is a requirement for selling and buying cloth, it is immaterial whether it is measured in centimetres or inches, as long as we are accustomed to the metric. Dressing as a merchant, a farmer or a judge is a different matter. Certainly, clothing has a common practical function in that it protects us from the cold, but, it also communicates meaning as it becomes part of the identity, or expresses the identity of the person wearing it. Indeed, the same might be said about language – it is a practical convention. However, it also creates identity. This double-sidedness is not curious. Internalizing a convention is likely to affect us, and changing a convention may cause conflict. Nevertheless, its basic rationale is to simplify coordination.

3.1.2 Norms

Norms also take a variety of forms. They may be distinguished from a convention since they combine a certain situation with a *required* act or solution which supports an underlying *value*. A norm typically says that you 'should not do x' or you 'should do y'. It is a prescription intended to support a certain definition of how we should treat others, what is a good life and so on. Typical examples are rules like 'you should greet people when meeting them'; 'you should not lie'; 'you are not allowed to cheat' and so on.

Norms are about developing and sustaining certain types of relations between people. They are the archetype of institutions in *civil society*. People live in societies and the number of interrelated acts is vast. It is here that the value aspect enters. Certainly, as already suggested in Chapter 1, there are overlaps between norms and conventions. The way the distinction is drawn here, we see that to greet when one meets is a norm. It signals respect and acceptance of the other. The way we greet is, however, a convention. Therefore we distinguish between the greeting as a norm to follow and the form as a convention.

In the case of norms, it is the creation of human character, human values and proper human relations that is foremost at stake. While coordination aspects are also involved here, norms go further. They define what is an appropriate or right act. As an example, we have the problem of contaminated water. If just one person emits a pollutant, there may be no problem – the level is below that which creates a negative effect. Conversely, if everybody emits, it does not help if only one person stops this activity. Following this logic, it becomes individually rational for each agent not to care about his or her emissions, while the effect for the whole group is detrimental since everybody 'participating in such a game' is motivated to think likewise. A norm saying that you should not dispose of matter x in waters of type y, is a possible way to solve this type of problem. It binds everybody to the collectively sensible solution by creating a norm – that is, an internalized motive for acting in a specific way. Not to pollute is a duty of the citizen, and we do not (always) need to resort to legal regulations – that is, state 'intervention' – to solve such collective choice problems. At the same time, there is often a latent conflict involved concerning normative behaviour.[1]

When norms are fully internalized, they work via a feeling of guilt and no external sanctions are necessary. People do the proper thing just because it is the *right* thing to do. In the extreme case, they see no available alternative. Normally there are alternatives, but the norm, if internalized, defines which alternative to choose. When someone considers deviating from the norm, then external punishments become relevant. Thus a norm may be supported, not only by the internalized feeling of guilt, but also by external sanctions. In the process of internalizing the norm, this is evident. Parents not only tell their child to avoid doing a certain thing; they may also have to punish that child by telling him or her that what was done was bad. The child may not yet have internalized the norm fully. If the issue is serious and violation is repeated, reactions may be more severe. At a community level, people who break the rule of 'we do it like this here', may be treated negatively, they become outcasts. A firm that is known for cheating will, for example, be 'blacklisted' in various ways.

Here we face the difficult issue concerning what norms a society should have and who should have the power to decide over them (see the discussion about value relativism in Chapter 2). I shall return to the implication of such questions later in this chapter. At this stage we shall merely observe that there is often a need for enforcing a norm in some way beyond that of personal sanctions. A special type of enforcement is to use the power of the law, which brings us to the realm of formally sanctioned rules.

3.1.3 Formally Sanctioned Rules: Legal Relations

Formally sanctioned rules, for simplicity just 'formal rules', are different from the above categories in two ways. They combine a certain situation with an act that is *required or forbidden* and which is governed by *third-party sanctioning*. Such a sanctioning system may be the law.[2] Violating what is prescribed implies formalized types of punishment such as being fined, imprisoned and so on.

As emphasized by Bromley (1989), legal relations are fundamental to creating order in societies, not least in the form of economic relations. They exist where interests are or may be explicitly *conflicting* and the collective finds it necessary to empower the regulation of this conflict by the formalized control of its collective power, like the authority of the court system of a state.

Wesley Hohfeld, a legal scholar of the early twentieth century, developed a structure of fundamental legal relations, which emphasized the dual character of any right (Hohfeld 1913). He furthermore distinguished between static and dynamic relations (see Table 3.1). Static refers to a given relation between, for example, individuals Alpha and Beta, while dynamic relates to the capacity or power to change a legal relation.

The first of the static correlates is right versus duty. If Alpha has the right to a certain good, let's say timber from a certain piece of land, then Beta is not allowed to cut down the trees. Beta is duty bound to let Alpha decide

Table 3.1 The four basic legal relations

	Alpha	Beta
Static correlates	Right	Duty
	Privilege	No right
Dynamic correlates	Power	Liability
	Immunity	No power

Source: Hohfeld (1913, 1917).

what to do with the resource. If Beta does not do so, the formal power of the collective is executed and Beta will be punished. This is also an expectation that is implied by the system.

The second static correlate is different in that Alpha in this case is free to behave in a certain way towards Beta, and Beta has no right to oppose this act. A privilege may imply that Alpha is free to cross land that is owned by Beta. In Scandinavia, as an example, it is 'every man's right' to walk in the forests, to pick berries and so on. The owner of the land has no right to stop this. In Hohfeld's terminology this is a privilege. We may also call it a liberty (Hahn 2000).

The 'right–duty' versus 'privilege–no right' correlates are distinguishable on the basis of how responsibility relates to action. Bromley (1989) uses the example of solar collectors. If Alpha is allowed to grow trees to a height where Beta's solar collectors become useless, Alpha is privileged and Beta has no right. If the law protects Beta, then s/he has a right and Alpha the duty to keep the trees low. The same issue can be dealt with by both systems. The type of problem and the definition of whose interest is to be protected defines which is logical.

The dynamic correlates are divided into power versus liability and immunity versus no power. Concerning the former, Alpha has the power to voluntarily create a new legal relation which affects Beta. Alpha may be the parliament of a state and Beta its citizen. Alpha may define a new law concerning the regulation of polluting substances. When this is set up, Beta must observe the regulation or accept punishment. At the fellow citizen level, we may have a situation where Beta wants to cross Alpha's land. This need may be created because Beta wants to cut down some trees on his/her land and it is impossible to get them out without crossing Alpha's land. In this case a contract may be established defining what Beta must do in order to be allowed to cross the land. As the property owner, Alpha has the power to define these demands, and Beta is obliged to comply. Otherwise there will be no contract. We observe how it is the right, the static term, to a specific piece of land that gives Alpha the power to set the conditions – that is, the dynamic aspect.

Immunity means that Alpha is not subject to Beta's attempt to voluntarily create a new legal relation which binds Alpha. Alpha may have a right to cross Beta's land protected by an immunity rule. Beta may want to sell the land, but s/he is not free to change Alpha's right. The land may be sold, but Alpha's right stays the same. Alpha has immunity and Beta has no power to change the relation.

As we have seen, to make a legal relation binding, to ensure that Alpha's right is observed by Beta, a third party must be instituted that has the power to bind Beta. Legal relations are in general triadic. They dictate what Alpha may or may not do towards Beta, and, in the event of non-compliance, some kind of reaction from this third party will follow.

3.2 A 'GRAMMAR' OF INSTITUTIONS

The logical differences between the above categories are replicated in language. This should not come as a surprise, since language is the (main) medium for formulating institutions.[3] From the above presentation, a legal relation may have the following form:

> Alpha's animals must not feed on Beta's cultivated land during the growing season or else Alpha will be fined.

This formulation consists of five elements (Crawford and Ostrom 1995):

A: An *Attribute* is the characteristics of those to whom the institution applies. In this case the attributes concern owners of animals.
D: A *Deontic*[4] defines what one *may* (permitted), *must* (obliged) or *must not* (forbidden) do. In our case the deontic is 'must not'.
I: An *Aim* describes actions or outcomes to which the deontic is designated. The formulation above implies that the forbidden action is feeding on others' cultivated land.
C: A *Condition* defines when, where, how or to what extent an *Aim* is permitted, obligatory or forbidden. In our case the condition is 'during the growing season'.
O: An *Or Else* defines the sanction for not following the rule – that is, a fine will be issued.

Crawford and Ostrom call this the ADICO format from the (first) letters of the different elements. Any legal relationship has this format. The 'grammar' of legal institutions contains all five components.

In the case of a norm the 'Or Else' is omitted. We are down to ADIC. The following formulation is a typical example of a norm: everybody must wash their hands before dinner. The norm thus consists only of an attribute (in this case 'everybody'), a deontic (must), an aim (wash hands) and a condition (before dinner). Following their 'grammar', a norm is not based on a sanction, it is just something that people are obliged to follow and when fully internalized as a norm, becomes part of what is natural to do. It is 'obvious' or 'self-sanctioning'.

The following formulation is an example of a convention: people in Scandinavia greet each other by shaking hands. Here both the 'Or Else' and the deontic are omitted. We have reduced the format to AIC. A convention just tells how something is to be done.

While we have used the concepts 'conventions', 'norms' and 'legal relations', Crawford and Ostrom use a somewhat different specification of their categories. They also use the concept of a norm. However, they call conventions 'shared strategies' and legal relations 'rules'. In the former case I find the meaning of a convention to be similar to that of a shared strategy. It is primarily a question of choice of words. Nevertheless, a convention covers more than actions or strategies if these are understood as acts. The concept should cover more than acts, that is, it should also cover measurement scales and so on. These artefacts do not fit well into the structure of Crawford and Ostrom's 'grammar' because it seems not to be part of their concept of an institution.

Also in the latter case – that of rules – I think there is a deviation. The concept of a 'rule' is not specific enough. In my mind the type of sanction, so important to this category, should also be signalled by the naming. However, while we use a somewhat different definition of the concepts, their 'grammar' is still useful for us in distinguishing between the categories.

There is another issue: the distinction between a norm and a legal relation, as defined above, is not as clear-cut as it may seem from the 'grammar'. As we have emphasized earlier, there may also be sanctions – that is, some 'Or Else' – involved in the case of a norm, even though it is not part of the defined norm itself. If a norm is not fully internalized – that is, not automatic – group pressure may still make people follow it. The unspecified threat may be reduced public standing or reputation. We may talk of an implicit, unformalized 'Or Else' in the case of a norm. So while fellow citizens sanction the norm, a third party with extended power to use force sanctions a formal or legal relation.

3.3 INTERESTS AND INSTITUTIONS

One important function of institutions is to protect interests. In a world of restricted and physically interrelated resources, there will always be conflicts over whose interests are to be protected. This is an important aspect of norm development, and it is the very core in the case of legal relations.

In the case of *restricted resources*, there is a question about who will have access to the resource, Alpha or Beta. What Alpha owns is not available to Beta. The situation may be changed if Beta owns another resource and Alpha agrees to do a trade. A resource may be restricted simply for physical reasons. The amount of water in a lake, the acres of land in a specific county and so on is given. Institutional arrangements define who has access, in what form and to what degree.

While scarcity is normally thought of as a simple relationship between the size of the resource and the number of users, one should note that scarcity might also be an effect of the institutional regulation *per se*. The size of the land, its productivity, may be sufficient to feed the whole population well. However, uneven distribution of the land and lack of purchasing power may still create scarcity for some. Thus, we observe that food is exported from areas where many people are starving (for example, Sen 1981) or water is scarce despite the fact that a better distribution system could avoid shortage (for example, Aguillera-Klinck et al. 2000). In the latter case the authors show how institutions may be used to create shortage to increase resource rents.

In the case of *physically interrelated resources* – that is, natural, or more precisely biogeochemical resources – use of different parts will influence the quality of others. The use of a parcel of land will influence neighbouring resources, such as a stream whose water partly comes from rain falling on this piece of land. The movement of wildlife may be influenced. Air quality may be affected over a wide region by emissions of, for example, ammonia. Genes from crops may mix with genes in the vegetation of the neighbouring fields, natural habitats and so on.

By treating land or other physical resources as property, one may certainly secure for the owner the 'fruits of own' labour (Locke). This is a positive and important aspect motivating increased productivity and quality of the owned resource for production purposes. However, there are also some problematic issues involved. First, the property solution is based on keeping other people out. This is no problem if there is an abundance of resources. If, however, resources are scarce, the very distribution of resources influences people's options. Second, the formal border established by the property institution does not necessarily constitute a physically strict demarcation. Rather, there will be many physical 'exchanges' going on, which the legal arrangement of ownership may not be able to cover or avoid. Gas emissions, soil erosion, nutrient leaching, moving organisms (macro and micro) and so on are difficult to regulate with the help of property rights. Conflicts may thus arise due to both immediate scarcities, and to 'spillovers' or 'external effects'.

The situations may vary substantially. In some cases the interrelations between people described above may be handled well by conventions. In other cases norms will act as regulators. Finally, it may be necessary to solve the problem by instituting a legal regulation.

To illustrate the above, consider an example of 'land development' in an area bordering a town. The city council has decided that the land should be sold to people wanting to settle there. Several people have handed in bids, and after a long process each plot, as demarcated by the authorities, has an owner. Certainly, much institutional development lies behind this defined starting point: a city council is set up; it is given the power to define how certain types of land should be used; it can buy land and sell it to people who want to settle there. Finally, those buying land must have the purchasing power to do so.

As the plots are distributed, people gather to decide how common issues should be addressed. Many questions need to be settled because of all the physical interrelationships involved: some produce little or no conflict; others are more difficult to handle.

3.3.1 Coordination Problems with Little or No Conflict

Let us assume that one of the first decisions to make is to choose street names. People realize that individual or personal naming is indeed impractical, and they very soon agree to a 'shared strategy' or convention, to a common system of street names and house numbering.

The debate may still not be easily settled. In a meeting, some may propose to use a system based on local, that is, old, names in the area. They refer to the fact that this is something that is often done. Others may argue that there are insufficient old names and they provide no common structure on which to build. Therefore it would be better to avoid a 'mess' by developing a new, more coherent system. These people agree to a suggestion that the roads should be named after the flowers in the area, which would signal a peaceful environment. Against this some argue that it is 'bureaucratic' to have one system – let the people settling in each road choose for themselves. However, the names have to be accepted by all, and there is no immediate agreement on this issue. In a second meeting, a large majority voice support for the 'flower idea', which is finally chosen by consensus. The main argument is that it is easier to remember each name if it is part of a larger structure; in addition, the 'flower idea' is supported as it creates a kind of identity that most inhabitants find they can identify with.

The issue of naming is a typical example of a coordination problem with little or no conflict. With other issues, the conflict might be greater. An alternative is then to move from the pure communicative solution resulting in conventions and perhaps rely on some kind of normative pressure, too.

3.3.2 Coordination Problems with Conflicting Interests, but Potential for Internal Solutions

Norms are positioned in the interface between self-restraint and coercion. As an example, the group of newcomers may be faced with the issue of accepting a particular style or colour for their house. This is taken up at one of the meetings, and arouses more intense conflict than in the case of naming roads. While the latter was about finding a practical solution to a common problem, people are now confronted with a situation where individual demands may be much more at odds.

Some argue that the houses should be in harmony. It would improve the quality of the area if the buildings were not only appropriate to the individual plots, but also fit together as a whole. They also refer to examples of specific, often older, villages where only a few materials, colours and forms have been utilized. In their mind, this constitutes important character and continuity. They also argue that there should be some specific restrictions on the height of each house due to the negative external effects of high houses on neighbouring properties.

Others argue that this is an issue which everybody should be free to decide for themselves. If someone wants a pink or a high house, why should someone else, who does not own the property, be allowed to influence that decision? By demanding a common set of rules, the opportunities for each individual are restricted. They also argue that if these issues are to be agreed by all, then it would be years before any construction could start and people would not be able to bear the cost of waiting.

The situation is quite serious. Let us envisage some possible paths to solve the problem: (i) 'preference alteration'; (ii) 'self-restraint'; (iii) 'side payments'; and (iv) 'coercion'. We shall start by defining the dilemma as set up in Figure 3.1. For simplicity, the newcomers are divided into two

		Individual type A: Cooperate	Individual type A: Defect
Individual type B	Cooperate	I — A: 100; B: 80	II — A: 70; B: 80
	Defect	III — A: 60; B: 100	IV — A: 65; B: 100

Figure 3.1 Preferences for cooperation versus individual solutions concerning housing rules with both cooperative and non-cooperative preferences

groups – those who want some common rules concerning the type of home (solution A) and those who support the right of each person to be free to choose (solution B). The reasons why people may support solutions A or B may vary. Some support A for aesthetic reasons, others for community reasons, still others do not want to make a choice themselves. Supporters of strategy B may be generally in favour of individual choice; they may more specifically support heterogeneity and so on.

In Figure 3.1 the positions of each group are presented via a typical representative. The figure gives the utility for types A and B – that is, individuals supporting solutions A and B, respectively. The type A position implies a preference for cooperation, but only if others also cooperate. If these do not do so, type A would also prefer to make individual choices: since the cooperative solution is not working, why not take personal advantage of that situation. Type B prefers individual choice independently of what the others might do.

If no common agreement is reached – that is, that everybody just chooses on the basis of the above payoffs – solution IV will be the result. Type B will immediately acknowledge that independently of A's choice, it is best not to cooperate. Type A will recognize this and realize that a non-cooperative solution is then best also for them. Solution IV is the outcome.[5]

Looking at the figures, however, we observe that total utility measured as the sum of A and B's utility, is greatest if solution I is chosen.[6] On the other hand, there is no way to move to this square without some type of action that goes beyond individuals choosing between the payoffs of Figure 3.1.

Changed preferences

One way to alter the conclusion is if preferences change. The group of people will normally not exist in isolation. We have already seen that they have meetings and discuss what to do. The result of these meetings may be a development towards some kind of consensus over what is best based on a process of argumentation as to what is the best solution. Type B realize that they do not really mind if they are not entirely free to choose themselves or if they have to refrain from making some choices. They may learn that the consequences are less problematic than believed. They may shift perspective and support the common solutions proposed, becoming aware that their position also implies that a neighbour may be free to erect a six-floor building, thus turning their own plot into a backyard. They may be persuaded that the others have a better argument. This situation is depicted in Figure 3.2.

Here, both types A and B in the end prefer the cooperative solution, and the move to this situation is obtained via communication and the associated learning. Type B have learned that the solution of square I is also best for them. Certainly, the opposite may also occur – that is, that the debate

		Individual type A	
		Cooperate	Defect
Individual type B	Cooperate	I 100 100	II 70 70
	Defect	III 60 90	IV 65 80

Figure 3.2 Preferences for cooperation versus individual solutions concerning housing rules where cooperation is preferred

in the group results in a situation where type A change preferences. However, this move does not result in a change in the solution compared to the one obtained in Figure 3.1.

Norms and self-restraint

The communicative process focusing on gains and losses may not necessarily result in changes in type B's preferences. The cooperatively minded people may have to accept that persons of type B maintain their prior preferences and vice versa. Given the structure of the problem, type A may, however, also argue that their gains from a cooperative solution are greater than the losses encountered by type B. This argument may produce derision from type B, whose interests are protected by the status quo. It may, however, also fuel an intricate debate over how to compare utilities across individuals.

If type A outnumber those of type B, the situation may change. There may be a majority norm to accept common rules concerning the height and colour of houses. In this case, type B may abstain from pressing forward a solution where everybody is free to choose. This may follow as a consequence of two different kinds of argument:

1. Type B may invoke or feel bound by a prior existing norm that they should not go against the majority. If most households prefer cooperation, they will abide by the majority decision.
2. Type B may become concerned about their standing in the community. People in favour of the cooperative solution may make complaints or

> type B may just sense that problems may occur. While type B's preferences concerning which choice is best for them is not altered – that is, they are as in Figure 3.1 – they expect the negative reactions to be such that it does not pay to stick to the original position.

While both arguments result in self-restraint – that is, to accept the cooperative solution – the type of arguments invoked in (1) and (2) are very different. The type (1) argument refers to a common norm of accepting a majority solution. The gains for the type B persons are not changed compared to Figure 3.1. They just accept that it is right to let this be a majority issue. What gives the highest personal utility is irrelevant for their decision. Another decision rule is invoked.

In the case of (2) it is the trade-off that is changed. The fear of being criticized or of becoming a 'bad neighbour' with lower social standing reduces the utility of defecting. If it is reduced to less than 80, we observe a switch to the cooperative solution on the basis of an individual calculation by type B persons. Note that there is an important distinction to be made here. While the fear of being picked on refers directly to own utility, the issue of social standing may go beyond the immediate perspective of a loss of individual utility. It may also refer to issues like self-respect or even the norm of social obedience, which goes beyond a simple utility calculation of pleasure and pain and actually takes us back to some of the reasoning around (1) above.

Side payments

There may be some who do not follow the majority preference. This may be accepted and we are back to the solution in Figure 3.1. There are, however, further options to pursue for the type A interest. Since type A gain more from a cooperative solution than type B lose, side payments may increase utility for both categories compared to the equilibrium in square IV of Figure 3.1. Let us simplify and not consider how many persons happen to be in each group – that is, let there be only person A and person B.

First, we encounter the problem of comparing individual utilities. So far we have implicitly assumed that a utility of 100 is the same across A and B. Such a comparison cannot be easily made (see Chapter 6). We are actually only able to produce a ranking of the options for each individual separately. If we want to compare them, we have to construct a numeraire into which both individuals can translate their utilities. Money is one such numeraire. Let us therefore shift assumptions and presume that the figures in Figure 3.1 are willingness to pay estimates.

If so, a person of type A is willing to pay a person of type B up to 35 monetary units to get from situation IV to situation I and still be as well off

		Individual A	
		Cooperate	Defect
Individual B	Cooperate	I *100* (65) 15 *80* (100)	II *70* *80*
	Defect	III *60* *100*	IV *65* *100*

Figure 3.3 The solution with side payments

(see Figure 3.3 where the italicized figures are equal to those of Figure 3.1). If A pays B 20 units to cooperate, B will be as well off as if s/he defected (see the figures in parentheses in square I). By doing this there are still 15 monetary units left which they could divide and both be better off than in the previous solution, square IV. This solution is what is normally called a Pareto improvement, that is, at least some gain and nobody loses from the new solution.

There are principally three very important and very different problems attached to this result, though: the first concerns the issue of utility measured in the form of monetary bids; the second concerns the rights distribution; and the third is about the effect of positive transaction costs. Concerning the first issue, B may argue that while the compensation covers what B loses by refraining from a non-cooperative solution, A is much richer, and it is still not fair that s/he can so easily buy the right to shift the rules. B at least claims to be compensated by being paid all the 35 monetary units that A gains by changed rules.

Concerning the second issue, A may not accept that B has the right to stick to the individual solution in the first place. Why should the non-cooperative solution be the reference point? S/he may claim that s/he has the same right to a cooperative solution as B has to the non-cooperative one. It should be B who pays A to move from the initial position which then is solution I in Figure 3.3. If B cannot come up with the necessary payments, solution I must be optimal. A may also consider it principally wrong to pay B because it would support, give legitimacy, to B's privilege as implicit in situation IV. This privilege has, however, not been granted. A concludes that to make a side payment is not wise. This reasoning may finally involve the thought that opening up for side payments will begin to spoil the community spirit. If everybody is free to claim payment for refraining from doing something that is considered a nuisance by others, it will pay to create nuisance and the solidarity of the community will erode.

Concerning the issue of transaction costs, if A accepts that the bargain has to start on the basis of the non-cooperative situation, s/he may observe that transacting with B may be so costly that the potential common gain of 15 is more than wasted. At least A has to consider whether this may be the situation and decide whether it will be worth while trying to strike a bargain. In a situation with two individuals, this may be fairly easy to figure out. If there are many actors involved, transaction costs increase. Uncertainty concerning their magnitude also increases. The chance of not obtaining a gain through bargaining becomes larger. If transaction costs are 20, they exceed the potential gain of trading. In other words, since A also has to cover transaction costs when approaching B, her/his maximum willingness to pay reduces from 35 to 15, which in the end is not enough to make B shift to the cooperative solution.

The first message delivered when studying our example was that institutional structures may influence preferences or motives and therefore which solution will be chosen. Studying the issue of side payments, we encounter the other basic message of this book. Both rights and transaction costs matter for what becomes an optimal solution. Side payments may work, but require prior acceptance concerning the rights distribution. In the case of environmental issues where so-called externalities are pervasive, this is the fundamental question. However, it is often overlooked, as rights are often thought to be implicitly defined by the status quo.

Transaction costs are very important in that they may block solutions which are otherwise sensible. What is costly for each individual to undertake, may be less so if done collectively. Thus, there are two reasons for supporting some kind of common institutional decision structures. First, we need someone to decide which interests should get the protection of the collective – that is, the basic rights distribution. Second, such structures may also be used to reduce transaction costs between rights holders considerably, implying as an example that many Pareto-irrelevant options are transformed into Pareto-relevant ones. This issue will be discussed more thoroughly, in Chapters 8 and 13.

3.3.3 Conflicting Interests: The Extended 'We' or Third-Party Solution

The situation described in Figure 3.1 is one of physical interrelationships where the choice of one by necessity influences the situation for and the well-being of others. If individuals of type B are free to do as they like, then type A will suffer a loss. If type A put pressure on type B so that they conform to the majority view, it is the latter who suffer a loss. An authority with the necessary power to define which rights should exist is indeed needed. If none of the above solutions work, the issue may be sent back to the city council for resolution.

The city council may decide in favour of the cooperative solution. In practice this may be instituted in the form of a *mandated solution*. This implies in this specific case that everybody has to produce a plan for the building they want to set up. The plan will normally be made public. A civil servant or a committee will evaluate it and check that certain predefined rules concerning the construction of new houses are met. Neighbours are given the opportunity to make formal complaints.

An alternative to the mandated procedure is to institute a system whereby people who do not want to follow the rules are taxed according to the nuisance they create. The right structure is the same in both cases. It is the mechanism that is different. Both solutions are anchored in the law.

The city council may also rule that everybody has the privilege of developing their own plot of land as they want. This turns the rights structure upside down compared to the previous solution and it supports the 'unregulated' result of Figure 3.1. Granting a privilege of this kind is still a rare exception if we look at what is practised in different parts of the world. This, I believe, follows from the fact that there are strong reasons why most city councils or national legal systems grant some rights to the collective in such cases. It is simply because the privilege of one in a case like this is also a privilege for everybody else. Then the privilege actually erodes since the other side of the 'privilege for everybody' coin is a 'no right for all' or 'open access'. If I do not want the neighbours to build a high house, I must by mere consequence undertake not to build one myself.

This is the fundamental logic underpinning zoning laws. By putting similar activities together, conflicts are reduced. Manufacturing may produce noise and much heavy traffic. Located together, and thereby detached from housing areas, the negative consequences are minimized. This is the same with shops, restaurants and so on.

How then do such institutions appear to us some time after they are set up? Well, mainly they appear as a given constraint! Since we, the latecomers, did not participate when the institutions were set up, we may not see that they *both restrict and liberate*. For example, we may move into the above described area of house construction many years after the rules concerning building homes were set up. We may buy one of the houses because we believe that establishing a restaurant here would be a good idea. There are many people around whom we believe would like to go out eating and dancing in their neighbourhood. Starting the process of rebuilding the house and setting up a car park we realize, however, that this is not going to be easy. The city council planning office informs us that we cannot do this. The area is regulated for housing. We argue fiercely that the area will benefit from having a restaurant and accuse the office of being a bureaucratic organization obstructing free enterprise.

We have returned to the point made by Berger and Luckmann (1967). When a system, an institution, is set up to solve a problem, it is the rule not the arguments behind it that survives over time. Those encountering the rule at a later stage will often be unaware of its history and rationale. The conflict between A and B was resolved by the city council and a regulated system was set up. What we meet many years later is only 'the system' and we are deluded into believing that it is the bureaucracy that is against us. Nevertheless, the basic conflict is between us and the others living in the neighbourhood into which we have moved intending to set up the restaurant.

We observe this in many situations. Typically issues like smoking regulations, reduced speed limits, stricter laws concerning driving and alcohol use and so on are often seen as the authorities versus the liberties of the common man. Certainly it is not. It is a conflict between those who want to drive as safely as possible and those looking for speed and excitement. The authorities are a sort of 'extended we', constructed as a third party with the power to adjudicate in conflicts among the citizens.

Certainly, regulation systems may fall out of step with the situations they regulate as these may change. Officials may also execute undue power. They are not an 'extended we', but are running their own agenda. In the case of our neighbourhood, the situation may have developed so that people now might accept the establishment of a restaurant. Peace and quiet is less important than it was in the beginning. Furthermore, giving planning permission for one restaurant does not imply a general acceptance that everybody can transform their property into a noisy business. Here we encounter another issue, the rigidity of institutions which in some situations may obstruct solutions that are acceptable at a later stage. However, this is something very different from claiming that the conflicts are basically between the individual and the state/city council. Rather, they are about which interests should be protected by the collective of citizens.

3.4 THE PROTECTION OF RIGHTS

In the above discussion we have learned that inhabitants of a certain neighbourhood may have been granted a right to be protected from the nuisance of high houses or noisy establishments. We have also seen that the authorities might have decided otherwise and given a privilege or liberty to erect whatever building one may like. We also observe that in our case it was the city council which had the power to decide in such issues. Others were obliged to comply with that decision.

A right to a piece of land or to sunshine or to a quiet neighbourhood must be protected if it is to function properly. This protection will normally work on different levels. A high level of local acceptance of rules and rights largely creates a self-policing environment. People will normally abstain from causing what is considered to be a nuisance. Those who still violate the rules will have to face the reactions of the people living there. Even legally conferred rights may, however, be broken, and a formal system for handling such situations is needed.

According to Bromley (1989), such a protection may take three different forms. First we have the protection given by a *property rule*. In this case the party wishing to contravene a right held by somebody must initiate a bargaining process with the rights holder before any interference occurs. In our case, those wanting to set up a restaurant have to negotiate with the people in the neighbourhood to see if they accept an offer to cover the nuisance that its establishment will cause. The person contravening the right must carry the costs incurred by the bargaining – that is, the transaction costs. A property rule involves *ex ante* acceptance by the rights holder.

In other cases the problem is of such a character that using a property rule may be considered very impractical. Transaction costs will simply be too high. As an example, constructing a building in the area implies transporting much material to the site. Doing this involves some risks. A truck may end up in the garden of someone because the brakes failed, or, in the process of building the foundations for a new house, someone may destroy an existing pipeline. In such situations we often observe that a *liability rule* is in place. This implies that the company transporting, in our case building material, has to compensate for any damage they may cause.

A liability rule is typically used in cases where there is a risk of some damage occurring, but it is impossible to say where and when. Transport, not least by sea, is a typical case. The number of property owners along a coast is normally very high, and *ex ante* negotiation would actually make such transport more or less infeasible. Transaction costs would be insurmountable, and *ex post* compensation has become the standard.

Finally, we have the *inalienability rule*. In this situation, transacting is blocked. One is not allowed to interfere with the owner under any circumstance. Similarly, the owner is not free to sell. Both *ex ante* (property rule) and *ex post* (liability rule) bargaining are prohibited. Bromley (1989) cites the ban of some toxic chemicals as an example. While transaction costs seem to be important for the choice of property or liability rule, the perceived seriousness of the problem influences whether an inalienability rule becomes established.

3.5 COERCION, FREEDOM AND INSTITUTIONS

There is a tendency, not least by many economists, to view market transactions as free and uncoerced while collective choices coerce people. Friedman (1962) claims that markets in themselves are free: they both constitute freedom and are an important base for political liberty. Bromley (1989: 65) remarks:

> This connection by the market and freedom is said to be established by private enterprise and the fact that individuals are free to enter into any particular exchange. Freedom of the individual to deny any particular exchange is seen by Friedman as insuring [*sic*] maximum freedom for the individual . . . A related position, most often espoused by Buchanan, is that collective action implies political externalities unless it is accompanied by Wicksellian unanimity. . . . A careful assessment will reveal, however, that there is no logical support for the familiar proposition that markets are coercion free while non-unanimous collective action is coercive. Both markets and collective action simultaneously constrain and liberate the individual.

There are several issues involved here. First, we have the fact that no institutional structure, be it a market or a system for collective decision making is coercion free. Second, given that markets exist, are people really equally free to enter into any particular transaction? Third, we have issues related to the fact that in a world of physical interconnections, we intervene by necessity into each other's lives via the choices we make. We shall discuss these issues in turn.

For a market to exist, several rules are needed. Markets are social constructs dependent on defining an initial distribution of rights not least over the physical and biological resources that sustain our way of life. This distribution is a coercive act in that the right of one implies no access by others. Certainly, the system of distribution may vary from society to society. In some countries everybody has access to necessary resources for a high standard of living, in others this may not be the case. None the less, the point made here is that whatever distribution there is, coercion is necessary to establish that structure.

Furthermore, many arrangements found in markets have to involve decisions that are rarely unanimous. For example, paying for the necessary court system and police force to handle rights violations; the construction of common product controls which some may want due to the high transaction costs involved in instituting the control individually; the establishment of contract law which is equally important, but which could be formulated differently and hence defend interests differently; and so on.

If we move to the second element, whether we are equally free to enter into any particular transaction, we observe that the initial distribution of resources or wealth is of great importance. Viewed formally, the Friedman position is right. That is, as soon as a market exists with all its structures, there is nothing in the institutional set-up that forces anybody to make a specific transaction. Everybody is equal in that respect. You are free to buy a blue or a red shirt, eggs from conventional farming or from organic production and so on. Nevertheless, formal equality does not imply actual equality or freedom for all. If the market for organically produced eggs is small, you may not be able to choose this product. Such eggs are not widely available due to the cost of their supply. If everybody else had the same preference as you, the situation would have been different. Thus, you are often not free to satisfy 'rare' preferences.

The basic issue still lies elsewhere. Bromley (1989: 66) emphasizes:

> The matter here concerns the logical ability to affirm individual freedom (the absence of coercion) by the mere fact that I can choose to avoid any particular transaction (the purchase of toothpaste). Macpherson would argue that freedom is present when I have the opportunity to avoid all transactions. Not just any particular transaction. To the extent that the rich have more choice in avoiding certain transactions – such as hiring their labour out to owners of capital – then they are less coerced than the poor.

We may take this even further and follow Commons ([1924] 1974) and Macpherson (1973), who maintain that freedom also concerns the ability to understand and develop those areas of opportunity on which one depends.

The formal equality of trade often makes people unaware of the coercion involved. The poor Indian farmer who every year runs short of rice some months before harvest and must borrow to be able to sustain his family is free not to borrow in formal terms, but not in real ones. Whether he turns against the system or accepts his situation to be a 'natural' one, depends not least on how well the initial distribution of access to land is legitimized, and what kind of understanding or consciousness prevails concerning the existing structure.

What we observe here is that coercion may become invisible because it is concealed in the structure of historically defined institutions defining rights and access to resources. No physical or open power is executed even in the Indian example. It is built into the structures. We may call this 'systems coercion'. Building the necessary coercion into the institutional structures is a necessary for any system not to be constantly exposed to violent conflict. Nevertheless, there is coercion. The question is rather: what, why or who do we coerce? These are issues that lend themselves to critical reasoning.

Finally, in a world of physical interconnections, the freedom of A is always a restriction on B's possibilities. This is exactly what we experienced in our previous example of the settlement area and the conflicts concerning what rights people should have when choosing building projects. If people do not have equal preferences in such cases, no unanimous consent can, as we saw, be obtained. Whether market or non-market, is immaterial. Informal social pressure or the use of formalized power by the city council constitute different types of visible coercion. However, the situation where everybody does as they like, is not free of coercion either. The physical interdependencies dictate that coercion will have to be involved. As in Figure 3.1, the 'non-regulated' situation gave type B the opportunity to coerce type A individuals.

The difference between the Hobbesian war of all and the state of ordered relations, as in a market or in our community example, is not a world of coercion and fear to be compared with a world of unlimited freedom. It is rather that the freedom we grant each other builds on restraint or coercive acts. The question is not primarily about coercion versus freedom, but about which coercive acts and which interests we defend, so that these interests may thrive.

There is a strong tendency in the literature to associate coercion with 'bad will'. As an example, Hayek (1960) makes his main distinction between coercion, which is a constraint put on somebody by someone else, and physical circumstances, which is not coercion since it is something we cannot avoid. In this dichotomy between coercion as acts of will and mere 'physical circumstances', there is a danger that the power relations built into the rules of a system – that is, the law of property, the market and so on – become associated with 'physical circumstances'. I shall close this chapter with a very instructive quotation from Bromley (1989: 67; original emphasis):

> If A *wills* some restraint on B then that would comprise coercion. If conditions are such that A can behave (in my terminology A has a *privilege*) in a manner that is seriously detrimental to the interests of B – but is oblivious to B's suffering, or absentmindedly harms B – that is not coercion. Those who defend markets and the status quo would suggest that when B seeks relief from this intolerable situation the *presence of will on the part of* B, coupled with B's necessity to seek some official sanction to be relieved (usually in the form of government action), comprises the essence of coercion. For if B had only the *will* to alter A's behaviour,

> and rather than relying upon the state had attempted to bargain with A over the interference and had failed, then the status quo would be reaffirmed as efficient and B would simply be out of luck; the freedom of the market would be confirmed. As a defence of minimal government and *laissez-faire*, Hayek's selective perception of coercion seems purposeful – if not very logical.

Thus, if A emits a pollutant and B, who suffers from it, is not able or willing to pay what is necessary to reduce or stop this activity, the situation is optimal and should continue. If B goes to the government and asks for relief, it is to ask for coercive acts. However, this is not a question of a coercive versus a non-coercive solution. Both situations involve coercion and the issue is which of the interests the collective chooses to defend.

3.6 SUMMARY

In this chapter we have presented the definition of institutions on which this book is based:

> Institutions are the conventions, norms and formally sanctioned rules of a society. They provide expectations, stability and meaning essential to human existence and coordination. Institutions regularize life, support values and produce and protect interests.

The various elements of this definition – the conventions, norms and formally sanctioned rules – are understood as responses to various types of problems. First, we have the fact that both the natural and the social worlds are complex. Second, we have emphasized that our actions have interrelated consequences. Actions by one person influence the possibilities for others. This has given rise to the analytical structure as shown in Table 3.2.

Table 3.2 Institutions as responses to different problem situations

Problem	Consequence	Type of institution
Complex world	→ need for coordination	→ conventions
Interrelated actions type I: interests can be harmonized	→ potential for creating common values	→ norms
Interrelated actions type II: interests cannot be harmonized	→ need to regulate conflict	→ formally sanctioned rules

The concept of a norm overlaps the other two. It resembles that of a convention in that it is dominantly developed from below – from within the civil society. It is, however, different, since it is not a solution to a mere coordination problem, but defines and supports a certain value that is a solution to a potential conflict. On the other hand it resembles a formally sanctioned rule in that a sanction is also a potential reaction if a norm is contravened. Nevertheless, to survive as a norm, this sanctioning from below is sufficient to keep the norm viable. In the case of a formal rule, the conflict potential is stronger, and/or the cost of sanctioning is beyond that of the civil society. Third-party regulations – that is, state regulations – are necessary.

We have illustrated the various dimensions in this with an example of a situation, the establishment of a new housing area, involving different problems from that of mere coordination to serious conflicts. In the case of conflicts, we have seen how changed preferences, invoked norms, side payments and public (formal) regulations can all be involved in defining solutions. In particular, in the case of side payments and public regulations, we also saw that issues concerning both rights and transaction costs are of importance for which solution becomes the chosen one. Hence the three basic issues raised in Chapter 1, concerning the effect of institutions on individuals' motivation, the effects of the rights structure, and finally the transaction costs faced by individual agents, are all shown to be important for the chosen solution to a resource allocation problem.

The definition and protection of rights is a core issue. The right of one is the duty of others. This means that any rights structure implies both freedom and coercion. This holds for markets as well as for collective action. Rights can be protected with a property, liability or inalienability rule. While transaction costs seem to be important for the choice of property versus liability rule, the perceived seriousness of the problem seems to play a significant role in the establishment of an inalienability rule.

The fundamental aspect of rights is how they distribute access to resources and which interests they protect. In the case of environmental resources – that is, of physical interconnectedness – the freedom of one person will always imply a restriction for others. There is another duality in this. As institutions regulate conflicts, they also tend to normalize them, or make them 'invisible'. The fact that some have much and others have little may seem to be the 'natural order of things', not an effect of the chosen and protected rules.

NOTES

1. The distinction between individual and social norms is sometimes made in the literature. Individual norms are rules that individuals formulate for themselves, while social norms

are those that are learned. I do not deny that individuals may formulate their own norms, but it is rare to observe norms that only one person holds. I do not focus on this distinction because to become an institution, a norm must be socially constructed – that is, reciprocally typified. Individual norms, if really purely individual, are not institutions.

2. One should be aware that while the law is the formal sanctioning system in societies with a state as the top political level, other types of third-party regulation may be observed as the 'council of the elderly' and so on in societies that do not have state structures. While the distinction from a norm may be less pronounced in this case, in principle that is also a type of third-party structure.
3. Certainly, if you are trained into a regular way of doing something, a habit, by just watching someone else, it is still an institution, but it is not internalized via the use of the spoken word. A typical example of this is the way an apprentice learns directly from the master. At least not all conventions are transferred via the use of oral mechanisms.
4. From deontic logic.
5. In game theory this solution is called a Nash equilibrium. It is the solution obtained if all players – in one-shot games – play the strategy that is individually the best.
6. This assumes, however, that the utilities of A and B can be compared. I shall return to this.

4. Institutional economics: different positions

After the neoclassical 'revolution' in economics ended, by the late 1930s (see Chapter 2), the interest in institutional phenomena waned. However, a strong revival of interest in such issues can be observed from the early 1960s, with substantial growth in the last 20 years or so.

In some sense, the focus on institutions is an old issue in economics. The writings of the classical economists of the eighteenth and nineteenth centuries contain elements of an institutional nature. This is due to the fact that these authors were much interested in the organization of the economy. With the development of neoclassicism and its more abstract schemes of costless exchange and maximization of individual utility, institutional issues became unimportant. American or classical institutionalism, the first tradition in economics specifically focusing on institutions, was developed as a reaction to the trend of neoclassicism. It seems to have started with Thorstein Veblen's famous paper 'Why is economics not an evolutionary science' (1898). Veblen challenged the contemporary tendency to make economics the study of abstract equilibrium ideas based on individuals with fixed preferences. In the decades following, institutional thinking attained a dominant position among American economists, a position it retained until the Second World War (Hodgson 2000). It was a somewhat heterogeneous tradition, though, which may explain some of its mixed success thereafter.[1]

From the 1960s, when there was renewed interest in institutional issues in economics, it is notable that people with a neoclassical orientation entered the scene. We observe the birth of 'new institutional economics' (see also Chapter 2). Basic to the neoclassical tradition is voluntary exchange. This exchange, however, has to be based on a set of predefined property rights, and some economists questioned how these rights have evolved and which rights structures are the most efficient. Second, many economists have observed that the economy is not costless to run. Transaction costs are pervasive and may explain the fact that not all transactions are undertaken in market institutions, as envisaged by the neoclassical model, but by command structures like firms and the state. Standard economic theory, however, had no explanation for this.

Possibly as a result of this trend, we can also observe a renewed interest in classical institutionalism, which is a reaction to the rather narrow understanding of institutions taken up by the 'new' institutionalists. These modern 'classical' institutionalists are not only influenced by Veblen and his contemporaries. They base their ideas on a more modern social constructivist perspective and have refined the thinking of the role of institutions based on this view.

There is also a third position that is of interest to us. The 'institutions-as-equilibria' stance has cultivated the idea of the independent individual to a greater extent than the new institutional economists have. They argue that even institution building is a market process. These authors reduce all institutions to mere conventions based on a kind of market selection. They try to build a theory of institutions where everything is really market, or with as minimal a role of third-party engagement, state control, as possible.

In Chapter 2 we defined the *core* of neoclassical economics consisting of: (a) rational choice as maximizing individual utility, (b) stable preferences, and (c) outcomes as equilibrium states. We furthermore defined the *standard application area* to be: (a) no information costs, (b) no transaction costs, and (c) private property rights for all goods which are exchanged in competitive markets.

Given these assumptions, the economy can run without any institutional structures other than private property rights. However, if transaction costs are zero – that is, in a world of full information (no uncertainty), with no costs of policing contracts and so on – it is impossible to differentiate between any institutional structures concerning their efficiency. As already emphasized, competitive markets, oligopolies, monopolies or even planned economies will under these circumstances give the same results concerning resource allocation (Williamson 1985). Private property can be replaced by other assumptions about the structure of the economy without changing the motion of the system. On the other hand, as soon as transaction/information costs are accepted as positive, institutional structures (such as property rights structures) matter. Then many problems concerning the consistency of the model also appear.

To structure the presentation of the different positions, I shall utilize the above definition of the neoclassical model and describe various institutional perspectives on the basis of the way they make changes in the core or application area of that model. I shall start by presenting the stance of the new institutional economists, with Douglass North and Oliver Williamson as representatives of important models (Section 4.1). I shall then cover the 'institutions-as-equlibria position' (Section 4.2), before I turn to the classical stance (Section 4.3). Here I shall cover both some 'old' and some contemporary positions within this tradition. Since the issue of institutions is very much about the theory of authority and the role of the state, I will close the chapter by discussing the main understandings of the state as they appear in the economics literature (Section 4.4).

I should like to emphasize that the landscape we are now entering is a complex one. There are many different perspectives appearing in the literature. I have put much effort into simplifying and structuring so that the 'core' positions become as clear and 'pure' as possible. This may have a cost since variations within positions, and the tendency of various authors to 'move across' stances, may become underemphasized.

4.1 THE NEW INSTITUTIONAL ECONOMICS

While we have already divided institutional economists into three main strands, I find it necessary to also split the position of new institutional economics into three different sub-branches: the property rights view, the transaction costs school, and the specific position of Oliver Williamson.[2] There is a lot of common ground covered by the three. They are all based on the idea that institutions are external constraints – the 'rules of the game'. They are moreover all strongly inspired by the neoclassical model. When differentiating between them, it is therefore fruitful to make distinctions on the basis of how each tradition positions itself in relation to that model.

The *property rights school* can be taken to simply claim that neoclassical economics is not consistent in using its own assumptions concerning the core and standard application area. It is argued that in a world of zero transaction costs, no public policy is necessary. None the less, neoclassical economics has developed several subdisciplines concerning different policies for the allocation of public goods – for example, health economics, and resource and environmental economics. The protagonists for the property rights school argue that this is unnecessary, since with zero transaction costs, all resource allocations can be made via individual bargains. The *transaction costs school* takes another route. It positions itself by studying the effects of accepting positive transaction costs. Finally, the *Williamson tradition* goes one step further and also makes a change in the core by suggesting that humans are not fully rational, only boundedly so. He actually takes on board the full consequence of accepting positive information and transaction costs.

4.1.1 The Property Rights Position: Accepting the Neoclassical Model as It Is

Much of the basis of the property rights position is found in the writings of Coase (1960), Alchian (1961), Demsetz (1967) and Posner (1977). Basically this stand can be viewed as an attack on neoclassical welfare theory for not treating its assumptions in a consistent way.[3] Normally, representatives of neoclassical welfare theory would support 'state intervention' to secure the production of public goods (defence, education and so on) and to correct for externalities in the economy (for example, pollution). Ever since the work of Pigou (1920) it had been standard for neoclassical economists to argue that if the activity of one agent influences other agents without these latter agents being compensated (externalities) resources would not be optimally allocated. Put into the neoclassical model as defined above: if some goods are not owned, if they are not commodities, then resource allocation will not be optimal, and some state regulation is needed.

Demsetz (1967) argued that this reasoning was flawed. If some transactions did not appear, it was because it was optimal not to transact (compensate). He states:

> [T]he emergence of new property rights takes place in response to the desires of the interacting persons for adjustment to new benefit–cost possibilities . . . property rights develop to internalize externalities when the gains of internalization becomes larger than the costs of internalisation. (Ibid.: 350)

Thus he actually accuses standard economic theory of not being consistent, given its own assumptions. When preferable, private property rights to resources will emerge. Given that these rights exist, individual resource owners will bargain over the effects of physical interrelationships, as in the case of pollution. If the gain of the factory owner by emitting is higher than the costs experienced by the owner of, say, a receiving river, emissions should take place. If the benefit–cost ratio is otherwise, there should be no (or fewer) emissions and the private agents will also reach this conclusion via private bargains. There is no need for 'state intervention'.

Actually, Demsetz's position merely emphasizes the various assumptions underlying the standard economic model. Assuming rational agents with given preferences and zero transaction costs there are no problems for society to handle. Private property will solve any allocative concerns. Bromley (1989) suggests that the property rights tradition has a strong ideological bent towards private property and, for example, wrongly associates the concept of common property with open access. For the issue of different property rights systems to become interesting, however, we have to assume positive transaction costs.[4]

4.1.2 The Transaction Costs School: Accepting Positive Transaction Costs

Changing the application area to include positive transaction costs is the dominant trend by neoclassically orientated institutional economists.[5] When I call these institutionalists 'neoclassically orientated', it is because they tend to accept the core of this tradition.

The basic issue is simple. If it is costly to transact, market exchange may not be the least costly way to solve a resource allocation problem – be it the problem of allocating inputs such as land, labour and capital for different uses, or the exchange of commodities. This idea goes back to Coase (1937), where he asked the following question: if markets are favourable, why do firms exist? Firms are command structures – that is, the negation of voluntary exchanges. The proposition made by Coase is that the costs of exchange may in some cases be so high that everybody is better served by a command structure. Producing a car on the basis of selling and buying parts among all producers/workers involved is more costly than to join the same firm and manufacture a car under the authority of the firm's management.

North and Thomas (1973) took Coase's ideas one step further, claiming that the development of an economy depends on its institutional structure and that the trade-off between transaction costs and the establishment of property rights structures is the core issue:

> Economic growth will occur if property rights make it worthwhile to undertake socially productive activity. The creating, specifying and enacting of such property rights are costly . . . As the potential grows for private gains to exceed transaction costs, effort will be made to establish such property rights. Governments take over the protection and enforcement of property rights because they can do it at a lower cost than private volunteer groups. (p. 8)

Thus the existence of the state is also understood on the basis of its ability to reduce transaction costs – that is, its capacity to protect and enforce property rights. North stresses the important role of the state as the ultimate source of coercion. His point is that a theory of institutions inevitably involves an analysis of the political structure of a society and the degree to which that political structure provides a framework of effective enforcement.

North strongly emphasizes that the major role of institutions as 'the rule of the game' is to establish a stable structure for human interaction. Institutions reduce uncertainty. They also make it possible to capture the gains arising from specialization and division of labour since, for example, contracts define who is to do what for what compensation. The building of trust implicit in these arrangements is also of great importance. There are many important insights in this reasoning. Nevertheless, the strong focus on efficiency and the fact that North looks upon institutions as mere constraints is a weakness (see also Chapters 1 and 2). Actually, new institutional economics sees no relationship between institutions and the constitution of the individual *per se*.

The basic idea is that building institutions is a way to economize on transaction costs. While these are assumed to be zero in standard neoclassical analyses, Wallis and North (1986) estimated the resource use of the private and public transaction cost sectors to cover over half the gross national product in the economy of the United States in 1970. The transaction costs borne by individuals when searching for information, doing shopping and so on were not included. Transactions within firms, however, were included: measured as a fraction of all costs, the amount had approximately doubled since 1870. This indicates that transaction costs economics is important. The increase in the level does not imply that transactions have become less efficient over time, rather the opposite. Since these costs are reduced per unit transaction, it is advantageous to undertake more transactions. The increase in aggregate transaction costs is rather an effect of the fact that economies grow and differentiate. As we specialize, we trade more and the costs of transacting become relatively more important than the costs of producing. We see that markets are not a free good.

Eggertsson (1990) put emphasis on three characteristics of the tradition that we here call 'transaction costs economics'. The basic idea is to involve positive transaction costs. The authors of this tradition also focus explicitly on constraints like rules and contracts that govern exchange. Finally the standard assumption that commodities have only two dimensions, price and quantity, is changed to allow for variation also in quality. Variation in quality increases information and transaction costs as it makes it uncertain what is purchased – compare the difference between buying a box of (standardized) nails and a piece of meat. The issue becomes even more important if it is, say, a health service that is to be bought: it may be very difficult for the 'customer' or 'patient' to evaluate the quality of the service. In the latter case it is also an issue that the quality of the good is not known until the service is given.[6]

The transaction costs theory has therefore been used not only to explain the existence of authority structures like the firm or the state, but also to study the economics of information, the various institutional structures around the markets for goods of various complexities (qualities) and so on.

There are certainly many insights to be obtained in studies of this kind. We have already stressed some of these in our focus on the importance of institutions in creating order and making coordination simpler.

There are two problems, though. First, there is a tendency to view any institutional structure as a solution to a pure coordination problem, and very often the prevailing institutional setting is regarded as an efficient solution to that problem. Otherwise it would not exist. This cannot be taken as given. Institutional structures may also be based on the execution of power and the protection of certain values or interests (Bromley 1989; Pitelis 1993).

Second, accepting positive transaction costs makes it difficult to defend the core assumption of rational choice as maximizing. If it is costly to gather information and to transact, it becomes impossible to define what is an optimal bargain. It is simply not known when the optimal amount of knowledge about the market is obtained. We shall return to this issue in more detail later, especially in Chapter 5.

Eggertsson (1990) strongly supports developing the research programme of transaction costs economics, leaving all core assumptions of the neoclassical model unaltered. He particularly emphasizes the importance of sticking to the rationality assumption. He argues that this will produce the most productive hypotheses. While I agree that building on the hypothesis of rationality as maximizing at least makes it easier to produce formalized hypotheses, I believe it is difficult to support a programme that starts out from internally inconsistent presumptions. While Eggertsson claims that North is supportive of such a programme, we observe that he, at least in his recent writings, accepts that people are not maximizers. They are rather boundedly rational (North 1990).[7]

4.1.3 The Williamson Position: Accepting Bounded Rationality

While North seems to have partly taken bounded rationality on board over the years, Oliver Williamson is well known for having built this assumption into the centre of his research programme early on. Turning to Williamson and his work in industrial organization, we therefore observe two changes from the standard set-up of neoclassical economics: the inclusion of bounded rationality (core) and the acceptance of positive information and transaction costs (application area).

Williamson's focus is mainly on different types of contractual arrangements and business structures under capitalism. Rather than viewing the firm as a production function,[8] as is standard in neoclassical expositions, it should be regarded as a governance structure (Williamson 1985). He suggests that the main purpose of the institutions of capitalism is to economize on transaction costs. He uses this to explain the development of the modern corporation, why we observe vertical integration in some sectors and not in others, why we find different franchise structures and so on.

Concerning transaction costs, Williamson's reasoning is quite in line with that of the transaction costs school. However, he takes the reasoning one step further. Given positive transaction costs, it is impossible to undertake all bargaining prior to the contracting, at the *ex ante* contracting stage. Many aspects of the good to be delivered or the costs of producing it may be unclear or impossible to define with enough precision at the time the initial contract is written – for example, the delivery of parts for the construction of a car or a computer network. The uncertainties involved may make it reasonable to guard against future problems by choosing governance structures such as vertically integrated firms.

Williamson's point that people are both boundedly rational and opportunistic is important in relation to this. Bounded rationality implies that decision makers do not optimize. They try instead to reach defined targets. This may be viewed as a way to circumvent the information problem inherent in neoclassical economics – that is, the problem of defining what is optimal to do when information is costly.[9] Furthermore, given positive information and transaction costs, opportunism may flourish.[10] Williamson also focuses on variations in the characteristics of different goods, what he calls their 'asset specificity', and the importance of the frequency of a certain transaction. High asset specificity, which for simplicity we can view as a low degree of standardization,[11] makes it more demanding to specify the qualities to be delivered and the higher the transaction costs will tend to be. Parallel to this, the greater will be the gain of a merger between firms as compared to transacting in markets, since the need for a specified contract is avoided. High frequency reduces transaction costs as it also develops trust through increased contact. Then ordinary market transactions may work, while other contract forms may develop if frequencies are low.

Concerning the concept of bounded rationality, Williamson draws on the work of Simon (1957, 1959, 1979). The position is a type of hybrid between the 'economic man' of neoclassical economics and the 'institutional man' of social constructivism. Institutions like 'rules of thumb' and other bounded decision algorithms simplify decisions and support decision making. A carpenter may not know the exact optimum for which nails are right to use for a specific part of a wooden construction. He just follows existing rules and thereby obtains a satisfactory result.

However, the various rules are not accepted as influencing the personality of the 'bounded man'. What the model of bounded rationality does, is to focus on institutions that can help humans to better handle the fact that they are not all knowing. I find this move, while limited, important not least because it is consistent with the focus on positive information and transaction costs. While Williamson in his later writings (for example, 2000), also touches upon issues such as culture and social embeddedness – the social capacities of institutions – his own focus is still on the bounded abilities of *independent* individuals.

4.2 THE 'INSTITUTIONS-AS-EQUILIBRIA' POSITION

Basically this position builds on the assumptions of neoclassical rationality, but it is even more individualist in that it denies a role for the collective and for any intentional creation of institutions. According to this stance, institutions are 'equilibrium strategies of the players in a game'. Institutions are *spontaneously formed.* Aoki (2001) is a core representative of the 'institutions-as-equilibria' position. He contrasts it with the 'rules of the game' theories – for example, North – as seeing institutions as consciously designed by, for example, the state. The 'institutions-as-equilibria' position, however, looks at institutions as a result of spontaneous emergence, 'a convention of behaviour [that] establishes itself without third-party enforcement or conscious design' (Aoki 2001: 7). It is assumed to be a solution supported by everybody, not 'forced' by the state, or any other third party.

Aoki thus defines an institution in the following way:

> An institution is a self-sustaining system of *shared beliefs about how the game is played.* Its substance is a compressed representation of the salient, invariant features of an *equilibrium path,* perceived by almost all agents in the domain as relevant to their own strategic choices. As such, it governs the strategic interactions of agents in a *self-enforcing* manner, and in turn reproduced by their actual choices in a continually changing environment. (Ibid.: 185; my emphasis)

Important references that can be positioned under this tradition in addition to Aoki are Hayek (1973, 1988), Schotter (1981), Sugden (1986) and Sened (1997).

It is standard for this tradition to focus on institutions as conventions. Sugden (1986: 132) defines a convention as 'any stable equilibrium in a game that has two or more stable equilibria'. Shaking hands has two possible equilibria, both individuals using the left or the right hand. The convention of using the right hand solves the problem by choosing one of the two. In this respect, the position is not much different from what is said elsewhere about conventions, and restricted to this level it offers interesting perspectives. The problem is that the authors of this tradition insist that *every institution is such a convention* or spontaneously developed 'equilibrium'. This is the source of several problems.

First, if there are issues where an agreement by all does not exist – that is, a situation with conflict – a legal regulation made by the state/a third person is needed (see the discussions in Chapter 3). The existence of such regulations is rather pervasive, not least in modern economies. How can the supporters of the 'institutions-as-equilibria' position explain this fact? Well, such observations may just be dismissed as illegitimate. Sugden (ibid.: 5) solves the problem by arguing that legal arrangements merely formalize 'conventions of behaviour that have evolved out of essentially anarchic situations . . . [and] reflect codes of behaviour that most individuals impose on themselves'. According to this, formal institutions also grow spontaneously out of tradition, supported if not by all, at least by 'most individuals'.

Aoki takes a somewhat different route, claiming that 'statutory laws or regulations may induce an institution to evolve, but they themselves are not institutions' (2001: 20). Removing the problem by just defining the law as non-institutional must be considered rather simplistic. He is somewhat more eloquent when he suggests that every researcher in his study has to construct a distinction between existing rules that are exogenous to the game and those evolving from the game. In Aoki's mind one should always formulate the inquiry so that the evolution of institutions – institutional change – is viewed as an endogenous process.

This could be meritorious, but not in this case since the definition of an institution as something evolving from below (or spontaneous) is secured by merely excluding 'institutions created from above' from the study. We need go no further than North to find the idea that the state is an endogenous and intended solution to the problem of establishing an efficient enforcement mechanism in a society. There is nothing implying that an endogenous solution demands something 'from below'.

A third alternative for the 'institutions-as-equilibria' position is to try to avoid the state or third party altogether. This is the route mainly travelled by Hayek. Hayek's stance[12] is strongly characterized by the idea of spontaneous order – that is, unconscious or unplanned order and subjectivity (for example, Hayek 1948, 1967, 1988). According to Hayek, all information is individual specific – that is, subjective – and cannot be fully communicated. Using a metaphor based on biological selection, he envisages institutions as selected on the basis of their capacity to foster human survival. While this is an interesting thought, there are again several problems involved.

First, there is no simple way to ascertain that the spontaneous development of institutions by 'necessity' creates order. There is a substantial debate about this, which is well covered in Hodgson (1996). Hayek makes many references to evolutionary theory, as developed in biology. In transforming these ideas to societal issues, he proposes that conventions or rules could be looked upon as equal to genes. However, he does not present any ideas about how such rules or 'genes' are selected apart from claiming that they are functional to the order. Hayek uses a functionalistic explanation (see Chapter 2, Section 2.4) without defining the selection mechanism. Hodgson suggests that Hayek's theory also produces inconsistencies since he does not accept that individuals change as a consequence of the evolution of new rules.

Second, there is no reason why tradition should be less coercive than institutions that are consciously defined by some actors, group representatives or the state. There is a tendency, also in economic positions beyond Hayek, to view some acts as coercive almost by definition, while others are considered free (see Chapter 3, Section 3.5). Rules that follow from decisions made by the state are generally seen to be coercive – that is, they do not normally have unanimous consent. Rules following from tradition are seen to have the opposite quality. One may ask, however, what it is that secures this? Traditions are also invented (Hobsbawm 1983). They are social constructs, and may also be coercive and built on inequalities. Access to commons may be established by custom. It may regulate the right to cut down trees or allow animals to graze and so on. Often the tradition dictates that only persons having this and that characteristic, owning this or that type of property, have access. There is no reason why there is less coercion or less consent involved in these cases than in cases where more formal collectives like the city council or the state make the decision.

The mistake made here is that of juxtaposing unanimity with tradition. It may appear natural that only those owning land in the valley also have access to the woods and the pastures of the surrounding mountains – at least only those farmers have animals and the equipment to cut down trees and so on. Some (maybe some hundreds of) years back there may still have been a conflict over this solution when the use of these common resources became a source of conflict for the first time. Those with little or no private land in the valley may then have argued that they were in greater need of the grass and wood of the commons. The alternative view was to distribute the common resources in proportion to the land that was owned in the valley. If the latter position became 'tradition', we may envisage that to survive, those with little land over time became labourers on the larger farms. They ended up with no animals, and ultimately, who can oppose a solution where only those with animals have the right to the pastures?

Basically, Hayek's problem is how to establish the neutral starting point from where everybody freely transacts. Actually, his ideas mirror an ideal market-type selection of institutions. The problem is then from where do the rules defining this market come? From what type of market is the selection of institutions made, and why and how do people settle spontaneously for the same solution? Hayek is well aware that market transactions presuppose non-market traditions. Markets are embedded in tradition. This is a sound observation – see also Chapter 2 and the discussion on contracts. The problem is to establish the (neutral) basis for these traditions. Michael Oakeshott remarks that Hayek actually sets up 'a plan to resist all planning'.[13]

There is a strong ideological drive in Hayek's writing. The idea of liberalism and a specific understanding of individualism as a goal in itself seems to form the basis for this. The idea of free individual choice is, however, confronted by the conflicts following from the need to distribute resources in a society. While the state can be used as an oppressive instrument, an authority of some kind is a necessary tool not least in complex modern societies where resource conflicts appear daily as a function of technological change, resource shortages and population growth. New resources continuously become scarce.

Therefore, simply claiming that there should be no authority or third-party solution cannot eliminate the problem of authority and power. As Polanyi ([1944] 1957) emphasized long ago, the extension of markets implied not less but rather a parallel extension of state powers. The state is fundamental to the very structure of private property. Furthermore, more conflicts appeared as markets grew and new resources were constantly commoditized. These conflicts have demanded regulation. Finally, there is nothing in Hayek's position that goes against the development of (common) authorities if individuals favour such solutions. However, this is a problem which he never seems to consider.

4.3 THE CLASSICAL TRADITION OF INSTITUTIONAL ECONOMICS

As laid out in the introduction to this chapter, the classical tradition of institutional economics was established more than 100 years ago in the United States as an explicit reaction to the neoclassical trend developing in Europe in the late nineteenth century, and it gained a dominant position among the US economists of the first half of the twentieth century.

While the new institutional economist and the institutions-as-equilibria positions both take on an individualistic perception of the problem, the classical view stresses the role of the collective and the effect institutions have on forming the individual. While the positions presented in Sections

4.1 and 4.2 tend to make at most one or two changes in the standard assumptions underlying neoclassical theory, the classical institutionalists tend to challenge the whole structure of both the core and the standard application area. Most important is the stand taken concerning the core assumptions. In our presentation we shall distinguish between the 'old' and the 'contemporary' scholars of the classical tradition.

4.3.1 The Old

The group of old institutional economists were all Americans led by Thorstein Veblen, John R. Commons, Clarence E. Ayres and Wesley C. Mitchell. We shall not cover the positions of all of them here. I have chosen to focus on Veblen and Commons in this short introduction. They are the most novel and cover the scope of this tradition well.

According to Mayhew (1987) Veblen was the first economic anthropologist, the first to study the customs of the American economy as it developed around the turn of the nineteenth century. He first of all focused on change, on the evolutionary and cumulative processes of an economy (Veblen 1898, 1919). He thus formulated his analyses very much in opposition to the position taken by his contemporary neoclassical colleagues who focused on equilibrium ideas. Instead he came to see continuous change as the characterizing aspect not least of market economies, and he saw 'both the agent and his environment being at any point the outcome of the last process' (1919: 75). He developed the position that humans are influenced by the institutional framework within which they live, which is very much in line with the perspective of Berger and Luckmann (1967) as presented in Chapter 2. Hodgson (1996: 126) writes:

> Veblen may have originally entertained a reductionist position in which explanations of human behaviour can be reduced to instinctive drives. However, he quickly moved away from it when he realized that institutions could be seen as not only being formed by, but formative of, such elements.

We should observe that Veblen's evolutionary ideas were very different from those observed in Hayek's writings. Specifically, he does not base his theory on methodological individualism and avoids the problems related to defining an evolutionary process where the individual him-/herself does not change.[14] Yet, Veblen must also be criticized for not defining the mechanisms of the institutional selection process clearly.

Veblen was critical of the concept of marginal utility so fundamental to the neoclassical stance (Veblen 1909). This followed from his view that institutions affect the preferences that individuals hold. He especially ridiculed much of the consumption he observed among the upper economic classes by calling it 'conspicuous'. He found it to be merely a flaunting of their economic status and position (Veblen 1899).

Veblen tended to focus more on the conserving or 'negative' side of institutions than on their liberating capacities. He contrasted institutions with technology. The former were the 'settled habits of thought' while technology was the source of change. He looked at institutions as 'ceremonial' while technology was instrumental, knowledge based and progressive. He specifically distinguished business (making money) with its 'predatory' habits of thought, from industry (making goods) with its 'productive' thought habits. He related this distinction to the contemporary growing class of 'absentee owners', the new class of capital owners, who did not work in the production themselves. In his mind these were predatory in their search for pecuniary gains. Their 'instincts' were very different from those of the engineers, and their 'workmanship' focused on production.

John R. Commons took a rather different route from that of Veblen. Commons thought that institutions were ways of supporting interests and handle conflicts. He focused on how collectives, organizations, the court system and the state, formed institutions to protect specific interests. He reacted against the tendency among economists in general to look at economic issues as harmonious exchange instead of conflicting situations. Resource scarcity made economic choices conflicting, and institutions were the remedy by which (some) harmony could be created (Commons 1934).

He also reacted to the tendency to focus on psychological features as a substitute for institutional ones. Neoclassical economics, or 'hedonism' as Commons tended to call it, deals with individuals and their relationship to material things or nature. According to Commons, the important relation is not between a person and an object; it is between that person and other people. Nevertheless, the focus of hedonism is on individualist concepts like marginal utility, time preferences and so on and not on social constructs like rights, duties and ownership. He emphasizes:

> Thus an institution is collective action in control, liberation and expansion of individual action. These individual actions are really *trans*-actions instead of either individual behavior or the 'exchange' of commodities. It is this shift from commodities and individuals to transactions and working rules of collective action that marks the transition from classical[15] and hedonic schools to the institutional schools of economic thinking. . . . The smallest unit of the classic economists was a commodity produced by labor. The smallest unit of the hedonic economists was the same . . . commodity enjoyed by ultimate consumers. One was the objective side, the other the subjective side, of the same relation between the individual and the forces of nature. (Commons 1931: 651–2; original emphasis)

To Commons, the smallest unit for the institutional economist was the transaction. Transactions intervene, according to him, between the labour of the classical economist and the pleasure of the neoclassical (hedonist) economist: 'simply because it is society that controls access to the forces of nature, and transactions are, not the "exchange of commodities" but the alienation and acquisition, between individuals, of the *rights* of property and liberty created by society' (p. 652; original emphasis). The very existence of any economic transaction rests in the institutional structures within which it is embedded. It is the working rules, the institutions, of any system which define who can do what and to whom, what a person must or must not do and so on. These deontics are specified by the collective group of people belonging to the actual society or 'going concerns'. The defined institutions both constrain and liberate, and we see how Commons assigned a much more positive role to institutions than did Veblen.

Commons distinguished three types of transactions: bargaining, managerial and rationing (Commons 1934). The *bargaining transaction* is typically what is observed between sellers and buyers in *markets*, undertaken under the existing rules of competition, fair or unfair, with equal or unequal bargaining power and so on. Therefore the economic issues arising out of a bargaining transaction are 'competition, discrimination, economic power and working rules' (Commons 1931: 653).

The *managerial transaction*, those observed in, for example, *firms*, is between the superior and the inferior. Here there are only two parties and the focus is on the character of commands – reasonable or unreasonable – and on obedience – willing or unwilling.

With *rationing transactions*, Commons had in mind decisions made by *governments* and *courts*, but also by *boards of corporations*. These differ from managerial transactions 'in that the superior is a collective superior while the inferiors are individuals' (ibid.: 653). These transactions flow from the fact that resources are scarce and the conflicts this creates. They form the 'working rules' within which bargaining and managerial transactions take place. The rationing transactions (perhaps today we would use the term 'policy formulations'), involve negotiation 'but in the form of argument, pleading or eloquence, because they come under the rule of command and obedience instead of the rule of equality and liberty' (ibid.: 654). An important point for Commons was how the state in modern democracies had developed institutions fostering public deliberation over the rationing transactions (Commons 1934).

Commons had much experience in conflict resolution. He was professionally involved in labour and public utility legislation, and programmes of industrial safety (Rutherford 1994). This certainly influenced him in his perspectives on the dynamics of the economic process and his focus on both power and communication. What was lacking in his writing, as I see it, was a treatment of how institutions influence the individuals *per se*. So where Veblen was strong, Commons was rather weak and vice versa.

4.3.2 The Contemporary Classical Institutional Economists

The 'old' institutional economists have been criticized for their complex messages, and because they focused more on empirical analyses than on theory building (Coase 1984). It is true that they were not able to develop a structure or model with the same stringency as the neoclassicals: focusing on the economy as a structure of institutions embedded in the broader society does not easily foster that kind of theory building. Yet, it is wrong to conclude that the work of these scholars was not theoretical in its orientation – its theorizing was, however, built on constructs based on empirical observation, not on 'axioms' like that of maximization and stable preferences.

The intention to start out from 'practice' is important and fruitful. In my mind the strong focus on relevance and representativity is important if one's ambition goes beyond that of a pure intellectual endeavour and is instead focused on understanding real phenomena. However, it is also true that the greater complexity one faces, the harder it becomes to build a complete system of theoretical ideas. The cost of increased relevance is to some extent a more heterogeneous body of thought. This is a 'price' that all institutionally orientated schools pay, including also representatives of the new institutional economics.

The visions of the old classical institutionalists have been taken up by several authors over the last 20–30 years.[16] Important contributions have been made by Schmid (1987), Hodgson (1988, 1999), Bromley (1989, 1991), Mäki et al. (1993), Groenewegen et al. (1995), Sjöstrand (1995), Tool (1995) and Samuels et al. (1997).

This book itself is written within the tradition established by the developments of the classical institutional economics in its contemporary form.[17] Thus, the central themes raised by the above authors are covered more generally by the overall text. Here I shall list the main issues raised by these authors, to clarify important links to the rest of the book. More details follow, especially in Chapters 5–8.

While presenting the position as a response to the core and application theorems of the neoclassical model, we first observe that in much of this literature the human is regarded as *multi rational* (Hodgson 1988; Sjöstrand 1995). The idea of maximizing individual utility as the only form of rationality finds little support. It is not irrational behaviour that is emphasized, or so much bounded rationality, as the thought that what is rational depends crucially on the institutional setting. The kind of rationality involved is defined by the meaning and expectations as given by each context. Considering what is right and wrong is an alternative form of rationality compared to the calculus of an individual gain. Which rationality is relevant is defined by the institutional context.

Second, and implicit in the above, we observe a strong focus on the importance of the *institutional context on preferences* and value expressions (Samuels et al. 1997) and thus price formation (Tool 1995). This issue was raised, as we have seen, by Veblen. The institutional arrangements also influence the evaluation of what becomes efficient (Schmid 1987; Bromley 1989). Efficiency is actually a reflex of the defined rights and the interests that are protected by the status quo institutions. This issue was manifest already in the writings of Commons.

Third, we observe a strong interest in the old Veblenian theme of *evolution* (Hodgson 1988, 1996, 1999) as opposed to that of equilibrating forces. The issue of internal theoretical consistency of the neoclassical model is also important (Bromley 1989; Mäki et al. 1993). Actually it is a common feature of the group of authors we are referring to here, that consistently taking care of the properties of institutions is impossible from a perspective which looks at individuals as maximizers creating equilibrium states.

The above points all relate to the neoclassical core as defined here. Concerning the application area, the issues of positive transaction and information costs have certainly been addressed by several of these authors – for example, Schmid (1987); Hodgson (1988); Bromley (1989, 1991); Mäki et al. (1993). Thus, the contemporary classical institutionalists also focus on many of the issues that are central to, for example, the transaction costs school. But the role of the transaction costs issue is not the same since it is understood within a model that is otherwise quite different. Questions that in the new institutional economics are understood as ways to reduce transaction costs – that is, command structures and hierarchies – are in this literature understood also as a function of power relations or expressions of power (Pitelis 1993). Therefore issues concerning *power* and the *protection of interests* play a more general role in the studies (for example, Schmid 1987; Hodgson 1988; Bromley 1989).

Finally, contemporary classical institutionalists are interested in a wide variety of institutional structures – for example, property structures beyond that of private property – and these again are discussed not only in relation to efficiency, but also in relation to the issue of power and interest protection (for example, Schmid 1987; Bromley 1989; Pitelis 1993).

Thus the contemporary classical institutionalists challenge all the fundamental assumptions of the neoclassical model. Each and every one has important consequences for the evaluation not least of public policies. This will be the recurrent theme of the rest of the book. Before we start on that endeavour, however, I shall make a short comment concerning the perspectives on the state and public policy as being implicit in the previous presentations of this chapter.

4.4 PERSPECTIVES ON THE STATE AND PUBLIC POLICY

While the terrain we enter here is again a complex one, we shall concentrate on three different stances concerning the role of the state and thus the role and characteristics of public policy. The three positions are the welfare theoretic extension of neoclassical economics, the public choice view, and the view following from the stand of the classical institutionalists. The presentation given will be very brief. The literature we draw on is not always directly overlapping with that of the institutional positions previously defined. Nevertheless, they can, as we shall see, be fitted in rather easily.

4.4.1 The Welfare Theoretic Position of Neoclassical Economics

In neoclassical welfare economics, policy making is understood as a technically rational procedure where goals are chosen and measures implemented in a consistent way. It is furthermore based on a division between two institutional structures: the market and the planner. The rule for the planner, or the state, is to maximize social welfare. Agents are assumed to be individually rational and pursue subjective goals, as is the general basis for neoclassical economics.

The planner plays an important role, especially in situations where markets seem to fail – that is, in situations like pollution where costs are external to the market. In these situations the role of the planner is to create solutions as if markets had existed. Hence, the same type of calculative rationality dominates both policy making and agent behaviour (for example, Boadway and Bruce 1984).

While neoclassical micro theory assumes people's preferences to be uninfluenced by the institutional setting, it is interesting to observe that the planner as invoked in the welfare theoretical extension of that theory, is thus influenced. So, while the economic agents that operate within the boundaries of markets are egoist, the planner is assumed to be 'benevolent'. How this characteristic comes about is, however, not clarified. There is no explicit discussion in this literature about the institutional prerequisites for the planner to have these capacities. S/he is *ad hoc* to the model and its logic. While neoclassical theory assumes that institutions do not influence actors' goals or capacities, welfare theory implicitly presumes this in the case of the planner.

4.4.2 Public Choice

The self-sacrificing planner in economic welfare theory has accordingly been criticized for being illusory. The *public choice* tradition is at least a theoretically consistent critique of welfare theory in that it generally assumes that institutions do not matter for behaviour. There is only selfishness. This is the case both for market actors and those participating in the political processes within the state. The model of calculative individualistic maximization behaviour is also transferred to the domain of policy making. Planning becomes mere 'politics'. Put the other way around, the market is extended to the arena of policy makers or administrators themselves (Niskanen 1971; Buchanan 1978; Dearlove 1989).

The theory of public choice was developed as a critique of the welfare theoretical ambition to use the state to rectify market failure (see also O'Neill 1998). However, the focus was shifted to that of policy failure – that is, the idea that greater loss than those probably created by incomplete markets would be produced via the political process to restore equilibrium. When state actors maximize their own interests, the interests of the public suffer. When bureaucrats maximize bureau budgets, taxes must be issued and the possibilities for individual market actors are reduced without any gains in return.

Concerning the institutional positions studied in this chapter, the greatest resemblance with public choice is found in strong individualist positions such as the property rights school, and the institutions-as-equilibria position – especially Hayek. The representatives of the transaction costs school are a bit more divided. North, as we have seen, acknowledges explicitly the role of the state in reducing transaction costs. On the other hand, he also acknowledges the negative effect of state involvement in economic affairs in many countries with manipulative or corrupt politicians and/or administrators (North 1981, 1990). This observation was, I believe, the main reason for the shift he made from the view that any institutional change is to be understood as efficiency enhancing (see note 7).

4.4.3 The Classical Institutionalist Position

Finally, the *classical institutionalist perspective* on choice represents a quite different type of solution to the dilemma of welfare theory compared to that of public choice. Instead of claiming that everything is a market solution, including the policy process, classical institutionalism generalizes the idea that all behaviour depends on institutions. As already underlined, institutions are also important when understanding motivation and knowledge itself, not only in describing choice sets and restrictions. Institutionalism, as understood here, accordingly rejects both the existence of a universal norm of rationality (welfare theory) and the assumption that behaviour only follows an individualistic rational calculus independent of the institutional setting (public choice). We are not only consumers ruthlessly maximizing utility. We are also citizens acting within the institutions of public decision making (Sagoff 1988).

In this view, meanings, values and what is considered proper behaviour will depend upon the given setting and culture. Institutions define both what is useful and what is right. This does not imply that it is a simple task to produce a loyal bureaucrat or a decent politician. Certainly, the existence of corrupt regimes in many present societies is a great challenge. To develop and sustain the institutions necessary for the political process to stay with the rules will be a continuous challenge for any society. Added to this, various parts of a state administration may tend to develop 'local' models of knowledge and proper action which will govern the performance of these different bodies. Hence we observe variations across the various responsibilities of the state (Vatn et al. 2002). This is exactly what the model of institutionalized behaviour would predict. The basic point is that by instituting responsibilities, procedures and controls, the roles of politicians and administrators are shaped and the difference in logics across spheres – for example, markets and policy arenas – is made tenable.

If one accepts and even institutes that politics and public administration are selfish types of activities, they will certainly become so. If votes can be 'bought', then market-like processes will also characterize politics. According to classical institutionalism, and understood as a branch of social constructivism, it is possible to influence behaviour via institutional processes. Individuals are able to accommodate their behaviour to a variety of logics or rationalities. The institutions define which ones will be emphasized.

Two important qualifications are needed at this stage. First, proponents of a more classical institutional position do not claim that the state is or should be neutral as to outcomes. The ideal is to produce institutional structures that are procedurally neutral, meaning that no interests or ideas are systematically kept out of the policy process. None the less, when deciding upon matters, the collective in the form of state bodies has to take a stand as to which interests should get its protection. This is the role of the citizen.

Second, the distinction made above between the consumer and the citizen is not strictly related to each sphere, the market and the policy arena. According to Etzioni (1988), elements of citizen concerns are also involved when people make choices within the market institution. Norms may exist that motivate people to let the concerns of others influence their private choices. This is typically the case of consumer boycotts, but it is observed more generally. However, a deeper study of this takes us into the issue of individual rationality and what motivates people's behaviour. This is the issue addressed in Part II.

4.5 SUMMARY

In defining the different positions within institutional economics, we have utilized the core and standard application areas of neoclassical economics. The core was defined as consisting of: (a) rational choice as maximizing individual utility, (b) stable preferences and (c) outcomes as equilibrium states. The standard application area contained (a) no information costs, (b) no transaction costs and (c) private property rights for all goods which are exchanged in competitive markets.

Three main traditions within institutional economics have been identified: new institutional economics, the institutions-as-equilibria tradition, and finally the position of classical institutional economics. These can all be classified as different responses to the neoclassical model (see Table 4.1).

The positions subsumed under the term 'new institutional economics' were loyal to the neoclassical model – its core and individualist perspective. The representatives of this position define both informal and formal institutions as 'rules of the game'. They are, however, seen as consciously designed. The dominant position among the 'new' is the *transaction costs school*. It bases its analyses on the neoclassical core, but changes the application area to include positive information and transaction costs. It is argued that the function of institutions is to reduce these costs, be it by the establishment of structures like the firm or the state. Williamson's position has much the same focus, but he also makes a change in the core through accepting bounded rationality. This is a rather radical change, which together with positive transaction costs is developed in order to understand various market and contract structures. The move to bounded rationality is partly justified as a logical consequence of accepting positive information and transaction costs.

The *institutions-as-equilibria* position understands institutions as the result of spontaneous 'games' without any conscious design or third-party enforcement. All institutions are conventions based on a kind of market selection. From one point of view, the authors belonging to this position are 'more neoclassical than the neoclassicals themselves' – that is, they take the idea of individualism further. Accordingly, knowledge is by some important representatives (for example, Hayek) understood to be purely subjective.

Table 4.1 A characterization of the main positions within institutional economics

Position	Definition of an institution	Process of institutionalization	Understanding of the individual	Differences from the neoclassical model
New institutional economics	'Rules of the game', common constraints	Mainly conscious construction	Self-contained	
• TC* school				• Positive TCs
• Williamson				• Bounded rationality, positive TCs
Institutions-as-equilibria	'Equilibrium strategies', individual constraints	Spontaneous	Self-contained	'Super-individualistic', evolutionary
Classical institutionalism	Both external, common constraints and formative of the individual	Mainly social and conscious construction	Socially formed	Plural, context-dependent rationalities. Preferences as socially influenced, evolution, positive TCs and information costs

Note: *TC = transaction cost.

The *classical institutionalists*, both the old and the contemporary, take a different route from both the above positions. Here the understanding is based on a social constructivist perspective and the model challenges in general all elements both of the neoclassical core and the application area. Most important is the position that rationality, what rationality means, is dependent on the institutional context itself. It is defined by this context. Preferences are also seen as socially influenced. The perspective of evolution replaces equilibrium. These scholars also emphasize the idea of positive information and transaction costs. The effect and importance of this is still somewhat different if we compare with the new institutionalists. The existence of institutions is understood not only as a way to reduce transaction costs, but also as a mode to protect values and interests and as expressions of control and power.

The ideas underlying the various positions defined above are also reflected in a different understanding of the state. The *welfare theoretic position* is based on the standard neoclassical model. It describes the choices made by individualistically rational economic agents, but adds a public sphere to that model, including a benevolent planner to make it possible to handle market failures – that is, public goods and externalities. The *public choice* tradition is an attack on this model, making the claim that all behaviour is individualistically motivated. Also the planner acts on the basis of a selfish calculation. This is often used as an argument against (any) state involvement. Correcting market failures creates policy failures instead. Finally, the *classical institutionalist* position stresses that all behaviour, both in markets and by state representatives, is institutionally influenced. This implies that the actions both of the individual and of the planner, in which rationality and responsibilities are involved, will be influenced by the institutions as they foster specific interests and ideas about what is a sensible or good society. Through the construction of institutions one can influence important characteristics of the policy formulation process.

NOTES

1. Important fora in the United States continue to be the Association for Evolutionary Economics and the *Journal of Economic Issues*.
2. Eggertsson (1990) uses the concept of 'new institutional economics' to cover only the position developed by Oliver Williamson. What is here called the transaction costs school, he calls 'neo-institutional economics'. To complete the picture, some of the contemporary classical institutionalists also use the term 'neo-institutional economics' to register their position as a modern version of classical institutionalism. To avoid confusion, I have totally abandoned the concept of 'neo-institutional' economics.

3. Therefore it may be argued that it is going too far to say that the property rights school is also based on a neoclassical perspective. However, the conclusion is right as far as the neoclassical core and standard application theorems are concerned. The property rights position attacks a specific part of neoclassical economics – that is, welfare theory – because it is seen to be redundant given the assumptions.
4. This issue will be extensively discussed in Chapter 8 and in Part IV of this book, especially Chapter 13.
5. By neoclassically orientated, I imply authors basing their analyses on the neoclassical core assumptions.
6. These relations are one set of reasons behind the fact that many countries have public health care, while those relying on private health care often have a large insurance sector added to it.
7. While the main production from North fits well into the programme of the 'transaction costs school', in his later writings he also advocates abandoning rational choice as maximizing and accepting bounded rationality. He therefore comes close to Williamson. He even abandons, at least partly, the efficiency view of institutional change (North 1981, 1990). Finally, he has accepted changing preferences, but not explained how this comes about (North 1990). These are all very positive developments. As suggested by Field (1994a) these moves have, however, made his messages inconsistent. Over the years North has accepted parts of the critique coming not least from people adhering to a more social constructivist position, but he is still sticking to a methodological individualist programme.
8. That is, a function with inputs as independent and output as dependent variables.
9. See Chapter 5 for a more comprehensive discussion of the information problem and the concept of bounded rationality.
10. If these costs were zero, every opportunist would be revealed up front. No problems of *ex post* contracting would exist.
11. Williamson (1985) elaborates the concept much beyond this, distinguishing between 'site specificity', 'physical asset specificity', 'human asset specificity' and 'dedicated assets'. It is beyond our aim to go into any depth here.
12. In my evaluation of Hayek I am especially indebted to Hodgson (1996), Streit (1997) and O'Neill (1998).
13. Oakeshott (1962), cited in Hodgson (1996: 183).
14. Hodgson (1996) is critical concerning Hayek as an evolutionary orientated economist. Hayek builds on methodological individualism. According to Hodgson, this position must be either redefined or abandoned if an evolutionary view is invoked. Evolutionary theory also demands that individuals change. According to Hodgson, Veblen understood this.
15. To avoid confusion, Commons is here talking about the tradition of classical economics – that is, the tradition of Smith, Malthus and Ricardo, which is to be clearly distinguished from the classic institutionalist position.
16. One should not forget that between the 1930s and the 1980s, there were also some important contributions in economics, keeping alive the institutional issues as perceived by the 'old'. Important names in that respect are in alphabetical order: John K. Galbraith, William K. Kapp, Gunnar Myrdal and Karl Polanyi.
17. I may be somewhat more influenced by the sociological tradition of institutional analyses than most of the authors listed above. Nevertheless, many of the same ideas are developed. What this book may offer in this respect is an integration of concepts from the sociological literature within the institutional economist model. I believe that it offers some opportunity for both expanding the model and producing a more consistent and generally applicable vocabulary.

8. [illegible]

10. [illegible] would be equal to optimal. No problems of [illegible]
11. [illegible]
12. [illegible]
13. [illegible]

15. [illegible]
16. [illegible]
17. [illegible]

PART II

From Institutions to Action

Part I developed the idea that institutions influence both individual and collective choice, and are themselves in turn influenced by such choices. In Part II the aim is to focus more thoroughly on the first category of influences – on how institutions form us and our actions. Part III will focus on the opposite type of dynamics – that is, how we form our institutions. While these dynamics continuously work both ways in society, it is none the less analytically sensible to handle them separately.

Part II will extend the understanding of how institutions influence action both directly and indirectly – that is, by defining which actions should be made and by influencing the motivation for action. The analysis will be divided in two. First, Chapter 5 will look at how institutions affect the rationality that is applicable to a certain problem. Thereafter Chapter 6 will concentrate on the context dependency of preferences and values. The issues of institutions and rationality, and of institutions and preferences are intimately connected. While Chapter 5 is mainly focused on the basic rationality concepts and what rationality may mean, Chapter 6 is more orientated towards the historical development of the various positions and the empirical verification of context dependency of preferences and values.

Again the neoclassical model of choice will form the reference point. We shall discuss the relevance and consistency of the rationality and preference concepts as developed within this tradition. Furthermore, we shall use this model to contrast and evaluate the competing position of social construction/classical institutionalism.

5. Rationality

Rational choice is the core concept of neoclassical economics. It seems reasonable to say that according to this tradition, economics is not about the development and functioning of economies. It is instead about discerning the consequences of rational choice as maximization. Neoclassical economics is defined by its method, not its object. Moreover, the idea is that rationality has just one form – that is, it is universally defined as maximizing individual utility.

The institutional perspective presented in this book sees this quite differently. It takes its departure from the object to be studied. Even more importantly, it perceives rationality as defined by the institutional setting within which choices are made. This implies that the rationality of, for example, the marketplace, the family or the policy arena, is different as to its basics. These social constructs represent different logics or rationalities which they are devised to support.

In disentangling this we shall divide this chapter as follows. First, we shall define what is meant by rationality when understood as maximizing (Section 5.1). Second, we shall look at the relationship between the rationality concept and the rest of the neoclassical model, especially the consistency problems that appear if we accept information to be costly (Section 5.2). Third, we shall look at the competing idea of satisficing or bounded rationality (Section 5.3). We shall define this position as a response to some of the consistency problems appearing when information is costly. Finally, Section 5.4 will present the position basic to this book, that what is rational is institutionally dependent and that the alternative or rather supplement to individual maximization/satisficing is social or cooperative rationality. This standpoint implies that rationalities may be plural and that different institutional settings support different rationalities.

5.1 RATIONALITY AS MAXIMIZING

According to neoclassical economic theory, rationality as maximizing implies that the individual maximizes her/his utility. Maximization is linked directly to the preferences of the individual. Maximizing is undertaken given the constraints the individual faces concerning what goods are available at what prices, and what budget s/he commands.

To act rationally then implies two things. First of all preferences must be rational – that is, follow a set of consistency claims. Otherwise maximization is not definable. Second, the individual must be able to make the necessary calculations and choose what s/he prefers.

Preferences are rational if they are *complete*, *transitive* and *continuous* (Hausman 1992):

- Preferences are complete if the person is able to rank all goods or bundles of goods. This implies that for all x and $y \in X$: $x \geq y$ or $y \geq x$.
- Preferences are transitive if the ranking is such that x is better than y and y is better than z then x must be better than z. Formally: for all x, y and $z \in X$ where $x \geq y$ and $y \geq z$ then also $x \geq z$ must hold.
- Preferences are continuous if x is preferred over y and z is sufficiently close to y, then x is also preferred over z. This implies that the consumer is able to distinguish between goods even though the difference in the utility they offer is infinitesimal.

According to Hausman (1992), to say that individuals are utility maximizers says nothing about the nature of their preferences: 'All it does is to connect preferences and choices' (p. 18). This is only partly true. It also says something about their form, not least that it must be possible to trade them off against each other.

The definition of rationality, as applied in economics, implies that *preferences are context independent*. This should be understood at two levels. First, it implies that the ranking of goods x and y is independent of the presence or not of a third good z. This is the kind of context independency mainly discussed among neoclassical economists. Second, and much more important to us, context independency implies that the choice is independent also of the social context – the institutional setting. Your choice of (that is, preference for) drinks served should be independent of whether the setting is a fiftieth anniversary or a dinner in relation to a funeral. More generally the logic of a market transaction and an act within, for example, the family, is principally of the same kind. This issue is less discussed among neoclassical economists. According to the view advocated in this book, it is not only individual preferences that count, but also what is considered *right* or *proper* behaviour given the situation. This is socially defined.

While not explicit in the model, right beliefs about how to accomplish what is preferred are also important for obtaining one's goals. This element is rarely focused on in the standard expositions of rational choice as maximizing, maybe because the standard application theorems of the neoclassical model include full information. 'False consciousness' still exists. It is not right to demand no errors in the understanding of cause–effect relationships in order to call acts rational. I shall return to this.

5.2 PROBLEMATIC EVIDENCES AND THEORETICAL INCONSISTENCY

How can we assess the utility function? If it exists, it can still not be directly observed. One way is to ask people what they prefer. To base the analysis on such expressions has been found problematic since people could say one thing and do something else. Samuelson, one of the most prominent neo-classical economists in the post-Second World War period, was very engaged by this issue. He (1948) offered a solution by looking at preferences as revealed by the choices people actually make. If someone chooses coffee when tea is available, this shows that coffee is preferred to tea. Samuelson's solution moved economics away from subjective introspection, and it was thought that he brought economics to a more scientific footing by building it on observations of actual behaviour. It made the theory testable or falsifiable, which was the claim of the day for what could be considered scientific.[1]

However, if inconsistent choices are revealed – for example, the same consumer chooses biscuits over bread one day and then bread over biscuits the next – does this necessarily imply that the person is irrational? Is the theory falsified by such observations? Maybe preferences have changed, or maybe the person is maximizing something other than the independent utilities of biscuits and bread?

The first answer, changed preferences, is problematic since it makes the theory in principle irrefutable. It cannot be tested. Moreover, it goes against the basic core assumption that preferences are stable, and we observe an argument for sticking with stable preferences other than the one presented in Chapter 2. There we looked at stable preferences as a defence for context-independent individuality – that is, the autonomous individual. Now we see that this assumption also seems to be necessary to make the theory of rational choice testable. The postulation of stability itself must still be taken on faith, though – that is, one cannot test both the assumption of rationality and the postulation of stable preferences on the basis of observed choices.

The second answer, that there is something else which is maximized, is equally problematic. Boland (1981) argues that if we observe, for example, intransitive choices, this cannot be used as an argument against maximization; we have merely misunderstood what the acting individual is making the most of. S/he may drink beer one day and wine the next but, what s/he is after may in fact be the alcohol. Then this shift may become perfectly rational. Certainly, this redefinition may make sense. Still, by making this move, Boland is actually pulling away the carpet from under Samuelson's solution. Observation can then be taken as evidence neither for, nor against the core assumptions of the model. There may always be 'something else' that is maximized. Actually, we are left with two alternatives: either to accept that the model is refuted or that it cannot be refuted/tested. Both positions are problematic, indeed.

Samuelson himself was well aware of the problem:

> Thus, the consumer's market behavior is explained in terms of preferences, which are in turn defined only by behavior. The result can very easily be circular, and in many formulations undoubtedly is. Often nothing more is stated than the conclusion that people behave as they behave, a theorem which has no empirical implication, since it contains no hypothesis and is consistent with all conceivable behavior, while refutable by none. (1947: 91)

One may doubt whether it is in any way possible to avoid the problem of circularity, given the structure of the model.[2] Bromley (2006) goes further and argues quite consistently that the model may offer mechanical causes for actions, but it is unable to understand or capture the reasons involved.

There is one more fundamental issue involved here. Recall that the standard version of the model includes an assumption that information is complete – that is, it is cost free. This is an assumption that is often changed, not least since the analysis of risk and uncertainty has become a very important part of economic analyses over the years.[3] Certainly, this makes the model much more realistic. The problem is, however, that as soon as one accepts that information is costly, the standard assumption of rational choice becomes indefinable.

In this situation the actor, when maximizing, will always have to decide whether resources should be used on conducting choices or on gathering more information as a basis for potentially better choices. At every point, gathering such information may result in choices so much better that it is worth the extra costs involved. None the less, it is impossible to know the answer to this question before the search is finished. Indeed, when should one stop searching? The information already gathered says nothing about the value of information not yet acquired, implying that there is no answer to this question. This is the basic characteristic of knowledge. You really cannot know before you know.

This self-reference problem was already acknowledged by Morgenstern (1935). Given its character, it cannot be solved other than by an arbitrary act or standardized interruption rule. According to Knudsen (1993), an inextricable problem arises as soon as the cost of optimizing becomes part of the optimization calculus itself. We enter into an infinite regress. At any point in the process one has to choose whether to allocate resources to improve the choice procedure (including information gathering) or make the choice. Diverting from the assumptions about full information (that is, zero information costs) is inconsistent with the core, or more precisely: the reasoning presented here implies that the *utility maximization* algorithm becomes indefinable when shifted to *expected utility*, since the latter is characterized by incomplete information. Actually, learning is not a logical part of the neoclassical model – that is, optimal learning could, in principle, have been – but cannot be defined.

Certainly, one may classify a choice as optimal or rational on the basis of a given set of information. It is rational relative to the information that the actor possesses. Boland (1981) retreats to this position. However, this only circumvents the problem. Moreover, empirical research shows that even in cases where the issue of optimal information is not relevant and preference changes cannot be involved, choices are observed that are counter to the rationality assumption. The phenomenon of preference reversals, documented especially in the psychological literature on choice, is possibly the most important case. It raises serious doubts about the human ability to rank alternatives even when all relevant information is available.

Lichtenstein and Slovic (1971), Slovic and Lichtenstein (1983) and Tversky and Kahneman (1986) refer to results that show preference reversals. These come from a series of tests where people are confronted with two gambles, one with high probability and low payoff and one with the opposite characteristics. While risk is involved, there is no uncertainty concerning the expected value of the gambles. Since it is a 'one-shot' experiment, preference changes cannot be involved. The gambles are set up such that the one with the highest probability of a gain has the lowest expected value. People confronted with the two gambles tend consistently to choose the one with the lowest expected value, but with the highest probability of a win. When asked, they still rank the one with the highest expected value as the best one. Their preferences are reversed.

This does not imply that market behaviour or choice experiments do not confirm that people in many situations can act in ways that fit the standard hypothesis of rational choice rather well given that information problems are not insurmountable – for example, Smith (1991, 2000).[4] Nevertheless, I believe that Simon's (1979)[5] summary of the debate on rational choice holds. He accepts that developments made in order to make economics less vulnerable to the kind of critique offered by, for example, Kahneman and Tversky – like rational expectations, game theory and Bayesian statistics[6] – have offered important insights. Simon still concludes:

> The axiomatization of utility and probability after World War II and the revival of Bayesian statistics opened the way to testing empirically whether people behaved in choice situations so as to maximize subjective expected utility (SEU). In early studies, using extremely simple choice situations, it appeared that perhaps they did. When even small complications were introduced into the situations, wide departures from the predictions of SEU theory soon became evident . . . the conclusion seems unavoidable that the SEU theory does not provide a good prediction – not even a good approximation – of actual behavior. (Ibid.: 506)

5.3 THE MODEL OF BOUNDED RATIONALITY

Simon did not just criticize the conventional wisdom. He is among those who have worked consistently on developing models of behaviour that fit better to what is observed than the model of maximizing (expected) utility. His work can in many ways be seen as a reaction to the many problems appearing in standard theory if information is costly and individuals lack the necessary capacity to handle all available information consistently (Simon 1947, 1957, 1959, 1979). Parallel developments are found in Cyert and March (1963) and March (1994).

The basic idea of bounded rationality is that the decision maker transforms complex or intractable decision problems into tractable ones:

> One procedure . . . is to look for satisfactory choices instead of optimal ones. Another is to replace abstract global goals with tangible sub goals, whose achievement can be observed and measured. A third is to divide up the decision making task among many specialists, coordinating their work by means of a structure of communications and authority relations. (Simon 1979: 501)

Of the three, satisficing is the one that has gained most attention. We shall restrict ourselves to this hypothesis.

March (1994) concludes that satisficing implies setting a target. All solutions falling short of the target are exempted. The first solution passing it will be the one chosen. It is satisfactory, and the decision maker does not consider going on to obtain something (more) 'optimal'. It is exactly a type of 'short cut' one would expect if information is costly to obtain and handle. Many economists interpret the writings of Simon and March as describing how agents economize on information costs. The target is a solution to the problem of optimal information search, which takes us back to the standard optimization calculus. This is not true of Simon and March's position. They do not look at satisficing as a way to optimize on information handling. From their perspective, the problem humans face is caused by the costs of gathering and handling information, but satisficing is not disguised optimizing. It is something else. It is about pragmatic, tractable solutions to 'intractable' problems.

There are several rather radical implications of this position. First, the idea that economies will establish equilibrium results has to be abandoned, since it is strongly linked to the idea of rationality as maximizing. If people satisfice, there is no way to establish the neat results from equilibrium theory (Simon 1979). Second, and more important to us, if people are more concerned with success and failure relative to a target than with graduation of success or failure, the status quo becomes important for people's evaluation of outcomes. This will not be the case if maximizing is the 'rule'. March suggests:

> If out-of-pocket expenditures are treated as decrements from a current aspiration level (and thus as unacceptable) and foregone gains are not, the former are more likely to be avoided than the latter. A satisficing decision maker is likely to make a distinction between risking the 'loss' of something that is yet not 'possessed' and risking the loss of something that is already considered a possession. (1994: 22).

This may explain the large deviations between 'willingness to pay' and 'willingness to accept' compensation measures consistently observed in the literature, be it for ordinary commodities or environmental goods. We shall return to this observation both in Chapter 6 on preferences and in the treatment of environmental valuation (Part IV).

Certainly, the challenge for the model of satisficing is to define how people develop targets. While no developed solution to that problem seems to exist, it may be a way to explain the existence of habits, 'rules of thumb' and so on, so often observed in real life. These concepts can be viewed as forms into which satisficing rules materialize. However, they are not targets concerning acceptable levels of goal attainments. They are instead regularized procedures that are seen as capable of producing satisfactory results.

From a neoclassical position, we can again envisage these rules as optimal ways to handle various information problems. They are just repeated types of actions sustained as long as the costs of changing them are perceived to be too high. Nevertheless, since these costs are unknown until one has tried to make the change, this type of explanation does not offer any reasonable response to the issues involved. The idea begs the question of how the persons involved can assess when to quit a habit or a rule of thumb. Should they continue to look for alternatives? If that is too demanding, when should they start looking? The answer to this question is impossible to make before one has tried, and it is only afterwards that one can (possibly) conclude whether it was a right or wrong investment to do the search.

This does not imply that habits and rules of thumb are not reasonable responses to the involved information problems. Certainly, it is reasonable to stick to procedures that have been shown to work. Hodgson (1988) lists several references concerning consumer behaviour, documenting that only a small fraction of the purchases we make is based on deliberation over costs and qualities. Buying a car – high stakes – may be deliberate, while using it may turn into a habit. An important point here is that habits, first acquired as ways to handle complexities, later tend to become 'valuable in themselves'. Behaviour which is repeated, tends to be reinforced by its effect on how we perceive that the actual problem should be solved. It tends to be transformed from a mere solution to also becoming valuable as an act in itself. This is similar to the perspective of social constructivism, and actually goes beyond the perspective of bounded rationality, which still takes the human as given (see Chapter 4).

Screpanti (1995) refers to another aspect with special reference to what is here called 'rules of thumb'. In a complex world where many solutions are open to us and it is difficult to assess their consequences, learning from each other may be a way to obtain better and more certain results. Screpanti thus looks at these rules not as individual responses, but as institutions in the form of rules which are socially tested – that is, they are common knowledge in practical form passed on between individuals. Farming is, as an example, a complex business, as is the work of a blacksmith. While it is impossible to assess and optimize all factors involved when producing a crop or forging a horseshoe, experience is condensed in a set of rules or skills concerning 'what to do when'. In the case of farming, this may involve issues such as when to plough, how to plough, when to fertilize, how much to fertilize, when to apply pesticides and so on. The answers to these questions are tested ways of behaviour – skills – and are passed on in the specific working environment, from 'father to son'. Polanyi (1967) focuses on this issue when describing the dynamics and relevance of so-called 'tacit knowledge' passed on as conventions (rules of thumb).

Finally, habits and rules of thumb narrow down the space of action. These constructs are also important to human coordination because they make it easier to form expectations about the behaviour of others. Indeed, in a complex world where information problems are pervasive and maximization unattainable, regularizing behaviour is an important way to create a firm basis from which to form expectations. This suggests that rules of satisficing behaviour, if settled in the forms of habits and rules of thumb, reduce the information and coordination problem in two ways. First, they reduce the need for information search for the decision maker. They are given solutions to the problem. Second, by their very existence, they then reduce the need for information search for other decision makers. Once such constructs are instituted, decision makers know what to expect from each other.[7]

5.4 RATIONALITY AND INSTITUTIONS

The response by Simon, March and others to the 'rationality as maximization' hypothesis focuses dominantly on the capacity of a human being to handle complex issues and large amounts of information. Moreover, it is stressed that the characteristics of the information problem make it impossible to optimize on information gathering, implying that some simpler rule of behaviour is advocated. The idea of satisficing is therefore helpful in sorting out the problems concerning information costs inherent in the theory of maximization. It has, however, less to offer concerning situations *where what is at stake goes beyond that of individual rationality*.

The alternative to individual rationality is not foremost irrational behaviour. It is instead to recognize that rationality can also be social. This implies that rational action – that is, reasoned action – may not be driven just by one logic. Behaviour can be said to follow different rationalities. While irrational behaviour implies that we do not act in accordance with what we prefer or have decided to do, the idea of plural rationalities is based on the observation that what is rational to do can be driven by reasons other than maximizing/satisficing individual utility.[8] Which rationality applies, depends then on the institutional context in which one finds oneself. This implies that in some settings it is considered appropriate to take only individual interests into account. Under other circumstances this is not so.

If the idea of maximizing is independent of institutional context, it must actually involve all aspects of life. If not, we have to define when this type of logic stops applying and another starts. However, then we must also reason over which type of logic or rationality this *second-order decision* itself should be based on and to accept that there must be at least two different types of rationality involved.

Becker (1976, 1993) takes a clear position. Maximizing individual utility is a universal human characteristic formative of any social sphere. It is the typical logic not only in market situations, but it should also be used to explain the existence of institutions like the family, the fact that people have children, the incidences of giving, of crime and so on. It is all the time about making a trade-off between (the present values of) individual gains and losses. The second-order problem emphasized above becomes irrelevant.

Etzioni (1988) adopts a position that is very much against this view. He also stresses the role of commitments and moral reasoning. There is not only calculation, but also involvement. There is behaviour motivated both by individual utility and there is behaviour founded on norms, on moral reasoning about what is the right thing to do. According to Etzioni, there is always a tension between the rationality of individuality and that of social belonging – between the logic of 'I' and that of 'We' – of individual maximization on the one hand and norms and moral reasoning on the other.[9]

People may, when operating in the public sphere, act in ways very different from what is normally their behaviour in markets. People may support public goods independently of what individual gains they may offer. They may cooperate in situations where free riding is individually preferable ('rational'). Specifically, people vote even though casting an individual vote will not influence the result of the election (see also Hodgson 1988).

5.4.1 On Social or Cooperative Rationality

The basis for thinking in 'We' terms is related to the fact that we influence each other's possibilities and are continually faced with the issue of how to balance own interests against the interests of the community to which we belong. The literature on what is here termed 'social' or 'cooperative' rationality is rather complex, and I find it relevant to divide the 'We' rationality in two: *reciprocal rationality* and *normative rationality*. While the two categories are to some extent overlapping, they also embody some characteristic features.

Gintis (2000) is among those suggesting that fairness and equal treatment is a basis for much of our behaviour. He uses the concept of *homo reciprocans* as a contrast to *homo economicus*.[10] The distinct feature of reciprocal rationality is a propensity to respond positively to sympathetic actions and negatively to unfriendly behaviour, despite individual losses in, for example, material rewards from such a response (Fehr and Falk 2002). Reciprocity can be viewed as a form of solidarity. Kind acts are rewarded and unkind ones are punished. In Chapter 6 much evidence of such behaviour will be documented. Typically, people are also willing to share in situations where this gives them a personal loss, and to punish others who do not share in a situation where sharing is expected. This will happen even though those retaliating will experience a loss.[11]

The position of Etzioni (1988) covers the other type of social rationality, that of humanly formulated norms about what is the right thing to do in certain situations. Following the reasoning of Chapters 2 and 3, norms are a required solution to a specific problem that furthermore supports a specific value. It is defined to solve a coordination problem in situations where the society has developed certain common views concerning how something should be done. If the norms are fully internalized, they are followed independently of whether others know and can punish those breaking the norm. In the process of becoming internalized, social control and punishment play a role, however.

Not least, the latter often makes it hard to draw a clear distinction between reciprocal and normative behaviour. Actually, we might argue that reciprocity is simply the norm of cooperating and showing others due respect. Normative behaviour is, on the other hand, not bound to just that. Norms can be based on a wider set of values, such as those defining various virtues. This is more about solving the wider issue of whom we should be as social beings than just how to act in cooperative settings. We could also argue that reciprocity resembles the idea more of a convention than of a norm. It is just the way things are done; compare the concept of institutions as 'reciprocal typifications' (Chapters 1 and 2). None the less, the retaliation part of reciprocal behaviour in particular resembles that of controlling a norm. I therefore believe that the distinction should not be exaggerated. The important observation is actually what these rationalities have in common – that people may act in ways that are unselfish.

There have been several attempts to explain behaviour that is here called 'reciprocal' or 'normative' by understanding them as special forms of egoistically motivated behaviour. The two most significant are the *Folk theorem* and the idea of *selfish altruism*. We shall look briefly into both kinds of arguments.

The Folk theorem (see Fudenberg and Tirole 1991; Romp 1997) is developed in game theory and is based on individual rationality as maximizing individual gain. In one shot prisoner's dilemma games (as illustrated in Box 2.1) the pay-offs are such that individual rational behaviour is deemed to end in results that are unfavourable to everybody. The solution to the game – the Nash equilibrium[12] – is Pareto inferior. The idea behind the Folk theorem is that if the game is repeated infinitely (or with an arbitrary stopping point), and people are sufficiently patient,[13] it becomes individually rational to sustain cooperation. While you may gain by defecting in, for example, round 1, the value of this defection is reduced in the future if also other players defect since this causes lower future pay-offs. If everybody instead cooperates, there is a sustained future gain for all as the prisoner's dilemma game is structured. If the game stops at a certain and known point in time, defection again becomes the individually rational strategy. This is so because it then becomes profitable to defect in the last round. Backwards induction leads to the conclusion that defection is also the individually rational act in the second last round and so on. On the basis of this it is argued that acts that look as if they were motivated by social rationality (reciprocity or norms) can be understood as based on pure individual rationality. It is the result of selfishness or strategic rationality, as individual rationality is often also called.[14]

There are two arguments against this reasoning. First, cooperation is observed also in situations where the above assumptions do not hold – for example, in situations where games are not repeated (Ostrom 2000). More information concerning the data on cooperation in one-shot games and so on is given in Chapter 6.[15]

Secondly, the Folk theorem cannot explain retaliation as is often observed. It implies that the individual experiences a loss in his or her effort to secure that everybody reciprocates. This observation cannot be explained by individual rationality, but is understandable if the rule of reciprocity is invoked (Fehr and Falk 2002; Ostrom 2000).

The idea of selfish altruism offers another kind of explanation based on individual rationality. It focuses on the possibility that the good act actually gives the agent a satisfaction which is at least as great as the offer involved. It is the 'warm glow' of giving (Kahnemann and Knetsch 1992), and it may therefore be viewed as an individual satisfaction entering a standard utility function. I do not deny that such an effect exists. Still, it would be wrong to label all non-egoistic acts as still being egoistic, while on a more 'sophisticated' level (see also Elster 1983b). One might certainly ask: how can we differentiate? Helping an old man across the street may certainly give satisfaction, even though you were already late for a meeting. Nevertheless, you might not have done it if you were not thus raised. You just did what was right.

Certainly, the act in itself cannot be used as (final) evidence (Paavola 2002). Even in the case where a person risks his/her life trying to save somebody, it may be argued that if the attempt had not been made, that person would have suffered a guilty conscience for the rest of his/her life. In the end it is a mere calculation of pleasure and pain that 'drives' an individual to jump into the cold river.

This is to trivialize the argument beyond reason. Certainly, going against an internalized norm will cause a sense of guilt that could be viewed as a cost to be compared to the gain. This is one way that such a norm works. This cost or pain is, however, the effect of a social creation, of internalizing the norm. The feeling of guilt is exactly a sign of that.

Moreover, many acts are obviously performed where what is done is simply to follow the norm for what is right in the actual situation without thinking about consequences, the possible upcoming level of guilt and so on. Instead, I think, most of us would rather consider a person who is always calculating the potential guilt feeling of not following conventions and norms of society to be rather abnormal. Acting socially is about adhering to something – following a commitment. It is not a calculative business.[16]

The above observations have specific importance for the environmental field to which we shall turn more systematically in Part IV. To suggest a line of reasoning picked up there, I shall offer a short remark. The environment, the physical and biological processes of nature, produces interlinkages between humans when they act. If one releases a poisonous substance into the atmosphere or into water, it will by necessity influence others through the web of interconnected processes. If one destroys a lake, others cannot enjoy it. This is a typical problem structure where moral considerations do appear (Etzioni 1988).

Put the other way around: is it sound to base decisions in such situations only on personal utility considerations – that is, to build decisions on market allocation principles? This implies building them (only) on individual calculation. One might rather ask: is it reasonable in such a situation to argue that preferences are really private? The preferences of one influence the opportunities available to others. If I prefer consumption to a clean environment, then through my acts I reduce the possibilities for others to live in one. These others may then want to reason with me regarding what are reasonable or defendable preferences. This kind of reasoning – that is, *communicative rationality* to follow Habermas (1984) – might be viewed as more appropriate than summing individual price bids. Communication about what we should aim at together is then an aspect of social rationality. Put more precisely, it is its *dynamic aspect*. It is about reasoning together about which solution should be sought for the collective sharing of the common good. It is about developing, criticizing and testing arguments concerning which norms or behavioural rules should be supported. This applies to existing norms as well as to norms being developed to solve new issues or conflicts. While in the case of individual rationality, utility or willingness to pay forms the basis, it is the *argument* that is the core of social rationality.

According to O'Neill (1998) the most profound element of the neoclassical revolution in economics was to remove from the equation any question about what is a good or right way of living. Due to the interdependencies constituting life, it is odd to individualize issues concerning the environment. On the other hand, if you think of the world only in market terms – that is, in terms where no physical interlinkages exist – it does not seem so strange. If I buy a pair of gloves, there are still others available to you. In an idealized competitive market our acts are not linked in the way we observe in, for example, the social or physical environment. The problem is that there are more institutional arenas; there are many other types of goods and values involved in life than just the marketable ones. Maybe these things are linked?

5.4.2 Rationalities Vary across Different Institutional Contexts

From a classical institutionalist point of view, what is rational to do is socially or institutionally dependent. Instead of searching for a logic that may be common to all spheres of life, we should just accept that there are different institutionalized spheres existing and ask: why have these come into being, and what characterizes their logic – their rationality?

In accordance with the ideas of social constructivism, Etzioni suggests that even individual rationality has a certain social aspect to it. Referring to Srole (1975) he argues that to create the 'I' a 'We' is actually needed. Specifically he refers to research showing that people who are left alone lose their ability to act rationally on an individual basis. Not only social belonging, but even individual rationality may seem to depend on the social sphere.

Next, as we have suggested, rationality is not only about what is 'best' for the individual. It is more generally about choosing the appropriate means to defined ends. These ends may be individually defined – as instituted in the marketplace. However, they may also be established to support, for example, the creation or provision of a common good. This brings the social forms of rationality on stage and consequently also different logics concerning the ways we go about attaining goals – that is, what are the appropriate means.

As a first introduction, recall the point made in Chapter 2 that both the agents and their goals are influenced by the institutional structure. Firms seek profits, bureaucrats seek promotion, politicians seek increased power/more votes and so on. The different goals emphasized, even what becomes self-interest (O'Neill 1998), are themselves influenced by institutional structures like the firm and the bureaucracy. Promotion demands a hierarchy, be it a private or public bureaucracy. This has to be institutionalized.

Box 5.1 exemplifies some characteristics of five important institutional arrangements concerning the type of rationality involved and the way it is instituted via the forming of characteristic roles. All five arrangements are here treated as 'ideal' types.

BOX 5.1 DIFFERENT INSTITUTIONAL SPHERES AND RATIONALITIES: IDEAL TYPES

Institutional system	Rationality and type of interaction	Roles
The market	Individual rationality Utility maximization or satisficing Exchange	Consumer and producer
The firm	Individual rationality Profit maximization or satisficing Command	Employer and employees
The family	Social rationality Caring Norms, reciprocity Communicative process	Parents and children
The community (the civil society)	Social rationality Norms, reciprocity Communicative process	Neighbours Friends Members of civil organizations Working collectives
The political arena (the state)	Individual or social rationality Communicative process	Politicians and voters Citizens

The market is the archetype of an institutional system fostering maximization. Maximization is an unattainable goal. Markets with their institutional structures can be viewed as a way to make it easier to approximate the ideal. The institution of money is one such structure. Advertisement, stocks and stock exchange are others. In real markets, information problems are pervasive, resulting not least in rules to counteract cheating, false or 'excessive' advertisement, insider trading and the like. Common norms of good business practices have become important, but are actually counter to the basic logic of the institution.

The firm is in the ideal form based on much of the same instrumental logic as the market. It is again maximizing/satisficing – in this case of profits – which is the fundamental logic instituted. We still observe a distinct difference compared to the market. This arena is one of command, not exchange. A basic issue becomes how to handle various control problems. This might be solved by systems of carrots and sticks, following the idea of individual rationality. However, we also observe initiatives to create processes for developing, for example, a worker's/employee's identification with the firm. Even here a wider perspective on rationality may come into play.

Moving to the sphere of the family, the basic logic is fundamentally changed. While some economists, as we have seen, may even view the family as a mere 'trade', the existence of concepts like loving and caring make most people realize that there is more – much more – to this story. The role of the parents, as instituted, is that of support, of raising children, and creating durable relationships. The internal logic of the family is that of fairness, norm building and reciprocity as opposed to exchange. Children do not pay for the various elements of the goodnight 'ceremonies'. While it may be tempting to give more to the nice child, parents may feel that it is not right to do so. The dominant norm is that any child deserves care.

Again, the types of relationships involved may vary across societies. 'Trade' may even sometimes be involved in the form of marriage portions. None the less, it is actually these variations which show that social construction is involved and that different logics may apply. There is nothing purely natural to any institution. Which values to support and which solutions serve our goals the best has always to be reasoned over. Yet, the fact that family institutions exist all over the globe indicates that exchange is not the typical logic for this kind of relationship.

The community arena, or civil society, is the one representing the most variable structures of the five arenas of Box 5.1. It may cover everything from rather loose relationships in a neighbourhood to quite formalized common property arrangements, organizations, and working communities. Nevertheless, many commonalities are observed, not least since reciprocity and norms concerning fairness seem to be of crucial importance for building the necessary social coherence for the relevant group to function.

Turning last to the policy arena it seems that this field can fit both perspectives – that is, that of individual or social rationality. We observe policy arenas governed dominantly by individual or interest competition (pure political games) to situations dominated by strong social rationality and communicative processes. Hence which logic is instituted becomes crucial. Certainly, no political arena could exist based on pure interest competition – the very basis for politics would erode. On the other hand, politics is about defining which interests or values should thrive. Believing that competition between interests and fights over solutions could be institutionally removed, would because of this seem rather naïve. Nevertheless, the institutional structures defining which are acceptable and which are not acceptable acts in this arena play a decisive role in defining its characteristics. Many 'policy failures' observed around the globe follow from either weak insights about theses issues, or lack of institutional capacity to do anything about it.

We therefore also observe great variation across societies concerning what is allowed for at this institutional level of a society. In some cases the political arena has degenerated to pure nepotism. Even in what is normally called well-functioning democracies there is great variation concerning what measures are allowed in election campaigns. It is especially interesting to see to what degree the 'market mechanism' is allowed to intrude into the arena. In some societies individual grants or grants from firms or organizations are accepted without limit to fund campaigns. In others this is considered a bribe. In some societies political advertisements dominate elections, while in others the focus is more on debate and the testing of arguments.[17]

As emphasized, the logic of the political arena is to decide which interests should get protection from the collective. Certainly, this creates a difficult balancing between the 'common' interest and the interests of specific groups. This issue will be addressed in Chapter 8. Here we shall just make the point that the way we institute the political sphere will influence what is acceptable behaviour and define what is allowed concerning the priority of own interests, the interests of specific groups and the polity more at large.

Based on the same line of reasoning as above, Sagoff (1988) makes a distinction between the consumer and the citizen. While the role of the consumer concerns individual utility, that of the citizen is quite different. It is about defining and protecting the common good, be it social security systems, distributional issues more at large, the natural environment or the capacity to form institutions. While in the case of the consumer good, ability or willingness to pay is the important allocative mechanism, it is the *argument* and hence communicative rationality which is the comparable concept for citizens in the public realm.

Certainly, consumers may also act to support the notion of the common good while acting in markets. Etzioni (1988) offers a series of examples showing that people may involve moral reasoning even when they act in the marketplace. They may boycott certain products depending on who the producer is (reaction to political systems, political conflicts and so on) and the way some products are produced (use of child labour, polluting practices and so on). They often do not cheat even if the cost of being detected is very low. None of these reactions is rational in the sense of maximizing individual utility.

These are important observations and could serve as arguments against the basic idea that rationality is dependent on the institutional structure. If some people act reciprocally or morally in markets and others selfishly in more social arenas, there seems to be no clear relationship between behaviour and the institutional arena. First of all, the argument is not that institutions determine 100 per cent what is rational to do. The individual cannot be reduced to just the structure. There is always an important element of individual choice and adaptation, as different individuals follow norms to a different extent. One should rather talk about the dominant type of behaviour that an institutional structure fosters.

Second, even markets are dependent on a set of norms – for example, norms against cheating – and comprehensive processes are involved in trying to build good norms of business conduct. The various scandals in the segment of corporate businesses over the last few years and the reactions to these serve as good examples. Third, norms are not just 'out there'. They become internalized. Through that process they grow to be part of the individual and her/his character or integrity. This implies that they not only vary between arenas, but they also have to vary between individuals across the institutional setting. Some act more 'citizen like' in any arena, while others are more individualistic.

Actually, as all institutional arenas exist simultaneously, on many occasions they overlap. Society is a network of institutional arenas. To illustrate, somebody operating in the market, such as the shopkeeper downtown, may also be our neighbour. Thus, caring for the local community may influence our attitude towards him/her when we meet in the marketplace. The difficulties we experience in this mixing of roles are not least observed when we trade with friends. Many solve these conflicts or indeterminacies through explicitly stating that it is, for example, purely trade and not an act under the code of friendship that is about to take place. However, we also observe many disappointments in such situations. Who has not experienced the reaction that 'he took advantage of me being his friend and paid less than the normal price'. Most of us have also thought that 'I could not demand the full price from a friend. I just let him cover my external expenses'. The first kind of behaviour is in the long run likely to ruin the very essence of that friendship, while in the latter it may strengthen it since the use of own labour was made a gift. The rather neat distinctions observed here will make the analysis, the understanding of what is going on, challenging, but do not undermine the idea of institutional spheres. They show that when spheres overlap, extra sensitivity for the interplay of various motives is demanded.

In any society, decisions have to be made concerning which institutional arena is best suited for treating which issue. Some goods are explicitly kept outside of the marketplace. Walzer (1983) talks about spheres of blocked exchange. In general this has to do with (how we perceive) the character of the goods, their multidimensionality really. If only exchange value is perceived to be of relevance to us, exchange will dominate. If other dimensions are of importance, it becomes rational to establish spheres based on a different logic. The choice depends on which dimensions are found important and should be assessed or defended (Little 1957).

Health care is a good example of the various issues involved. Concerning this good or service, we observe very different solutions both across societies and across various dimensions of health. The basic question here is whether good health is a right for everybody or not – that is, whether the allocation of health-care services should be governed by market solutions or by reciprocity. The problem of information asymmetry is also of some importance and may influence the institutional structures chosen. The doctor knows more than the patient, and the patient is dependent on the doctor's decisions in a way that goes far beyond 'trading' a standard commodity. This is explicit in that the quality of what is offered is first observed after the treatment is finished.

If health is considered to be a right, the community will pay for treatment – from public budgets. Along the continuum from cancer and heart treatment – that is, saving lives – to plastic or 'cosmetic' surgery, we may, however, observe different 'switching points' from public to private payments, depending on what is considered a public responsibility. Some countries offer public treatment of a minimal kind, taking care only of basic needs for those not able to pay themselves. Others have the same system both for those capable of paying and for those less able. The costs are public – that is, covered by the tax system.

Even though treatment is paid for from the public purse, it may not necessarily be undertaken in public hospitals. Private hospitals may deliver while the bill goes to the public authorities. The choice societies make here will depend on several factors – for example, the costs involved, how the control or quality problem is envisaged, the basic values concerning what health is and so on. Creating a competitive environment – that is, private production – may be thought to reduce the costs. Control problems – for example, the development of 'luxury' treatments in the interest of both the hospital and the patient, or the problem of detecting low-quality products delivered at high prices – may still imply that public production with its internal control systems and professional codes is favoured. Finally, public and private delivery seems to evoke different expectations concerning whether the good should be of the same quality independent of the method of payment – that is, public provisioning – or the quality should be relative to what is paid – that is, private provisioning. The rationale activated may also depend on these types of signal.[18] Anyway, as this example illustrates, the decision must be made on evaluating arguments concerning which solution is considered the best or most appropriate. It is communicative rationality at work.

Other areas where trade is restricted or not allowed concern personal integrity/the human being itself, education, retirement schemes, public offices, criminal justice, freedom of speech, friendship, human body parts and so on. Walzer (1983) lists a total of 14 spheres from which the logic of exchange is exempted either totally or to some degree. The value dimensions fostered in this way are not only those of equality and redistribution, the above points concerning the inherent qualities of the good are also of great importance – that is, the interest in protecting qualities which would be destroyed or perverted if trade is accepted.

The fundamental logic of an institutional system as described in Box 5.1 may also be perverted. Walzer discusses the implication of this. He concludes that such occurrences do not violate the existence of various spheres. Rather they are strong proofs for the existence of different types of rationalities: 'Dishonesty is always a useful guide to the existence of moral standards. When people sneak across the boundary of the sphere of money, they advertise the existence of the boundary. It's there, roughly at the point where they begin to hide and dissemble' (ibid.: 98).

The above discussion is of great importance to the issues we shall later raise concerning which type of institutions should be used when making choices about the environment. Basically, this question will be about which type of rationality we find logical or reasonable for this sphere. One may argue that individual preferences should govern. In that case, markets – or, due to high transaction costs, perhaps simulated markets – and individual willingness to pay should govern. The cost–benefit analysis is a response to this logic.

An alternative to this would be to favour institutional structures built around social rationality, the role of norm building and of the argument. This would favour various types of deliberative, participatory procedures like consensus conferences, citizens' juries, political decision making and the like. While we shall not evaluate the various arguments here (this has to wait until Part IV), it is crucial to acknowledge that the basic issue is about what type of rationality we want to foster when treating the environment.

5.4.3 Irrationality

Defining rationality also implies defining implicitly what is irrational. Based on the preceding text, three observations should be made. First, inconsistent preferences may be an important form of irrationality. Second, if a person does not do what s/he prefers/has decided to do, s/he also acts irrationally. Finally, according to the perspective of social rationality, holding preferences for which no account or argument can be given is irrational.

The second and third types warrant some comments. While a person characterized by 'weakness of will' (see Chapter 2, Section 2.4.2) must be considered irrational according to the second point, what then about those having 'false consciousness'? Obtaining what one intends also depends on right beliefs concerning the effects of different acts. Yet, since information is costly, it would be wrong to describe acts based on mistaken cause–effect relationships as a sign of irrationality, since it is impossible to make a rational individual calculation of the gain of efforts directed at increased insight/knowledge. Rather it is an argument for accepting bounded rationality as rational and not as an example of irrationality. Nevertheless, the fact that people may deny the existence of certain information or rely on information that is considered dubious by most other people, casts some doubts on how to draw the exact line here.

The third type concerns the argument behind holding a certain value or preference. For the individualistic model this issue makes no sense. In the case of social rationality, it is the core question when evaluating what is rational – that is, when discussing what is good or reasonable. Being able to give an account of the held value/preference or the chosen act is necessary to call it reasoned. This does not mean, however, that everybody must accept these reasons. Certainly, one may hold different perspectives on what is a good life or a good society. The existence of interest and value conflicts[19] cannot be used as an argument against the acceptance of the very concept of social rationality. It would be the same as saying that the cost of information rules out any form of (individual) rational choice.

What then about internalized norms? I have already emphasized that the reason behind norms tends to be 'forgotten' while the practice still prevails. While the basis for the norm is/may be reasoned – that is, based on an evaluation of what creates a good community – the persons replicating the norm may not be aware of or be able to give any arguments for it. This point is not easily settled due to the social and historic characteristics of norms. One might say that it is irrational to follow a norm if, after being given time to think about reasons for its existence, one is still unable to come up with an account that seems reasonable. What then about the argument that 'I choose to stick to the norm despite the fact that I can find no good reasons for doing so. I have trust in the evaluation of the larger community that has developed and supported it'? Or what about the statement that 'It seems to work. Why not stick to it?'. Are these irrational conclusions? Given positive information costs, I think not. Nevertheless, at this point there are unresolved problems.

As emphasized, the first and second accounts of irrationality refer directly to individual rationality while the third refers to the perspective of social rationality. I shall close this section with some comments on their internal relationships. First, the issue of consistency not only applies to individual rationality, but in the case of social rationality, consistency must also play a role. However, since logics are different in different spheres of life, the perspective of social rationality must accept that people do things in one sphere that they would not find acceptable/preferable in another. While appearing to be inconsistent from the perspective of the neoclassical model, from the broader perspective this is rather signalling the existence of different spheres of rationality. It must be admitted, however, that to draw the line between such spheres is not a simple task.

The issue of 'weakness of will' is especially interesting when comparing the individual and social accounts of irrationality. If weakness of will implies that a person does something other than what s/he has decided, it must be termed irrational both under the individual and social perspectives. At the same time, there are situations where the model of individual rationality will define something as weakness of will while under the social perspective it appears to be something else. Here it may be understood as rational in the sense of being a 'self-sacrifice'. Holland (2002: 19–20; original emphasis) catches this very well:

> A first criticism of the belief/desire model[20] is that it appears incapable of explaining certain common phenomena such as weakness of will and self-sacrifice. There is a formal similarity between the phenomena. On the one hand we can describe weakness of will as someone's (apparently) sacrificing his or her own best interests, where this is perceived as an *ignoble* thing to do; and on the other hand we can describe self-sacrifice as someone's (apparently) sacrificing his or her own best interests, where this is perceived as a *noble* thing to do.

As we see, the same act can be given very different interpretations depending on how we understand the context. What is seen as rational, and what is to be termed irrational, may shift dramatically, depending on how we interpret the social sphere – what is noble or ignoble. This cannot be determined on the basis of the physical act itself.

5.5 SUMMARY

The reference point also in this chapter has been the neoclassical model and its specific understanding of rational choice as maximizing. It implies that preferences must be rational or consistent and that the individual chooses what maximizes her/his individual utility. Rationality is understood as context independent. There are some fundamental problems concerning the ability to test the core assumption of rationality as maximization. First, it demands stable preferences, since preferences cannot be observed independently of choice. What may seem to be irrational behaviour, may just be the result of changed preferences. However, if inconsistent preferences are observed, we do not know what is refuted: the maximization hypotheses, the idea of stable preferences or both.

Another problematic issue is the fact that if information and transaction costs are positive, rational choice as maximization becomes indefinable. The observation that information gathering and processing is costly, has therefore motivated the development of the model of satisficing and bounded rationality. Here it is argued that the individual shortcuts the information problem by defining a 'satisfactory' result in the form of a target. The solution that first passes the test is then chosen.

While even the school of bounded rationality, as applied to economics at least, looks at the self-contained individual and 'target reaching' as universal forms, the alternative is to look at rationality as a plural concept influenced by the kind of institutions in place. This is the idea basic to social constructivism and developed also by the classical school of institutional economics. It follows naturally from how social constructivists view institutions and behaviour. According to this position, the alternative to rationality as maximizing is not foremost irrational choice, but the existence of distinct types of rationalities. Specifically, the role of normative and reciprocal behaviour is important.

Thus, we can actually divide rationalities into two main types – the individual and the social/cooperative. The first class covers individual maximization of utility/profits and the idea of bounded rationality/satisficing. It is 'I' orientated and fostered not least by the creation of markets. The second class covers normative and reciprocal rationality. It is 'We' orientated and fostered by different types of communities. These distinctions are supported by a lot of empirical evidence. People share even though they experience individual losses by doing so. They retaliate even though it is costly for them. People follow norms even though they cannot be disciplined for breaking them.

We see that the degree of 'I' or 'We' logic varies across institutional arenas. These may actually be viewed as constructed to serve various types of rationalities since they offer certain structures supportive for the specific logic. Nevertheless, the fact that institutions in the form of, for example, norms become internalized and in this way part of our personalities, implies that the 'I'–'We' divide is not clear-cut across institutional arenas. While the market supports individual maximization, there is also reciprocity or normatively motivated behaviour in this sphere. Conversely, while the family or the community supports care and involvement, strategic or purely selfish acts are also part of these arenas. None the less, the basic logic of this kind of institutional structures is fundamentally different from the market.

Seeing rationality as institutionally dependent implies that the choice of, for example, 'which action is most efficient' is not only a technical issue. It depends on which rationality and therefore which values one wants to protect. The choice of institutions defines this. Some societies decide to create, for example, health care as a market good, while others build public systems to finance and also produce these goods. In some situations a market logic is applied when making environmental choices, in others it is not. The choice made here is about how we define the good involved and which rights individuals should have in relation to it. These issues go beyond that of technical efficiency. They are about which values we as a society want to foster and which interests we decide to protect. These are issues that will be given ample space for discussion in later chapters.

NOTES

1. We shall return to this issue more fully in Chapter 6.
2. In a more recent publication, Postlewaite (1998) makes a similar remark. He specifically emphasizes that models can have predictive power only to the extent that some behaviour can be inconsistent with the model. He continues: 'The assumption that agents choose those actions that maximize their self interest, however, puts no restriction on what might be agents' self-interest. If a modeller is free to specify what constitutes an agent's self-interest, he or she can simply posit that an agent's self-interest is such that any particular behaviour gives the most satisfaction; that behaviour is then consistent with maximization of self-interest' (p. 782). Thus no behaviour can be found to be irrational.
3. Recall, if information is cost free, then no uncertainty will exist. Knowledge about what will happen can be acquired for free. Then everything is by definition known with certainty.
4. Both Kahneman and Smith were awarded the 'Prize in economic sciences in the Memory of Alfred Nobel' in 2002 – which can be interpreted as a nice balance between competing positions.
5. Simon received the 'Nobel Prize in economics' in 1978. The reference is to the lecture he gave upon receiving the prize.
6. Bayesian statistics or techniques combine prior information sample data to produce estimates for future events. The technique provides a way of including subjective impressions and theoretical elements in quantitative analysis. It seems to be used as if this technique solves the Morgenstern self-reference problem. I do not find this to be reasonable.
7. Nørretranders (1991) offers a series of interesting observations, ranging from thermodynamics to neuro sciences, that are highly relevant to the above as he writes on the understanding of consciousness. He concludes – much like Georgescu-Roegen (1971) – that consciousness is establishing order. Or put the other way around, it is about getting rid of disorder/entropy in the form of information overload. While the amount of information reaching the human body is about 11 million bits per second, the amount we can

consciously register is between 10 and 30, depending on our skills. This implies that we need some way of sorting out what is important information. As Nørretranders also contends, institutions – not least in the form of conventions – help us to handle this. Thus, we 'get help' in sorting out what are the most important things among those that we have been exposed to.

8. Actually, defining irrationality this way covers the issue only in a partial way. I shall return to a more complete definition later in this section – that is, defining irrationality cannot be done before the concept of social rationality is explained.
9. It should be mentioned here that there are also developments within the more individualist perspective on rationality that take on board the question of moral motivation in a search for explanations of the same kinds of observation as those made by Etzioni – for example, Brekke et al. (2003). This is a growing field, although it is yet too early to say if any common conclusions can be drawn concerning the 'I' and 'We' perspectives.
10. There are other parallel concepts. Söderbaum (2000) advances the idea under the name of 'the political economic person. Sjöstrand (1995) develops the concept of 'homo complexicus' very much along the same lines as those referred to here.
11. The distinction between selfishness and reciprocity is well captured by the following familiar story: Peter and John are given a plate with two pieces of cake – one larger than the other. Peter offers the plate to John, who takes the larger piece, but recognizes that Peter is unhappy. Thus he asks: 'Which piece would you have taken?'. 'The smaller one', Peter announces proudly. 'There you are then' is the immediate response.
12. See Chapter 3 and the discussion around Figure 3.1.
13. Implying that the future is not discounted 'too much'.
14. A simpler model explaining cooperation in situations like the prisoner's dilemma is the so-called 'tit for tat' strategy (Axelrod 1984). It is based on bounded rationality instead of the 'hyper-rationality' of the Folk theorem. Here the logic is that everyone observes what the others do and cooperates if others cooperate and defects if others do so. Observe that this strategy may be interpreted as a form of reciprocity.
15. For an argument against interpreting the empirical results as refuting the standard rational choice model, see Binmore (1994), who in my mind goes quite far in his defence of the model.
16. In relation to the above discussion, see Crowards (1997) and his distinctions between 'selfish', 'reciprocal' and 'selfless' altruism. The last two concepts resemble what is here called reciprocity and normative behaviour. This reminds us again that the literature has many different concepts that cover much the same issues or ideas. This results from the fact that many equal observations are made by representatives of different disciplines, while they still anchor them in different traditions and conceptual structures. Nevertheless, the basic logic behind the various wordings follows much the same kind of logic.
17. At the present time there is an ongoing debate in Norway, on whether to accept political advertisements on television. A dominant argument behind the proposal to allow this is 'freedom of speech'. Moreover, the question is asked, why should political parties not be allowed to advertise when firms may do so? One wonders whether it is the general increase in advertisement-based channels that really drive this. However, I am quite confident that if accepted, it will further ruin the quality of the public space as an arena for debate and the development and testing of arguments. It will certainly make access to the public sphere more dependent on monetary resources too.
18. This is an issue to which we shall return in Chapter 6.
19. I shall return to a more specific definition of these concepts in Chapter 12.
20. That is, the neoclassical model as outlined here.

6. Preferences and values

The perspective offered in Chapter 5 – that of plural rationalities and the institutional dependency of rationalities – will in this chapter be taken a step further into analysing the role of institutions in forming preferences. As we have seen, the neoclassical understanding of rationality entails that preferences must satisfy certain demands like completeness and transitivity. However, they must also be institutionally independent – that is, they should be stable across institutional contexts.[1] The latter is crucial if we define choices as strictly individual. It is a necessity for a model built on methodological individualism of type 1 to be consistent.

The alternative to stable and given preferences would be to accept social circumstances to influence them. This would be to acknowledge that building norms is to define what it is to be a person – to be human. Further, socializing individuals to value and respect these norms, as they grow up, is to create this human being. The social constructivist position, as formulated in this book, sees this social dimension as creative and liberating. The issue is wider than that of forming preferences for various items or goods. It is about developing and supporting values that a society or group want to adhere to.

This chapter will provide a systematic treatment of these issues. It will be divided as follows. First, we shall go a bit deeper into the neoclassical understanding of preferences (Section 6.1). We shall look at (a) what it means, according to this theory, to prefer something, and (b) the importance of the core assumption of stable preferences. Second, we shall develop the distinction between preferences and values as the concepts are used in this book (Section 6.2). Third, we shall turn to the classical institutional position – the view that preferences and values are also social and that they may change according to social circumstances (Section 6.3). I shall both offer a theoretical foundation for this view and present a set of empirical evidence.

Fourth, we shall discuss the issue of plural values or preferences (section 6.4). Implicit in the discussion about plural rationalities in Chapter 5, lies the suggestion that preferences are not only context dependent. There may be no common denominator across them. They may be incommensurable. This has important implications not only for choice theory, but also for how to assess what are preferable states for society.

Section 6.5 comments on some of the implications of an institutionalist perspective on preferences/values and their formation. The most wide-ranging issue here is that, given the social dimension of preferences and values, the choice of institutions cannot be made only on the basis of instrumental considerations. It is not only about their capacity to reduce transaction costs. While reducing such costs is also an important question, a much more fundamental choice is involved – the choice of whom we want to become.

6.1 THE CONTEXT-INDEPENDENT PREFERENCE OF NEOCLASSICAL ECONOMICS

6.1.1 Preferences and Utility: From Substantive Content to a Ranked Order

The aim of this section is to briefly describe the development of the preference and utility concepts in economics from the perspective of the classical economists[2] to that of the modern neoclassical position. The process began with an understanding of utility – as usefulness. This was the way the concept was comprehended by the classical economists. It next took on the meaning of 'happiness' – that is, the measure of pleasure and pain – as understood by utilitarian philosophers and later taken over by the pioneering neoclassical economists. The latter, despite relating utility only to mental states, still accepted that it could be cardinally measured – that is, measured by a metric similar to those of weight and length. The development within neoclassical economics was finalized in the 1930s with the so-called 'ordinalist' revolution. Here the conclusion was that utility was just an index of preferences, and could only be measured ordinally – that is, preferences could only be ranked. Hausman makes the following observation concerning the meaning of utility and preferences in contemporary neoclassical economics:

> Good economists sometimes speak misleadingly of individuals as *aiming to* maximize utility or as *seeking* more utility, but they do not or should not mean that utility is an object of choice, some ultimately good thing that people want in addition to healthy children or better television. The theory of rational preferences or choice specifies no distinctive aims that all people must embrace. Utility is just an index of preferences. (1992: 18; original emphasis)

It took a long time to get to this position. If we go back to classical economists like Adam Smith, the perspective was very different. He distinguished between use value and exchange value. Utility flowed from the value an object had in its use. This was a concrete or substantive understanding of the value of the good. It captured its real ability to cover human needs like food (calories, vitamins), shelter and so on. Smith was in many ways an objectivist concerning the substance of well-being. To live well included, according to him, covering a set of objectively definable needs like physical requirements, but also intellectual and moral character (Smith [1759] 1976).

Furthermore, the classical economists looked at exchange value as different from the ability of a good to create utility. Its value in exchange was, as they understood it, relative to the amount of labour invested in producing it. This was the labour theory of value, allowing an objective measurement of exchange value: the hours involved. These economists were bothered, however, by the 'paradox of value' also observed by the ancient Greeks, the fact that goods having high immediate use value, like water and air, had no exchange value (Smith [1776] 1976).

Parallel to the development within classical economics, we observe the establishment of quite a different perspective, the theory of utilitarianism.[3] From the beginning, utilitarianism was developed by philosophers. It was based on a so-called 'hedonist' interpretation of utility. It gained much of its position through the work of Jeremy Bentham (see also Box 6.1). According to his interpretation, utility was defined as 'happiness' – as the gain of pleasure and the avoidance of pain.[4] Thus also for Bentham, utility had a concrete meaning, still clearly different from that of the classical economists. It was the capacity of actions or goods to create happiness that was the essence of its value. The perspective was changed from the ability a good had to cover (objective) human needs (use value) to its influence on the (subjective) mental state (happiness). The importance of intensity and duration for the level of pleasure and pain was accentuated (Bentham [1789] 1970). Bentham also stressed the possibility of calculating the amount of happiness. It was homogeneous and thus quantifiable as he saw it.

Bentham's work also played a role in the process of establishing liberalism and individualism, the development from the mid-eighteenth century to free society from the powers of not least the aristocracy. Bentham identified law and custom, the Common Law, with tradition, and recognized that it was supported by an authority system based on the rule of aristocracy. This he was against. He wanted to turn the focus away from the principles of *duty* embedded in the existing traditional order and towards the new ideal – the greatest happiness for all. Hence, in pursuing liberalism, Bentham went against tradition. This is a good example of the fact that law built on 'tradition' is no more neutral than law built on, for example, a parliamentary process.[5]

BOX 6.1 THE 'WONDERFUL YEAR' OF 1776

In a sweeping passage, John R. Commons describes the changes implicit not only in Bentham's work, but also in the general 'spirit' of the time within which he lived – the era of enlightenment and burgeoning individualism. This change took the form of an attack not least on custom and traditional (aristocratic and religious) power and establishing a basis for 'rational' development of society and its productive forces to increase wealth:

> It was Bentham who separated Economics from Law and Custom. The 'wonderful year' 1776 produced Bentham's *Fragment on Government*, Smith's *Wealth of Nations*, Watt's steam engine, and Jefferson's Declaration of Independence. The first was the philosophy of happiness, the second the philosophy of abundance, the third the technology of abundance, the fourth the revolutionary application of happiness to government. Eleven years before, Sir William Blackstone had published his *Commentaries on the Laws of England*, agreeing with Smith's Divine Origins but finding their earthly perfection in the Common Law of England. Jeremy Bentham's *Fragment* was a critique of Blackstone, substituting Greatest Happiness and Legislative Codes from Divine Origins and Common Law.[*] This was followed in 1780, by his *Morals and Legislation*, revised in 1789, wherein he eliminated *duty* and derived ethics from happiness. (Commons [1934] 1990: 218; original emphasis)

Thus, the focus on individual happiness explicitly took the form of an opposition to the existing normative order of society – a dislike of the aristocratic rule was generalized to an aversion against anything but individual utility. The mistake made by the utilitarians was to equate the existing norms with norms *per se*. It is perhaps easy to understand that this could happen in the given political context, but it was still a fundamental error.

Commons also comments on Smith's idea of abundance. In Smith's mind nature was rich. He adhered to a theology of 'divine beneficence'. According to Commons: 'If there is abundance of nature's resources, no person can injure any other person by taking from him all he can get, if he does this by exchanging his own labor for that of the other. The other has abundance of alternatives to which he can resort if he is not satisfied with the terms of exchange offered' (ibid.: 161). Smith saw abundance as a creation of God.

Note: * Common law – law based on tradition.

The hedonistic understanding of utility was introduced into economics not least by the work of John Stuart Mill ([1861] 1987), who embraced Bentham's perspective that utility was a measure of mental states. He went one step further, however, by emphasizing that pleasure had not only a quantitative, but also a qualitative aspect which was absent in Bentham's writing (ibid.; see also O'Neill et al. 2005). Following from this, Mill was both engaged in and bothered by the problems involved when trying to measure utility.

The neoclassical revolution in the 1870s[6] is characterized by the invention of the term 'marginal utility'. This shift was inspired by the utilitarian ideas about utility as a subjectivist feeling of happiness. By making this turn, Jevons claimed to have solved the 'paradox of value'. A relationship between price – that is, the exchange value of a good – and marginal utility was established (Jevons [1871] 1957). Water and air was not demanding a price simply because it was abundant. While its total value was great, its marginal value was zero. Edgeworth[7] (1899: 602) could conclude similarly that 'the relation of utility to value, which exercised the older economists [the classical economists] is thus simply explained by the mathematical school [the neoclassical economists]. The value in use of a certain quantity of a commodity corresponds to its total utility; the value in exchange to its marginal utility (multiplied by its quantity)'.

While neoclassical economists such as Jevons and Walras, and later also Marshall and Pigou, thought of utility in the form of pleasurable sensations, they still generally favoured a cardinal measurement of it. The utilitarianism of the early neoclassical economists had, moreover, a rather egalitarian flavour. Based on the idea of declining marginal utility, some concluded that the 'utility for all' would increase the most if more were given to the poor. One kilogram of vegetables in the hands of those having little, would increase utility more than giving this kilogram to those who already had plenty. This position was endorsed by most neoclassical economists up until the 1920s – for example, Marshall ([1890] 1949) and Pigou (1920).

Some economists of the late nineteenth century still did not support this egalitarian idea, though. Edgeworth ([1881] 1967) was rather disturbed by the conclusions drawn by his colleagues. He specifically argued that individuals differ in their capacity to enjoy happiness. Egalitarianism would not increase aggregate happiness. Rather, the most should be given to those able to enjoy it the best.[8]

The perspective that utility could not really be compared across individuals took hold gradually among economists. Edgeworth developed the indifference curve as a way to handle what he interpreted as a lack of a common denominator across individuals. Pareto ([1906] 1971) took this idea further, defining what was later called a 'Pareto optimum' – that is, the idea that an optimum is reached if it is impossible to further increase the utility of somebody without reducing the utility of someone else. This definition of an optimal level of 'happiness' circumvents the issue of cardinalization and interpersonal comparison by demanding that nobody should be worse off. We observe a move from thinking in maximum aggregate happiness or utility to changes which demanded that none loses. This new rule of optimality was dependent on the initial distribution, though, and it is criticized for implicitly supporting the status quo distribution.

Thus, Pareto rejected the idea of cardinal utility and moved to the concept of ordinal utility instead. This implies that utility can only be ranked, not compared on a proportional scale. We can say that something is better for somebody – for example, as revealed through choice or a higher willingness to pay – but we cannot, according to this view, say how much better. It was some time before this position became dominant within the discipline, though. What finally seems to have persuaded the profession was a debate in the 1920s and 1930s about what constituted a sound science. This debate had its roots in the ideas of the so-called 'logical positivists' of the Vienna Circle. Their central doctrine was the verification principle – the principle that the only valid knowledge is the knowledge that is verified by sensory experience. The fact that a conscious sensory experience depends on a preceding development of concepts that can capture this experience – see our discussions in Chapter 2 – seems to have been ignored by the logical positivists. Instead the *distinction* between 'facts' (as 'pure' observations) and 'values' (as 'non-scientific' entities) was underlined.

However, utility could not be observed. Ever since Bentham it had been considered an 'inner' experience of each individual. Thus, when the neoclassical economists of the 1930s finally adhered to the criteria of facts and science as developed by the logical positivists, the conclusions of the cardinalists had to be judged unsound. In the process of 'freeing' economics of value statements of the kind the cardinalists made, the work of Lionel Robbins was finally very important. In line with the ideas from the Vienna Circle, he stressed the need for distinguishing between *positive* and *normative* economics – that is, between 'what is' and 'what should be'. Value judgements – normative conclusions – should not be made in economics (Robbins 1935). The ordinal concept of utility was in his mind a way to secure this. Finally, Samuelson (1948) argued, as we saw in Chapter 5, that preferences defined this way could be observed or revealed via the choices made. Thus, utility had become just 'an index of preferences' established by the rankings revealed when individuals made their choice. It said nothing about its value content, either subjective or objective.

We have now reached the state described by Hausman in the introduction to this section. However, Hausman also emphasized that there is a strong tendency even in contemporary literature to load the concept of utility with different meanings, not least the meaning that utility has real content – that is, in the form of mental states (compare Bentham) or even usefulness (compare Smith). This is easy to understand – most of us believe that what is called basic human needs are to a considerable degree common to the human species as such. However, this is not the position embraced by today's version of neoclassical economics.

6.1.2 The Stable – Context-independent – Preference

In neoclassical economics the issue of preference stability is as important as the issues related to their form and content. There are actually two different reasons why neoclassical economists look at preferences as stable and given – that is, as exogenous to the (economic) system. First, and most fundamentally, the position follows from the insistence on the primacy of the individual. If the individual is a product (also) of social forces, this primacy vanishes. Second, there is a narrower issue involved. Assuming preferences to be stable makes empirical studies of economic behaviour simpler, while not necessarily more valid. This was an issue discussed in Chapter 5.[9] Here we shall therefore concentrate on the first issue.

Gary Becker has been among the most explicit on advocating the position that preferences are stable: 'The combined assumption of maximizing behaviour, market equilibrium, and stable preferences, used relentlessly and unflinchingly, form the heart of the economic approach as I see it' (1976: 5).[10] Becker views the individual utility function as both unchangeable and beyond dispute (see also Hodgson 1988). Changes in tastes may be observed. These are still just apparent and run from a single, basic utility function. The problem with this position is not least that it cannot be tested. It is more like a truism. One can always claim that the change we observe is just an apparent one. The 'basic' utility function is still intact. Yet, one can also claim the opposite.

In practice, not all neoclassical economists follow the assertion that preferences are stable. There is a tendency to accept that changes may occur. Why this is so and the consequences of accepting such changes, is emphasized much less. The effect of ageing is a typical example. That age influences physical needs is an important observation, while still keeping the analysis to the level of the individual only. Social explanations are rarely discussed or considered as a possibility (see also note 1 to this chapter). This is probably because it would conflict with the basic ideas underpinning the model. Hence, preferences are normally taken as given in the analysis – they are exogenous to the economic study. If explanations are sought, there is a clear tendency to seek them in other disciplines that are also based on methodological individualism. Hayek is, as we saw in Chapter 2, clear on this. He thinks explanations – if at all tenable – are to be found in the psyche of the individual. We observe that Hayek is a consistent methodological individualist of type 1. Social forces are of no interest to the understanding of preferences.

Another relevant aspect is the question of how individuals know their preferences. If given, are they immediately known or uncovered? Do we experience all our preferences from 'day one', or do we become aware of them as we go around choosing? There is an increased tendency in the literature to recognize preferences as constructed or 'found' by the individual in the process of choosing. Hanemann (1994) acknowledges this trend both in social psychology and in market research. Not least in rather unfamiliar choice situations, people construct their attitudes in the process of choosing. Unfamiliarity is often the typical situation encountered when people become involved in environmental valuation. It is no surprise that there was a trend in the 1990s to take account of this constructive element in valuation studies – for example, Gregory et al. (1993); Fischhoff et al. (1999); Payne and Bettman (1999).

Hanemann points out that the constructive perspective is not really a challenge to economics: 'The real issue is not whether preferences are a construct but whether they are a *stable* construct' (1994: 28, emphasis in the original). Hanemann is right in this. Whether individuals must learn about themselves, does not challenge the core of the neoclassical model. However, stability in the meaning of context independence is important.[11]

In this literature, 'construction' is thus viewed as an issue for individuals. They learn about themselves. It is not also part of a social process as we have formulated the position of social constructivism. The differences in perspective are rather fundamental. But before we can engage more fully in that discussion, we need to look at another aspect of the issues raised so far: the question of the relationship between the concept of preferences, values and welfare.[12]

6.2 PREFERENCES AND VALUES: CONSEQUENCES OR RIGHTS

There is a tendency to use concepts like preferences, values and utility almost interchangeably in the economics literature. In the interlinked historical process of establishing economics as a discipline and the economy as being the marketplace only, we observe a gradual shift in language, in particular the content of words. In the short history of the concepts of utility and preference given in Section 6.1.1, we also implicitly observed a change in the meaning of the concept of value. As a first step it was reduced to meaning the same as utility, second it was equated with the price of a good – that is, its 'value' as exchange value in the marketplace.

Over the years the concept of 'value' has gained a specific meaning in neoclassical economics – as the ability of a good to satisfy one's innate desires or wants. It is individual and subjective. Apart from the labour theory of value, this is not the only understanding of the word. It is rather a very particular one. Instead, one could define value to cover views about what is a good life and a good society, principles concerning what is important and right to do (for example, Sagoff 1988). This is the way the concept has been used in this book. It is therefore important to make a distinction between preferences as expressed in the marketplace and values as describing ethical or moral beliefs. As earlier emphasized – especially in Chapter 3 – it is these values that are supported by social norms.

By saying this, I do not imply that values, defined as principles for a good life and a good society, may not influence markets. First, they will influence the demarcation of the market – that is, what may be traded or not. As mentioned in Chapter 5, some goods and/or services are defined as non-tradable due to ethical reasons, distributional consequences and so on. Second, the individual may make choices in the marketplace that are influenced by held values, not just immediate desires. The willingness to pay higher prices for goods produced under certain circumstances – for example, eggs from free-range hens – is an expression of a value to the degree that it is concerns about the hens' quality of life and not just the quality of the eggs to be eaten that is at stake.

The distinctions made here are partly parallel to the distinction made in the literature about how to evaluate certain outcomes in a society – that is, the *consequentialist* and the *deontological* positions. According to the consequentialist stand, an action should be evaluated as a means to some end. Only the consequences in that respect hold importance. Deontologists claim that there are constraints on performing certain kinds of acts even though they may produce a better result in the form of higher individual utility. There are values involved that should not be compromised.

Utilitarianism or hedonism is a type of consequentialism. In this specific case it is the maximization of individual welfare that is the focused consequence, explaining why the position is also called welfarist. The best option is the one that creates the highest welfare or well-being for the agents involved. This may seem very innocent, but it has created a substantial debate, not least among philosophers. With reference to Bernard Williams O'Neill et al. (2005) offers the following case:[13]

> George is an unemployed chemist of poor health, with a family who is suffering in virtue of his being unemployed. An older chemist, knowing of the situation tells George he can swing him a decently paid job in a laboratory doing research into biological and chemical warfare. George is deeply opposed to biological and chemical warfare, but the older chemist points out that if George does not take the job then another chemist who is a real zealot for such research will get the job, and push the research along much faster than would the reluctant George. Should George take the job? (O'Neill et al. 2005)

From a consequentialist point of view George should answer yes to the question. He, his family and the society would all be better off. However, taking the job would undermine a held value that is important to George. Thus he would be demoralized and his integrity challenged.

It is especially in the issues related to personal integrity that consequentialism falls short. The deontological response is to hold that there are certain moral standards that are fundamental to individual integrity. This we see from the writings of Immanuel Kant (for example, [1785] 1981) to the writings of John Rawls (for example, 1971). The human being is an end in itself, not a mere means to some overall welfare:

> Now I say that man, and in general every rational being, exists as an end in himself, not merely as a means for arbitrary use by this or that will: he must in all his actions, whether they are directed to himself or to other rational beings, always be viewed at the same time as an end. . . . Beings whose existence depends, not on our will, but on nature, have none the less, if they are non-rational beings, only a relative value as means and are consequently called things. Rational beings, on the other hand, are called persons because their nature already marks them out as ends in themselves – that is, as something that ought not to be used merely as means – and consequently imposes to that extent a limit on all arbitrary treatment of them. (Kant [1785], 1981: 35)

According to Kant, it is the ability of human beings to make reasoned choices that makes them rational beings. It is from this that their dignity flows. It gives them a set of rights as a consequence of being able to reason. Beings that are not rational in this sense are mere objects or things.

O'Neill et al. (2005) draw attention to both the importance of this understanding and its restrictions. While it is important to understand the limitations of the consequentialist/welfarist position, it is also problematic to endorse the stand that individuals have an identity or a set of rights that predate or are independent of their membership of communities. From where comes the idea of 'man as an end in itself'? It cannot come from our nature. It must come from humans discussing what it means to be human. The point is that nature has given us the ability to reason, including the capacity to reason about who we should become. This capacity does not by itself grant us the right to be an end in itself. It is rather so that this capacity has given us *the opportunity to define* what it is to be rational and to do reasonable things. Recall that people have been kept as slaves, have been treated as objects. The conclusion that slavery is inhuman came exactly from the process of socially defining – even fighting over – what it means to be human. In this we see the role of the social context in the process of shaping the qualities of being human, the identity of individuals, and the meanings imprinted in various actions.

It is also important to be aware that it is not consistent to treat deontology as unaffected by consequences. In the case with the chemist, the deprivation of his self-esteem and integrity is certainly a consequence. It should rather be viewed in this way: George has been 'exposed' to the norm of not taking part in any activity that is supportive of war affairs. The development of such a norm may historically come from reasoning about people's moral obligations towards others. George has internalized this norm, and then it influences what he should do in the concrete case where it came to rule out the arguments pointing towards taking the job. From this short discussion we also see that there is some clear resemblance between deontology and normative rationality as described in Chapter 5.

6.3 THE CONTEXT DEPENDENCY OF PREFERENCES AND VALUES

6.3.1 Different Understanding of Context Dependence

There are several issues involved when we raise the question of social influence on preferences and values. Following Berger and Luckmann (1967), these are largely social constructs. They are learned through the process of socialization and embedded in our cognitive perspectives and social norms. They are supported by the various roles that we have been trained to perform. However, this evokes several questions. First, does the social process define all of our preferences? Second, if it influences at least some of them, is it actually the preferences that are affected, or is it just that certain institutional contexts make us emphasize a specific set of preferences which have been in our repertoire from the very beginning?

If we start with the first issue, Durkheim is famous for suggesting that the 'human mind is merely the indeterminate material that the social factors moulds and transforms' (Durkheim [1895] 1938: 106). Taken literally, this statement leaves much with the social and equally less with the individual. As we have discussed earlier, the methodological holism implicit in such an interpretation has some problems attached to it. There are two reasons for this.

First, humans have a number of physical needs that no social construction can 'do away with'. We need food, shelter and care to be able to function at all. Hence homo sapiens has some specific physiological and psychological characteristics. None the less, the way such needs are satisfied can take a variety of forms which are socially or culturally contingent. McCauley et al. suggest:

> [H]uman beings come into the world with certain likes and dislikes, such as innate dislike of pain, bitter tastes, and many types of strong stimulation, and an innate liking for certain types of touch and sweet tastes . . . Almost the entire adult ensemble of likes and dislikes is acquired, presumably in the process of enculturation. (1994: 27)[14]

So even these authors put most emphasis on the social side of the coin. The Japanese may value a 'perfect apple' at $10 or more, a Muslim would not eat pork, a westerner would not eat dogs, a Scandinavian would love aquavit while 'most Americans' find this to be a curious drink. All of these examples concern physiological needs, but their satiation is still culturally defined or moulded since they materialize as group phenomena.[15]

Second, people may choose to move from one culture to another, either locally – that is, to another social grouping – or by moving to another place – that is, becoming a member of another society. Here we may again draw attention to Screpanti (1995) and his point that people may choose to make such shifts because they find the institutions of the other culture to be better. There is room for individual choice and adaptation at this level, too.

Nevertheless, in the case of shifting from one culture to another we observe that both the culture we leave and the one we move to are existing social 'facts' and that the one moving accepts the norms and values of the people s/he joins. In this sense, individuality is contingent on the systems between which it is possible to choose. So, while there is always room for individual adaptation and choice, the social element still plays a basic role in that it delivers a necessary platform for our orientations and values. We simply could not do without some social structure.

Concerning the economic sphere more specifically, Bowles (1998) reviews much of the growing literature on the relationships between institutions and preferences. He starts by stating that '[m]arkets and other economic institutions do more than allocate goods and services: they also influence the evolution of values, tastes and personalities' (p. 75). After presenting results from different studies, he concludes that institutions influence preferences. More specifically, he accentuates that:

> [These studies] are consistent with the view that market-like situations induce self-regarding behaviour, not by making people intrinsically selfish, but by evoking self-regarding behaviours in their preferences. Thus, the hypothesis that market situations induce self-regarding behaviour does not imply that those living in non-market societies would be intrinsically less self-regarding. (p. 89)

According to this, the institutional context is of importance since it influences which preferences or values become mobilized. Torsvik (1996) takes a similar position. However, Bowles also suggests that preferences may be learned.

One may interpret this to mean that we have two (or more) sets of preferences. Which set to use is triggered by the context – the institutional setting. Much of what is said in Chapter 5 on the variation in rationalities between institutional contexts is consistent with such a position. But, we have also presented examples where people acted in an other-regarding mood when operating, for example, in markets. Hence, it seems reasonable to argue that some norms are so basic to us, so ingrained in our personality, that they apply independent of the institutional arena.

In the following we shall look at a set of empirical studies that can shed more light on these issues. I shall draw attention to three important findings. First, data from various experiments support the view so basic to this book, that humans are not just selfish beings. They also act in 'We' terms, act reciprocally and/or hold other-regarding values. Second, the degree to which people act selfishly or hold other-regarding values varies between individuals. Finally, the tendency to act selfishly, reciprocally or in other-regarding terms also depends on the actual institutional context. Expressed values and preferences vary across institutional structures.

6.3.2 Verifying the Social Being

Certainly, humans are not only self-regarding beings. We see over and over again that people act reciprocally or make sacrifices; to a small degree as in the street when we let someone pass before us; larger as when we do not strike a deal because it is against our principles even though the gains from an economic perspective are obvious. However, are these observations just a set of anecdotes, not changing the overall picture, or do they get more general support? Over the last 10–20 years an increasing amount of empirical evidence is building up not least within modern behavioural and experimental economics showing that such observations are typical. One important part of this literature concerns the so-called 'ultimatum' and 'dictator' games. We shall look at some of the results from this research to document that much behaviour cannot be explained by simply invoking the assumption of self-regarding behaviour.

Ultimatum games were first studied by Güth et al. (1982). These games are undertaken in the following way. The so-called proposer obtains a sum of money – for example, 10 dollars – which s/he must split between her-/himself and a respondent. The players do not know each other, and the game is not repeated. If the respondent accepts the division, both players get the money on the basis of the split made by the proposer. If the respondent turns the offer down, the two participants get nothing, however.

According to the standard theory of rational choice, one should expect the proposer to give away as small a sum to the respondent as possible, and the respondent should accept. Some is always better than none. Over the years a large number of studies have been published within this field. The results are quite consistent. Gintis (2000) sums up by concluding that a 50–50 split is the dominating offer, that most proposers make a positive offer to the other, and that respondents often turn down offers less than 30 per cent.

The fact that one is willing to share does not necessarily imply that people are strongly concerned about fairness or act reciprocally. It may be the result of some fear on behalf of the proposer that the other will reject the offer. However, implicit in that thought lies an element of fairness evaluation or reciprocity on behalf of the other. The thinking of the proposer must then imply that the respondent may reject a low positive bid, a behaviour that can hardly be explained as anything other than a rejection of something that is perceived as unfair, with a subsequent wish to punish the proposer. It is exactly such arguments that are made when the participants are asked about why they act as they do (Gintis 2000).

The dictator game is developed to eliminate the effect of potential strategic behaviour on the part of the proposer – that is, here the aspect of reciprocity and the willingness to give is accentuated, while the possibility of retaliation is ignored. The rules imply that the respondent now cannot turn down the bid. The result depends only on the choice of the proposer. In such situations the 'offer' goes down. The proposer is less willing to share, although a large number still make positive offers. As an example, a study by Forsythe et al. (1994) shows that 80 per cent of the participants in dictator games want to share. In this case the modal offer was 70–30. This suggests that a fraction of what is given in ultimatum games follows from the fear of punishment, but some must also follow from the fact that people genuinely want to share or find that they have an obligation to share.

A third type of 'game' is the so-called 'public goods' game. This is a type of experiment where a multiple prisoner's dilemma situation is created – that is, a structure where the best results are obtained if everybody cooperates, while it is beneficial for each individual to defect. The results are especially beneficial for the single individual if all the others cooperate. Ostrom (2000) shows that even in such situations, 30–40 per cent of the participants rank the cooperative result as better than a situation where they themselves defect and all others cooperate. Some 25–30 per cent of the participants were indifferent between these two outcomes.[16]

The argument has been made that if public goods games are repeated, it will become individually rational to cooperate – compare the Folk theorem described in Chapter 5. The hypothesis of self-regarding motivation can thus be defended. However, it demands that the game is not viewed as consisting of a finite number of rounds. Fehr and Gächter (2000) show that participants are willing to cooperate even if the game is not replicated, the composition of the group changes if replicated and so on. They also show that cooperation takes place even in the last round of a sequence of finite games.

Contextual conditions for the various games influence the results, as is also indicated above. Based on the literature in this field, the following examples of contextual variations can be given for the case of ultimatum and dictator games:

- The results are influenced by emphasizing 'divide' as against 'exchange' concerning the split, where the latter gave lower offers to the respondent (Hoffman et al. 1994).
- If the proposer gets the money without any effort or it is earned as a result of some activity – for example, a quiz – the willingness to give is influenced. The latter results, as expected, in less being given (ibid.).
- The tendency to reject an offer among the respondents goes down if the bid is made by a data-machine as compared to a person (Blount 1995). However, bids that are very uneven are often turned down even in a situation where they are generated by the machine.

In public goods games the following is observed:

- The possibility for communication between participants results in increased cooperation (Frohlich and Oppenheimer 1996, referred to in Ostrom 2000). Even seeing the other participants influenced the result towards more cooperation.
- If a prisoner's dilemma or public good game is called a 'Wall Street Game' or a 'Community Game', this influences the degree of cooperation even if the games are identical concerning the gains and losses (Ross and Ward 1996).

In relation to the last observations, it is interesting to see that there seem to be some elements of community relationships directly at work. Game theorists have focused predominantly on so-called 'non-cooperative' game structures, where behaviour is presumed to be only self-regarding and instrumental: 'If I cannot gain, I will not do it'. Communication is not allowed for. It is as if one ignores the fact that individuals engage in other than instrumental exchanges.

From these short glimpses into a growing and rather substantial literature, we observe both an individual variation and a variation following from the institutional context. Concerning the individual variations, I refer to different persons choosing differently under the same game structure and context. Some act more reciprocally or according to other-regarding values than others. It seems that we are confronted with two phenomena – one that concerns individuals' basic willingness or capacity to cooperate independently of the institutional context and one that reflects the actual effect of which institutional context frames the choice. We shall look at both issues in turn.

6.3.3 Cooperative Will and Cooperative Capacity

The will to cooperate may vary because of both cultural and individual aspects. The norm of acting in 'We' terms may vary due to different socializing environments both within and across societies. If the primary socialization favours individuality, this is likely to become a permanent characteristic of that person. However, we also observe that people who grow up under very similar conditions develop different capacities regarding their degree of selfishness, cooperative will and empathy.

A basic issue in this concerns our *ability* to learn norms. Norms and other institutions can be viewed as 'software' that can only be imprinted if the 'hardware', the ability to learn norms, is in place. The question has been raised whether both individuality and the capacity to learn to act cooperatively are fostered by evolution. Ostrom (2000) summarizes much of the discussion about this issue. She suggests that in the long period when human beings mainly operated in small groups as hunters and gatherers,

> [Survival was] dependent not only on aggressively seeking individual returns but also on solving many day-to-day collective action problems. Those of our ancestors who solved these problems most effectively, and learned how to recognize who was deceitful and who was a trustworthy reciprocator, had a selective advantage over those who did not. (p. 143)

Evolutionary psychologists have documented that the brain has a structure that fits well to internalizing norms. Humans have more skills than that of logic problem solving. The brain seems to have developed different domains for different types of logics: 'the human brain appears to have evolved a domain-specific, human reasoning architecture. . . . For example, humans use a different approach to reasoning about deontic relationships – what is forbidden, obligated, or permitted – as contrasted to reasoning about what is true and false' (ibid.).[17] The conclusion Ostrom draws is that the brain is structured so that it has the ability to be imprinted by social norms in the same way that it has the capacity to understand and master grammatical rules.

This provides evidence that evolution has enabled us to internalize norms. Certainly, this is in a way a tautological statement, given the perspective of social construction. Since this position is still disputed, and even rejected as in the case of neoclassical economics, the observation is important because it delivers independent support for the hypothesis of social construction. Together with the capability to develop complicated and rather precise languages, this is probably the most fundamental capacity characterizing the human being. It has not least been basic to our ability to build advanced civilizations.

As in the case of ability to learn languages, one may envisage that people have different abilities to learn and respect norms. While these issues cannot be compared – they are in quite different categories – it may explain some of the individual variations appearing in the 'games' referred to above. It is not only about the kind of socialization the individuals have been through. They may also be differently conditioned physically or genetically concerning their ability to learn and follow norms.

6.3.4 The Institutional Variation

While there is individual variation, there is also variation across institutional contexts. Such variation has already been observed in the material from the different 'games' referred to previously. Ostrom (2000) takes the issue further and refers to several findings concerning variations also across cultures. In the same vein, Henrich et al. (2001) show that offer and acceptance rates in games such as the ultimatum game have a cultural variation. Gowdy et al. (2003) emphasize that in many 'traditional' societies, fairness considerations explains economic behaviour well, while punishment does not: 'Instead non-cooperative behavior elicits cooperative response' (p. 470). They thus claim that *homo reciprocans* as described by Gintis (2000) and Fehr and Falk (2002) is not a universal, institution-independent model either. This may indicate that no behavioural typology is imprinted in the human concerning economic behaviour. If so, all economic behaviour has a societal or culture specific dimension.

In the economic literature the issue of context dependency is also gaining attention under the concept of endogenous preferences – for example, see Hodgson (1988), Lane (1991) and Bowles (1998). Endogenous preferences means preferences that are influenced by the economic institutions. Bowles (1998) discusses several relationships between preferences and the structure of the economic institutions. While he puts weight on the importance of institutions in the formation of individual preferences, norms and behaviour, he also underlines that our knowledge about the involved mechanisms explaining these relations is still restricted.

In his paper, Bowles also offers interesting insights because he bases his conclusions more on material from existing societies/economies than experiments. One might always argue that experiments are not 'real life'. Summing up the experiences from this material he specifically stresses that '[m]arkets and other economic institutions do more than allocate goods and services: they also influence the evolution of values, tastes and personalities' (ibid.: 75).[18]

Bowles suggests that in the choice among different economic institutions, one has to take into account how these institutional structures influence preferences. This is a second-order problem which is invisible if one takes the individual preferences as given or consider them unchangeable. Bowles is especially concerned with how the building of institutions that foster individual solutions can even erode the very same institutions because it erodes the norms or the trust that is necessary to avoid self-interested beings from descending into opportunistic behaviour. He concludes:

> Moreover, the analysis . . . suggests that approximating the market ideal by perfecting property rights may weaken non-market solutions to problems of social coordination . . . *approximating idealized complete contracting markets may exacerbate the underlying market failure (by undermining the reproduction of socially valuable norms such as trust and reciprocity) and result in less efficient equilibrium allocation.* (ibid.: 104, added emphasis)

A similar conclusion is offered by Etzioni (1988). Given that information or knowledge is costly, trying to increase 'efficiency' by transforming everything into market relations may also result in increased opportunities for opportunism 'to thrive' (Williamson 1985), which then results in increased coordination or transaction costs – simply more private and public control.

Literatures other than the above-mentioned give similar insights. One is the literature on the concept of 'crowding out' (for example, Frey 1997a; Frey and Oberholzer-Gee 1997; Frey and Jegen 2001). Here situations are observed where public policy instruments give the opposite effects of those expected due to the existence of what is termed 'intrinsic motivation'.

Following the perspective of this book, I would prefer to call such motivations 'internalized norms'. Most typically we have cases where paying for a good actually decreases supply. This is generally observed in situations where payment is not found to be adequate since the supply is considered to be an obligation. Blood donation is the classical example, where a shift from free donation to invoking payments has been observed to result in less supply (Titmus, cited in Frey 1997b). In a Swiss study, Frey and Oberholzer-Gee (1997) found that compensating people for hosting a nuclear waste facility actually reduced the 'willingness to host'. According to these authors, problems appear when the incentive mechanism used does not conform well to the logic of the concrete situation as perceived by the respondents or suppliers. The same type of literature also gives examples of how the institutional setting influences which preferences are expressed. Romer (1996)[19] documents how income transfer programmes shape preferences. Frey (1997b) shows how different constitutional arrangements affect predisposition to tax avoidance.

Finally, the discipline of social psychology has also delivered a series of examples which complement the picture. In a study by Gneezy and Rustichini (2000a), a fine to reduce latecoming by parents to a day-care centre in the afternoon, actually increased it. The authors explain the observation by the following dynamics. Before the fine was installed, a norm against arriving late motivated parents to try to get to the centre on time even though they might not always make it. The fine shifted the logic, so that 'stressed' parents now felt they were free to arrive late since they paid compensation. As a result, the fine was abolished. The level of late-coming was reduced again, but not to the original level.

Studies among social psychologists of the provision of different services that can be supplied both privately and publicly – such as transport (van Vugt et al. 1996; van Vugt 1997) and day care (Eek et al. 2001) – illustrate that the way such goods are accommodated influences which rules of distribution are expected. Public provisioning creates expectations about equal treatment, while private provisioning creates demands about treatment according to the level of payment. What is considered 'right' is again a function of the institutional context.

6.3.5 Individual versus Institutional Variation

The different literatures referred to in the above sections give a rather consistent picture of the mix of institutional and individual variation. Figure 6.1 illustrates some core aspects of this. To simplify the figure, I have equated acting reciprocally with other-regarding. The figure depicts individual I as more self-regarding than individual II as observed at a certain point in time.

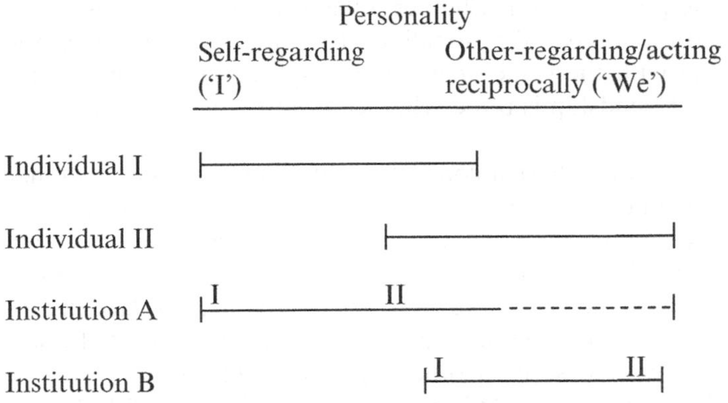

Figure 6.1 Institutions and personality

The variation between the two may be due to their prior history – how they are raised, life experiences – and factors like genetic disposition. Exposed to institutions A and B, one might expect the two individuals to position themselves differently. Institution type A is developed to foster individuality, but still does not deny other-regarding behaviour.[20] Institution type B motivates other-regarding behaviour or restricts the ability to act in a self-regarding way. The figure also indicates where the two persons would be expected to position themselves in the two contexts.

Although this is a simple illustration, it captures important aspects of the variations observed both in practical case studies and in the various experiments referred to above. How 'narrow' a certain institution is becomes important. A type I individual would, taking the figure literally, be unable to live well or be accepted in a society with many institutions that have strong other-regarding features. While being raised in such a society would probably have made that individual less self-regarding in the first place, it may explain why some people move to other societies, why there is crime and punishment and so on.

6.4 THE PLURALITY OF PREFERENCES AND VALUES

The above reasoning is not only counter to the standard neoclassical position that preferences are purely individual. It is also counter to the idea that they are one-dimensional. People seem not to have a single preference ordering which is complete and continuous. Instead, competing value orderings may exist which have no common denominator. Values are plural. This just restates the point made in Chapter 5 that different situations or contexts support different types of rationalities. The value of friendship is not commensurable with the value of a commodity like bread. One cannot buy friendship. Friendship is rather characterized by the fact that one cannot pay to obtain it.

There are many different conceptualizations of this observation. Smith ([1759] 1976) talked about moral sentiments. Sen (1977) makes a distinction between individual preferences and commitments when he emphasizes that we also hold values that are not tradable with preferences for ordinary commodities. Some goods are kept outside the marketplace just for this very reason. Etzioni (1988) uses, as we have seen, the concept of norms and involvement when describing other-regarding behaviour.

The philosopher John O'Neill (1993) presents a useful classification scheme for the degree of comparability between preferences or values. First, we have what he terms 'strong commensurability', which is the same as cardinality:

> To hold that values are strongly commensurable is to hold not only that the measure ranks objects, but that there is a particular single property that all objects possess which is the source of their value, and that our evaluation measure indicates the amount or degree to which that property is present. (p. 103)

The next class is that of 'weak commensurability' or 'strong comparability'. Here an ordinal scale is at work. This value structure implies that the individual is able to rank goods, but not make a direct trade-off: 'To hold strong value comparability is to hold that while there may be no single value in terms of which all states of affairs and objects can be ranked, there does exist a single comparative term of which they can be ordered' (p. 104).

The third class is then 'weak comparability'. Here we even lack a comparative term under which comparison can be made. We are clearly in the realm of value pluralism. This is a type of preferences that goes beyond the economic model. So while economics is based on the idea of value commensurability, Chang (1997) comments that among philosophers you really have to explain yourself if you take such a position. Here weak comparability is dominantly assumed to be the most typical.

O'Neill finally defines the term 'incomparability' as the situation where one holds that rational or reasoned choice between different states of affairs can be made without holding that there is a comparative term that orders them.

The above distinctions are important, not least for us since we shall later study the field of environmental decision making more in detail, a field where it may be problematic to assume strong commensurability. Thus, we shall follow O'Neill's reasoning in a bit more detail to capture the essence of the two positions – that of neoclassical economists and that of philosophers. O'Neill begins by referring to the Ramsey Centre Report (Attfield and Dell 1989), which focuses on the ethics involved in environmental decision making. This report supports the idea of commensurable values:

> Let us take just one typical case: comparing enjoyment of art or natural beauty with saving lives. It may look as if one could not say that a certain amount of aesthetic enjoyment was more or less valuable than one life: that such a comparison did not even make sense. But governments do make such comparisons, and it is hard to deny they make sense. For instance, the government of the United Kingdom has decided that it is justified in subsidising the Covent Garden Opera, even though it knows perfectly well that the money it spends could save a certain number of lives if it were transferred to the N.H.S. cancer screening programme. (Ibid.: 30)[21]

O'Neill's response is the following:

> This argument assumes that the claim that we can and must make choices between different objects and states of affairs, and that we can do so sensibly and rationally, entails that we are committed to saying that one state of affairs is more valuable than another. It is far from clear that this is the case. One might refuse to accept the statement 'X is more valuable than Y' while choosing X over Y where choice is required. Thus refusal stems not from moral squeamishness – that one does not want to accept that one really does find so much art better than so many lives – but, rather, from the vacuity of the comparative given a plurality of values. To say 'X is more valuable than Y' is to invite a response 'in what respect', and given value plurality there may be no respect in terms of which the comparative statement can be grounded. (1993: 104–5)

In relation to this discussion – especially the different views on the category of weak comparability – we should remind ourselves that market values are one-dimensional. Markets are created to foster comparisons concerning the exchange value of a good. This value does not necessarily say anything about the use value of the same good. The issue raised by philosophers like John O'Neill is the mistake that is made from (a) observing that all goods sold in markets demand a certain price, to (b) concluding that this must imply that all other goods can also be sensibly measured by way of the same one-dimensional measurement scale. One could turn the Attfield and Dell argument around and say that if there was nothing more to culture and life than trade-offs – their value in exchange – all these 'goods' could also be traded in markets. The reason why they are not traded supports O'Neill's position. Instead one may argue that by forcing commensurability on issues that concern not least ethics, we are making a fundamental categorical mistake. Justice is – as an example – a matter of what is a right order. It is a matter of principles, and belongs to a class other than commodity choices.

Douglas (1986: 124) makes this very clear as she also argues that issues of great importance like those concerning life and death, ethical principles, the possibilities of future generations and so on are not solved by individuals:

> It is a sort of problem that is insoluble if it is given to individuals as an intellectual puzzle . . . Individuals normally off-load such decisions to institutions. No private ratiocination can find the answer. The most profound decisions about justice are not made by individuals as such, but by individuals thinking within and on behalf of institutions. . . . Choosing rationally, on this argument, is not choosing intermittently among crises or private preferences, but choosing continuously among social institutions.

The literature on environmental valuation is, typically enough, full of examples where people may face a dilemma when asked to trade off an environmental good for a sum of money. Observing this dilemma, their response may appear as a refusal to pay even though these respondents show highly positive attitudes towards the good. They actually hold the position that these goods are characterized by weak comparability. In this literature, preferences are then classified as lexicographic (Stevens et al. 1991; Spash and Hanley 1995; Spash 2000). They belong to classes between which there are no trade-off possibilities. Respondents may view choices in the realm of the environment as 'citizen' issues as opposed to 'consumer' issues. They may also attribute rights to other species, restricting trade-off possibilities.

6.5 IMPLICATIONS

The observations made in Sections 6.3 and 6.4 are of great importance. In cases where value dimensions are incommensurable, institutions other than markets or market imitations are warranted when allocating resources. If preferences are affected by context, by the institutional setting, the issue concerning which contexts to prefer becomes crucial. The latter question is certainly the most fundamental.

First, and most basically, this last question makes us ask which preferences we should hold. Who do we actually want to become? Or in the words of Page (1997: 591): 'So the basic question is not "How do I satisfy given and fixed preferences" but "What sort of society do we want to become"'. This conclusion is clearly counter to the neoclassical position and its understanding of efficiency. None the less, if preferences are influenced by context, it becomes impossible to draw any conclusion about what action is optimal without making comparisons across contexts. This issue becomes, however, invisible if we think that there is only one logic and one value dimension. As in neoclassical welfare theory, the market is the only context applied. What falls outside of markets is measured in market terms to correct for the 'market failure'. We experience here the crucial role which the fixity of individual preferences plays in economic theory. If it is given up, the supremacy of both the market and the consumer is lost. The consumer of the day then offers no set point from which to make the calculation.

As suggested by Sunstein (1993), if preferences are affected by context, one cannot base policies – that is, the choice of institutional contexts – just on some aggregation of individual preferences. This produces mere circularities: 'When preferences are a function of legal rules, the rules cannot be justified by reference to the preferences. Social rules and practices cannot be justified by practices that they have produced' (ibid.: 235). While we cannot easily resolve such dilemmas, communication about which contexts we find to be the most relevant for expressing our different values is a very important second-order issue for society. From such processes a development of typifications or rules concerning which institutions have merit in which situations may follow. This is how societies have treated these issues all along (Walzer 1983; Douglas 1986). It is only in modern societies that the idea has developed that markets can be both 'the judge' and 'the defendant'.

The position taken here is by 'necessity' both 'perfectionist' and 'objectivist'. It is about how we perfect (making better) the society and its members, and it is about how we evaluate the values on which we build our institutions. Hence, while it is observed that there is something specific to being a human, 'objectivist' does not mean that values are objectively given by nature. The point is rather that values can be described and evaluated across individuals. A communicative process over what it should mean to be human and what is good for humans is both possible and important. This is very different from the modern neoclassical and also Austrian positions claiming that values are individual – that is, purely subjective – and cannot be socially evaluated.

According to O'Neill (1998), the idea of perfectionism is perhaps the strongest argument in defence even of the market itself – that it creates autonomy, entrepreneurship and so on.[22] Nevertheless, there is a danger for perfectionism to end in some sort of paternalism. To avoid this, it is crucial to build institutions that facilitate an open public debate concerning the choice of institutional structures. The values we want to foster must take form in open communication. This is the only way by which the standards we apply when making evaluations can be evaluated themselves.

The position developed here illustrates well the problems of the relativism found among many social constructivists. Since no objective or external judge exists – it is all constructed – one can as well resort to the position that 'any construct goes'. As was emphasized in Chapter 2, this is an erroneous conclusion. The constructs we make can be evaluated by reasoned debate. The rule is not that 'anything goes', but 'what goes is that which can be supported by reason'. The issue then becomes one of choosing institutions which foster a process of communication over what are important values to defend. This is an issue that will be taken up on several occasions in the rest of the book.

6.6 SUMMARY

The basic divide, the one between the individualist and the classical institutionalist or social constructivist understanding, also stands out clearly when we look at the issue of preferences and preference formation. The former position takes preferences as given and stable – that is, context independent – while the latter sees them as socially dependent. The divide concerning the definition of rationality (Chapter 5) is consistently carried further by how the origin of preferences is understood.

We have seen that the idea of the self-contained and independent individual was developed from the eighteenth century onwards as a reaction to the customs and power relations typical of European aristocracy. The development of ideas concerning individualism and liberalism went very much hand in hand. The focus was on maximizing the sum of individual utility as opposed to subordination to some divine normative order.

While the old utilitarians viewed utility as a concrete measure which could be aggregated across individual members of society, modern neoclassical economics looks at preferences as merely a ranked order. While the old utilitarians believed that one could compare different distributions in a society and say which gave the most total utility, the modern ordinalists accept only Paretian types of comparisons.

An alternative to the neoclassical stand is to look at preferences and values as plural and as context dependent. This links directly to the idea of plural rationalities as discussed in Chapter 5. Preferences are both self- and other-regarding. They may concern the 'I' as they may also concern the 'We'. Preferences may furthermore be incommensurable, implying blocked exchanges not least between spheres emphasizing individual utility satisfaction on the one hand and social or cooperative logics on the other. In addition, not only consequences, but also the issues of moral integrity and rights are important in themselves. This also creates restrictions as to the making of trade-offs.

Concerning the issue of context-dependent preferences, a large set of empirical studies support this. The following observations are made:

- The process of socialization or enculturation influences the preferences and values we hold. It affects both our preferences for various goods and the values we hold concerning the balancing between, for example, self- and other-regarding preferences.
- The institutional setting – as understood by the agents – influences which preferences and values in the continuum from 'I' to 'We' are found to be acceptable and/or relevant. This context 'mobilizes' certain sets of values and acts.
- There is individual variation too, exemplified by the fact that some people are consistently more self-regarding while others are more other-regarding across institutional contexts. This variation may be explained by genetic differences, personal evaluations and by the fact that individuals are raised differently.

The analysis undertaken here suggests both a plurality and a certain 'plasticity' of preferences. These can change, partly as an effect of socialization and partly as an effect of moving between institutions and thus (expected) behaviour/rationalities. The most egoistic market agent may also be a caring father.

According to our analyses, an important choice for society is to decide which institutional system should be in place for which type of problem. This is a second-order question which cannot be decided on the basis of the logic of any of these institutional systems themselves. A meta theory is needed. While no complete such theory exists, we shall consider this issue in a more comprehensive way in Part III – especially Chapter 8 – and in Part IV.

NOTES

1. That is, some authors accept that they are changed as an effect of learning about one's self, ageing and so on. Nevertheless, the reference always seems to be to the individual. McFadden (1999) emphasizes that both users and critics of the model sometimes formulate this understanding in unnecessarily restrictive ways: 'For example, immutability of preferences does not imply that consumers are unaffected by history or incapable of learning, but only that preferences develop consistently following a "rational" template' (p. 76). What this latter implies is not explained, though.

2. For example, Adam Smith, David Ricardo, Thomas Malthus.
3. Here we shall not discuss all the components of utilitarianism – just those related to the understanding of utility. A more complete presentation of utilitarianism as a theory about welfare is given in Chapter 8.
4. Bentham was not the first utilitarian theorist. Several philosophers, both English and French, had developed similar ideas before him. The 'greatest happiness principle' on which Bentham's concept of utility was based, was, for example, first defined by Priestley (Commons [1934] 1990). Nevertheless, Bentham coined the term by which the position would later be known – utilitarianism.
5. See the comments made on the institutions-as-equilibria school and Hayek in Chapter 4.
6. This revolution has been associated with the almost simultaneous publication of work by Jevons, Menger and Walras – that is, Jevons ([1871] 1957), Menger ([1871] 1981) and Walras ([1874] 1954) – which all had various and independent formulations of the issue of marginal utility. However, work from both the French engineer A.J. Dupuit and the German economist H.H. Gossen, in the 1840s and 1850s, respectively, included similar ideas.
7. Edgeworth is the creator of the 'Edgeworth box', so common in economic textbooks.
8. Despite the anti-aristocratic visions of neoclassical economics, Edgeworth still seems to carry with him some 'aristocratic norms' in making such distinctions between people. He furthermore argued that the 'capacity for happiness' could not be increased by, for example, education. I am indebted to Douglas (1986) and MacKenzie (1981) for this information on Edgeworth.
9. See the discussion of the ideas of Samuelson (1938, 1948) on revealed preferences.
10. McFadden (1999) is equally clear. It should be mentioned that Becker in his later writings (such as Becker 1996) seems to have abandoned his previous position. He now emphasizes that preferences are endogenous to the economic system and states that 'modern economics has lost a lot by completely abandoning the classical concern with the effect of the economy on preferences and attitudes' (ibid.: 18–19).
11. There is another problem involved though – the one about how a maximum can be defined if it is costly to learn about one's self. This issue is identical to the question of information costs and rationality discussed in Chapter 5.
12. It should be emphasized that social constructivism does not deny the existence also of purely individual preferences. Certainly, to what degree a preference can be moulded by culture or the social context will vary. We shall later discuss this in relation to what may be termed 'basic physical needs'.
13. O'Neill et al. (2005) refers to Williams (1973). For a more comprehensive discussion of the issues raised here, and their implications for environmental policy, see O'Neill et al.'s book.
14. Cited from Bowles (1998: 80).
15. Recently the theory concerning our liking for sweets has been given a functionalistic explanation, since eating sweet food increases the build-up of fat (energy storage) and thus the capacity of the individual to handle periods of food shortage better. In modern societies this is a minor problem, while access to sweets has increased dramatically, causing problems with obesity. What was functional has become 'dys-functional'. This in turn has resulted in various public campaigns to draw attention to the excessive intake of – that is, preference for – sweet food.
16. Ostrom (2000) refers both to own work and the work by Ahn et al. (1999).
17. Ostrom refers to work by Andy Clark and Annette Karmiloff-Smith, Ken Manktelow and David Over, Mike Oaksford and Nick Chater, plus Denise Cummins.
18. There actually seems to be some inconsistency in Bowles (1998). The main message of the quotation in Section 6.3.1 was that institutions do not influence the degree to which people become intrinsically selfish, they just evoke behaviour that is already there. The citation given here points in another direction – that institutions also influence the evolution of values and personalities. The latter resembles my view.
19. Referred to in Bowles (1998: 104, footnote 37).
20. Certainly, institutions may exist that force people to act selfishly in the sense that other-

regarding behaviour is seen as unacceptable. However, this is much less typical than the other way around, as captured by the figure.

21. Cited from O'Neill (1993: 104).
22. According to O'Neill, Mill was both a liberal and a perfectionist. He quotes Mill when he says that 'the first question in respect to any political institutions is, how far they tend to foster desirable qualities, moral and intellectual' (O'Neill, 1998: 17).

PART III

From Action to Institutions

We move now to the issue of forming or choosing institutions. As previously emphasized: while institutions influence human behaviour, humans also construct the institutions that subsequently play important roles in forming their lives. These are just two sides of the same coin. While the reader may have obtained the impression from Chapters 5 and 6 that the most important aspect is to understand how institutions form people, it is as significant to acknowledge that institutions are themselves human creations. Certainly, we are much more often engaged in reproducing existing institutions than producing new ones. That is the nature of institutions as durable structures. Changing institutions is, moreover, a demanding and complex task.

In Part III we shall study institutional change. Again we shall divide the subject matter into two chapters. Chapter 7 will look at different explanations of institutional change. The analysis will be mainly descriptive. We shall distinguish between theories about 'spontaneous' as opposed to 'designed' change. This relates to a core theme in the literature concerning what role conscious design plays as opposed to institutions as the unintended result of many uncoordinated acts. Chapter 8 will look more systematically at the normative issues involved when changing institutions. What are good institutions? How can we evaluate institutional change?

The division made between Chapters 7 and 8 is largely built on distinguishing between two aspects that are too often mixed up in analyses and debate. First, we have the question of what actually *causes* institutional changes to appear: a question that concerns mainly which interests have had the power to effect a change. It is about 'victory and defeat in the political battle', to quote O'Neill (1998: 1). This is the descriptive issue. Second, we have the issue of which institutional solutions can be supported by reason. It is about the 'victory and defeat in political argument . . . a question of truth and validity' (ibid.). This is the normative side of the question.

7. Explaining institutional change

The issue of institutional change has already been visited on several occasions. In Chapter 1 we contrasted the institutional structures surrounding the 'man of the forest' with the 'man of Manhattan'. Between the situations thus pictured lies a tremendous development in institutional structures. In Chapter 2 we looked at the social constructivist perspective, presenting a rather simple and general formula for how institutions come into being. In Chapter 3 we looked at the various institutions that could be established to solve coordination problems and regulate conflicts in the creation of a new housing development. In Chapter 4 we finally drew distinctions between different schools of institutional economics partly on the basis of their perspectives on how institutions change. However, these treatments were not systematically focused on the explanation of why institutions are altered. This is the topic for the following chapter.

Institutional change covers both the process of changing an existing institution and the establishment of an institution in a field where no institution has existed before. Both the move by many European Union (EU) member countries from national currencies to the euro in 2002 and the creation of the very first currency are examples of institutional changes.

Just as there are many theories of what institutions are and what they do, there are certainly also many different ways of explaining their development and change. There are theories about spontaneous development of institutions as against designed institutions, theories about unintentional versus intentional change, and theories about institutional change from below (for example, civil society) and from above (for example, the state). There is a certain overlap between these three groups. Hence, we see a tendency to put 'spontaneous', 'unintentional' and 'from below' together in one set and 'designed', 'intentional' and 'from above' into another – for example, Sened (1997).[1] While intentional emergence or change implies that the institution is built on conscious design, spontaneous change is seen as the unintended result of a series of uncoordinated acts. It is not planned in the sense that some collective has deliberately created it.[2]

While there is certainly some logic to this distinction, there is also a fundamental problem. The creation of institutions from below may certainly also be the result of intentional design. While people 'below' may not have the power of a formalized collective like the state, they may communicate and agree on certain conventions, norms and rules that they would like to institute. Indeed, such intentional creation from below is very important. So when I utilize the distinction between 'spontaneous' and 'designed' change, I shall also include institutional change that is still based on intention under the heading of 'spontaneous'. However, in this case such changes are characterized by the fact that they develop from below and are not part of a more comprehensive structure of a conscious design of institutions (Section 7.1).

Theories about designed institutional change will be divided into two groups: first, designed change that is driven by *efficiency considerations* (Section 7.2); and second, designed change as driven by the intent to *protect specific interests or values* (Section 7.3). This distinction follows a core idea of this book – that efficiency and interest/value factors are both crucial aspects of institution building. The internal relationship between the two will be further elaborated in Chapter 8.

Both spontaneous and designed institutions may fail and thus create crises which themselves can be countered by new institutional changes. I shall therefore briefly include a fourth group of theories that cover unintended effects of institutional structures and the change of institution as a *reaction to crises* (Section 7.4).

All the above types of explanation have some credit. While spontaneous creation and change may have a lot to offer when explaining the development of many conventions and perhaps also norms, it is the last three explanations that are relevant when studying the emergence of various formal institutions. By definition, these need to be based on design – that is, on some kind of formal collective choice. Efficiency- and interest-based explanations are often competing explanations. As will be argued here – and developed more fully in Chapter 8 – it is not possible to draw a clear distinction between pure efficiency considerations and the protection of interests. Rather, what *becomes efficient* is defined by the interests protected by the collective via the formulated institutions.

7.1 SPONTANEOUS INSTITUTIONAL CHANGE

The concepts of 'spontaneous order' and 'spontaneous institutional change' are, as already indicated, not always used in a well-defined manner in the literature. The issues of 'non-intentional', 'non-designed' and 'change from below' all seem often to be implied. However, changes from below may certainly involve some intentional creation. To clarify the various ways of thinking, I shall thus distinguish between 'theories of change from below' and 'pure spontaneous order theories', where in the latter case the emphasis is on changes as explicitly unintended or not designed. This is a restriction that does not necessarily follow from the fact that the creation is from below.

7.1.1 'Spontaneous Change' as Change from Below

Berger and Luckmann's model for social constructivism, as presented in Chapter 2, is a typical example of institutions as emerging from below. Such institutions are spontaneous in the sense that they are not the result of any collective design, meaning that elected boards and so on create them. They are rather built on the idea that institutions emerge as solutions to practical problems in everyday life. As such they may still be, and typically very often are, created intentionally by some. Next, the institution is expanded to other people who copy or reproduce. The meal – in the simple anecdote – became an organized act via an intentional choice in the first place, which next became an institution because it was copied and reproduced by others.

This process of copying can be explained in three different ways, where two of them also involve intention at this stage. First, the act may be copied because of some conformism – that is, based on some tendency by humans to just do as others do. In this case no intent is explicitly involved in the copying phase. Second, it may be picked up because – after we have considered it – we also find the solution sensible or good for us. Hence, intention may also be involved in the phase of copying. The solution is chosen since we 'like it' or think it 'functions well'.[3] Finally, the reproduction may be the result of some authority relation – for example, the parents of Berger and Luckmann's anecdote 'forced' the children to participate in meals since they were offered just this option. This explanation is based on the intention of the parents and the power they possess to make others comply.

Screpanti (1995) develops a reasoning of the above kind, accentuating the complexity of any social situation and the great uncertainties involved. According to him people will have to *simplify*, they will have to *rely on each other*, and they will not find it rational to give up something that already works. We therefore observe that people tend to 'stick to the options which have been tested socially' (ibid.: 67). While not a result of a plan, both intention and collective processes are involved.

At any point in time the durability of existing institutions is questionable, and the institution might be given up. If the more overall institutional structure is challenged, we may observe radical changes as in the case of the French revolution in 1789 where the whole existing aristocratic regime was overthrown. While it was initiated 'from below', this process soon developed into a collective design of new institutions – the building of a regime for the bourgeois society.

Changes that are purely from below are more gradual – step-by-step changes. They are evolutionary. Thus, Screpanti – using the language of biological evolutionary theory – picks up the concept of 'mutations' to describe the creation of an alternative behavioural pattern. This formation of an alternative solution becomes an institution when others imitate it.[4] Learning and copying replace the biological type of selection. Screpanti talks about 'artificial selection' as distinct from 'natural selection'. These changes may also, according to Screpanti, often originate in changed attitudes or preferences.

A typical example is the changing dress codes which can be observed over time. No one seeing movies from different epochs of the twentieth century will be in much doubt as to whether they are from the 1920s, the 1950s, the 1970s or the 1990s. The codes are so distinct. How these dress codes evolve and develop into a specific identifiable set is often difficult to trace. This is typically the case for processes from below. That they may influence whole generations is equally evident. How spontaneous these processes are, however, is an interesting issue. Klein (2000) argues that they are very often either co-opted by industry or even created by manufacturers in the continuing fight for new markets.

It would be wrong to say that the theory of social construction, as presented by Berger and Luckmann, can only be applied to institutional change from below. There is nothing in their model to indicate that, for example, parents cannot be replaced by a collective like the state or the local council, and children by citizens. So, while Berger and Luckmann were interested mainly in studying the evolution of informal institutions – specifically conventions (see Box 2.3 in Chapter 2) – their model can also be applied to explaining formal institutions by just redefining the agents and the form and content of the externalization, objectivization and socialization processes.

7.1.2 Pure Spontaneous Change Theories

Theories of spontaneous change as unintended – pure spontaneous change theories – are advocated foremost by the institutions-as-equilibria position and often cast in a game-theoretic language. It is a type of so-called evolutionary game theory. The focus here is on how equilibria are spontaneously developed and changed into new types of equilibria. This specific position looks at the individual as boundedly rational and considers knowledge to be subjective.[5] Preferences are, however, considered stable (Weibull 1995; Young 1998). Therefore this school is closer to standard neoclassical economics than, for example, Screpanti, who explains institutional changes as a result of changes in preferences.

Furthermore, evolutionary game theorists tend to avoid intentional explanations altogether. There seem to be strong links with the public choice position and its negative view of collectives and the political arena. By pursuing evolutionary game theory, they suggest that individual *behavioural deviations* ('errors') are the 'mutations' which create options for new institutions. Thus, they come closer to the biological model of evolution than Screpanti. The mutated gene, which is also an error, is replaced by the 'mutated', 'deviant' or 'erroneous' type of behaviour.[6] Whether in the end it becomes widespread, depends on the number of other persons repeating the act. The chance is low as for ordinary gene mutations, but in a few cases, what was originally an error ultimately becomes the standard. This is the same as for gene mutations.

While there are some merits to this position, there are also problems. First, it can only be used to explain the appearance of informal institutions – most typically, conventions. It may also to some extent explain the appearance of certain norms, not least if these give a competitive advantage to the group following this norm – for example, dietary norms, the incest taboo. However, there is no reason why one should not accept that at least some institutions are also purposely invented. As already mentioned, what becomes 'tradition' is often intentionally created. More specifically there are cases where conflicts are involved and then the question of intentionally constructing systems of third-party sanctions is of immediate importance.[7] The position of the Austrians is of great interest when evaluating this issue. I shall start with a short discussion about the creation of money as understood by Menger.[8]

Menger ([1871] 1981, [1883] 1963) saw money as a spontaneous social institution. As such it was similar to language. Menger stated that, 'the origin of money can truly be brought to our full understanding only by our learning to understand the *social* institution discussed here as the unintended result, as the unplanned outcome of specifically *individual* efforts of members of society' ([1871] 1981: 155; original emphasis).[9] Items that are especially saleable at a given time and place become money via a customary process. Hodgson (1996: 110) clarifies Menger's position by concluding: 'Hence the process begins on the basis of subjective evaluations, and becomes progressively reinforced through action and the perception of this action by other individuals'. What becomes money is spontaneous as defined above.

Menger compared the creation of money with that of language. These processes are equal in his mind. They are both understood as typical examples of spontaneity. However, there is one important difference overlooked by Menger. While in the case of language everybody has an incentive to follow the rules, this is not the case concerning money. When we talk to others it is (normally) in our interest to conform to the linguistic conventions so that we are rightly understood. Language is the archetype of a spontaneous institution since its rules are in fact self-policing.[10]

In the case of money, the situation is different. Here there is an incentive to cheat if the quality of the money is not controlled for. The commodity used as money may not be homogeneous in quality – be it spices, copper or silver coins. The possibility for individual control of the quality of each money item is so limited that there is an apparent need of an authority – the state/central bank – to secure the money standard. For Menger, the role of the state is to just declare by law what is already spontaneously – that is, customarily – acknowledged as money. This obscures the effect of 'state intervention'. The state or collective heavily influences the reliability by securing the quality of the currency and then also by substantially extending its potential use. It is actually fundamental to the role it plays, not least in modern market economies.

As mentioned already in Chapter 4, there is a strong tendency among many who view institutions as spontaneous – that is, the institutions-as-equilibria position – to look at the state as something that should only acknowledge what is already established via tradition (for example, Sugden 1986). While cast in the terms of a descriptive model, this is still a normative position based on the idea of a minimal state or a minimal formalized collective.

Hayek has developed the position of spontaneous institutional creation the farthest. In doing so, important clarifications, but also some important internal inconsistencies have become visible. His basic idea is that one should always let the markets do the job. They will produce great variety and via various trial and error processes, as in nature, the most functional or competitive society will evolve, which is a market-based selection of institutions. The problem, as we saw in Chapter 4, is then: from which market is this 'market for institutions' to be selected? In other words, there is a limit to spontaneous processes as a way to develop institutions. There is only a certain subset of institutions which can evolve in this way. Most fundamentally they will have to be restricted to (a) those that are self-policing and (b) those which individual agents can produce. In his plea against any type of collective action, Hayek throws out a substantial body of solutions to real-world problems, not least of great importance to those engaged in searching for constructive solutions to environmental problems, for example, local and national political bodies, state structures and international institutional structures.

The restriction implicit in Hayek's view has made Hirschman (1982) turn the argument around and assert that there is a potential for stagnation in individualist market economies. In all its consumer diversity, there is no real diversity. It is diversity of things but not of ideas. Instead there is only one idea – to be competitive. However, this is only one dimension of life, and according to Hirschman it may even be that the ability to compete in markets stagnates since the extra market forces from where the diversity of ideas come are fundamentally eroded.

Hence, there is an important inconsistency involved in relying only on spontaneous processes. And it goes further. Hayek is not really a *laissez-faire* advocate or advocate of spontaneity of all kinds. He supports diversity concerning economic agents, but not diversity concerning types of societies, that is, structural diversity. At one specific and important point he therefore becomes 'interventionist', very supportive of creating a specific type of society – the all-pervasive and individualist market society. Hodgson (1996: 183) writes:

> [T]his interventionist outcome creates still further problems for his system of thought. . . . [It] is not any spontaneous order that Hayek has in mind. It concerns just one type: The Great Society. What happens if the foundations of the Great Society are yet unbuilt or under threat? Rather than a faith in evolution towards perfection, Hayek believes that socio-economic intervention must be pushed down a particular track precisely by the creation of institutions and 'general rules' which are necessary for the formation and sustenance of the liberal utopia.
>
> The interventionist temptation in Hayek's thought is masked by the fact that the capitalist market systems are actually dominant in the modern world. In such real-world circumstances the advocate for free markets can then declare: when in doubt, do nothing. Accordingly, by placing the 'burden of proof on those wishing to do reform' (Hayek, 1988, p. 20), most proposals for state intervention can easily be opposed.

The breakdown of the command economies of Eastern Europe highlights the dilemma in Hayek's thinking. The question for the Austrian position became: should one rely on spontaneity or on the deliberate construction of market institutions in this case? Should one let new institutions form freely on the basis of what was falling down, or should one instead use the forces of collective bodies like the state to form private property and a market type of exchange structures? Should one therefore oppose the reforms of the Soviet Union, which were 'interventionist'/not spontaneous, or should one go for a construction of a market economy? Whatever stand is taken, it contradicts some basic features of the Hayekian model.

7.2 INSTITUTIONAL CHANGE AS DESIGNED: THE CREATION OF EFFICIENCY

There are two reasons for supporting the view that institutional change may also be intentionally created. Both follow from the above. First, some types of change become possible that will be beyond the reach of spontaneous or 'from below' processes as defined here. These are typically changes involving some sort of collective decision, which can transcend not least the large transaction costs involved in individual bargaining. This is the efficiency argument for institutional change. Second, any institution regulating conflicting interests depends on the intentional creation of the law. This section will address the view that institutional change is driven by the will to create efficiency. The issue of institutional change as interest or value driven will be covered in Section 7.3.

The idea that institutional change is efficiency determined is advocated mainly by the new institutional economists. Actually there are two types of issues that are dominantly focused on when arguing that institutional change is efficiency driven. We have the point above that institutional change enhances efficiency by economizing on transaction costs. However, we also have the idea that institutional changes occur as a response to technological change. They are necessary to make it possible to harvest the potential gains from this change.

7.2.1 Institutional Change to Reduce Transaction Costs

The idea of transaction costs reduction is used to explain many different types of institutional structure. We shall look briefly at the three focal ones: the existence of *property rights*, the creation of *firms* and the existence of *the state*.

Property rights imply a guarantee for the acquisition of benefit streams from a specific resource. This institution gives this benefit to the rights holder, and by creating such an arrangement there will in principle be no uncertainty about the distribution of the benefits. This reduces costs since the property holder does not need to physically protect what s/he defines as hers/his. Instead the collective/the state, after having acknowledged the exclusive right, protects it by the law. This considerably reduces the cost of protection borne by the individuals – their transaction costs.

Using a familiar example, the first gold miners of the early European settlements in California faced the problem of protecting the precious metal, first when found and then when extracted.[11] The use of threat or physical force from each individual became necessary. Many resources went into protection instead of production. Lives were also lost in these fights.

The establishment of rights both in land and in what was extracted, and the establishment of a necessary court system and police force were institutional developments that all reduced the involved costs of protection.

Following the same type of reasoning, Bromley (1991) argues that while property rights may reduce transaction costs, it does not follow that it is efficient to always opt for private property solutions. Instead one should acknowledge that it is also costly to institute private property. From a strictly economic point of view there is a trade-off between the costs of establishing the right to private property and the gains thereof. If the benefit streams are low in value compared to the costs of establishing exclusive rights, private property may not pay or be possible. Bromley (ibid.) views common property as a way to shift this trade-off point. As an example, pastures may not be productive enough to carry the costs of fencing individual plots and so on. Instead a common pasture – that is, a common property regime – is constructed, implying rules concerning both who has access and under what conditions. This reduces transaction costs even further as compared to the private solution. If costs of establishing a common property regime are too high, state property or open access may be actual regimes.

One should not confuse this with the idea that the resource involved – even in the case of open access – is necessarily of low value. We talk of costs relative to gains, not their absolute values. The value of fresh air is very high for each of us. Nevertheless, the establishment of individual rights in the involved benefit streams may be far too costly to make this solution work. A common property or some sort of state regulation is the only viable solution. Until recently, open access has dominated concerning the issue of air. During the last part of the twentieth century, various regulations – both national and international – have been put in place. The sulphur protocol, the CFC (chlorofluorocarbon) regulations and the Kyoto protocol on climate gas emissions are typical examples (Young 2002). We can view all of these as a type of common property regime where the participants in this case are the involved states. More on the above issues will follow in Chapter 10.

Moving to the issue of *the firm*, we may recollect a point already made in Chapter 4, that firms might be efficient organizations compared to markets. They economize better on transaction costs. It is cheaper to coordinate production within the firm than via market transactions (Coase 1937). Thus, the gradual establishment of the firm and the differentiations into various types may be seen as a response to high transaction costs in societies experiencing increasing division of labour.

Eggertsson (1990) clarifies by comparing a standard firm with two alternatives. First we have the alternative where it is the consumer who negotiates with several separate producers of inputs and assembles the final product him – herself. Wanting a bicycle s/he makes contracts – shops around – with producers of wheels, frames, brakes and so on. This is time consuming and demands that the consumer has extensive knowledge of how to assemble the final product. It also demands that the different components are produced according to common standards so that it is possible to construct a final product. According to the second alternative, the producers of inputs agree to make one of them responsible for negotiating with the consumer over the final product. They are still individual producers, but invest in a common marketer.

In reality, we can observe all combinations from traditional firms producing and delivering the whole product, to situations where the consumers actually assemble the final good themselves. A typical example of the latter is the homemade meal based on purchased inputs from a variety of producers. Modern car making falls somewhere between the second form mentioned above and a classic firm since many parts are bought from more specialized producers – that is, subcontractors. The car factory in the end is here not much more than an assembly line.

The main point for the 'efficiency explanation' is that the reorganization of production systems are responses to the opportunities evolving to reduce transaction costs. Williamson (1975, 1985) has been instrumental in defining which factors may explain the varieties of forms observed in both organizational structures and contractual arrangements. His dominant focus is on asset specificity. The more specific a good is to the transaction, the more costly contracting becomes. In such situations, gains may be obtained by undertaking the production within the boundaries of the same firm where contracting is reduced to the condition for wage payments only. Cheung (1983) offers a set of examples not least concerning the form that such payments to workers may take, for example, per hour or per unit of output depending on the type of product to be delivered. If the product is easily observable, payments will most probably be per unit of output, if not, ordinary wages are paid.

Within the efficiency perspective of institutions, the existence of *the state* also is understood in transaction costs terms. It economizes on costs of enforcement. North (North and Thomas 1973; North 1990) has been the clearest proponent of this view. According to him, state control of the quality of money, for example, has evolved because it is more efficient than other solutions. The gains in efficiency by creating a state appear at two levels. First, the state is central to the very establishment of the contract institutions since it offers a third-party form of enforcement. This is in itself an explanation based on efficiency arguments. The trustworthy contract makes the gains from trade possible. It is the state monopoly of coercion that creates this ability. Second, the same monopoly force gives the state a greater capacity to handle conflicts over contracts than any other body. Hence, the existence of the state reduces transaction costs at both the individual and collective levels, and thus enables the full utilization of the gains from specialization, a capacity otherwise thought of as a gift of the market.

The role of the state in creating and securing markets is important. However, it is also of importance for us to acknowledge the potential role of the state in creating institutional structures in situations where markets are too costly to use, as is often the case if we think about the allocation of environmental goods. The cost of transforming these goods into commodities may simply be too high. To demarcate air into sections so that it can be traded is virtually impossible. The state can, however, be involved in setting up regulations concerning the use of air, such as emission taxes or permits. It has the capacity both to make decisions about what to do concerning such common resource dilemmas and to enforce the solution – a capacity no market agent or private organization has. We shall return to this issue in much more depth in Part IV.

7.2.2 Institutional Change as a Response to Imbalances Created by Technological Change

The second issue concerning the creation of efficiency is that of technological change demanding institutional change to become practice. According to efficiency theory, this mechanism mainly comes about due to changes in relative prices of input factors that follow from technological change. Institutional change may be necessary to restore equilibrium in input markets.

However, there is more to the story. The potential of the new technology may not become available without some changes in the institutional setting. This was the case with the assembly line and firm organization structures. This is the case concerning the introduction of genetically modified organisms in present markets followed by necessary legislation concerning property rights of genes, the role of patent laws and so on. This is a necessary change if one wants private firms to be able to make profits from the new technology, which otherwise may be copied by others for free. Finally, new technology may also result in the loss of some income or benefit streams, a situation which may be opposed by the groups that are hurt. Agricultural policy in modern western countries can be understood as a response to such losses (Vatn 1984).

To illustrate some of the core mechanisms involved, let us look at the effects of the 'green revolution' of the 1960s and onwards. Basically, new varieties of crops, fertilizers and pesticides were introduced in developing countries to combat poverty. To make this transition possible, systems to distribute both the goods involved and the necessary knowledge had to be set up. Furthermore, the system of credit institutions had to be established or further developed to help farmers to raise the necessary capital. We also observe changes in the rules governing the distribution of the costs and the gains following the new technology.

These points seem reasonable and are all typical of many changes in the institutional settings. Studying the effect of technology on institutions demands a great deal of caution, though. Not all that is observed is – as an example – an effect of creating efficiency. To illustrate this, we shall look briefly at a study of the green revolution in the Philippines by Hayami and Ruttan (1985). They observe that this process increased yields and that this increase was split in accordance with the rules of the existing share tenancy institution. In the view of the authors, this created disequilibrium. The introduction of new technologies had produced disequilibrium between marginal returns and marginal costs of factor inputs. According to Hayami and Ruttan, workers received more than their marginal product.

They argue that a shift from share tenancy to sub-tenancy was one way to restore equilibrium. Other changes also appeared. Traditionally, landless labourers had been paid to weed the fields. Now, suddenly, weeding was no longer paid for. It became instead a prerequisite for being allowed to take part in the harvest and getting paid for that operation in the form of a defined part of the harvest. Hayami and Ruttan concluded:

> To test the hypothesis the [new labour payment] system was adopted primarily because it represented an institutional innovation that permitted farm operators to equate the harvesters' share of output to the marginal productivity of labour, imputed wage costs were compared with the actual harvesters' shares . . . The results indicate that a substantial gap existed between the imputed wage for the harvesters' labour alone and the actual harvesters' shares. This gap was eliminated if the imputed wages for harvesting and weeding were added. (Ibid.: 208–9)

The difficulty with the type of explanations implied by the Hayami and Ruttan study is the belief that distribution is the result of some natural forces – the correction of some externally forced disequilibrium. The problem with this reasoning is that it is not 'nature', but institutional arrangements that define what is income and what is a cost and for whom. The explanation of the institutional change is based on concepts (for example, economic rent) that are themselves defined by the given institutional setting. No neutral point exists and, as we shall soon see, the change can be better explained with reference to the power implied by given rights structures.

7.3 INSTITUTIONAL CHANGE AS DESIGNED: THE ROLE OF INTERESTS, VALUES AND POWER

The second perspective on institutional change as intended and designed, focuses on the issue of interest protection. Turning more systematically to this view, let us again start with the topic of *property rights*. As we have seen, such rights make transactions possible, as they also reduce transaction costs. However, this is only a part of the story. The most basic issue is still: who gets the right to the resources in the first place? While economic theory tends to take distributions of endowments as given – they are thought to be outside the realm of scientific enquiry – this distribution is at the heart of institutional change. As so strongly emphasized by Bromley (1989, 1991, 2006), institutional change is foremost about protecting interests. This is a view generally held not least by classical institutionalists.

Basically, it is the distribution of rights that defines the opportunities faced by different people. In a situation where some own capital and others own only their labour,[12] there exists a very uneven distribution of power and potential for consumption. Owners of capital can rely on this resource for their sustenance – that is, they are not dependent on continuous operation. Those who only own their own labour depend on a continuously running wage to survive. Adam Smith was among the many acknowledging this.[13] The asymmetry has consequences for the ability of the parties to influence the distribution of the net result of their joint operation, as in a firm. Moreover, the ability of the rich to continuously invest and increase their capital may tend to increase the uneven power relation since those with little cannot afford to set much aside. So while the parties to a labour contract are formally equal, they are not equal in reality. They simply have different capacities to handle a conflict.

Platteau (2000: 15) suggests that there is 'ample evidence that rules and institutions can be selected for distributive rather than efficiency reasons, or that institutional change can be redistributive rather than efficiency-improving'. In substantiating this he refers among others to Allen's (1992) study of the second wave of enclosure in eighteenth-century Britain. This move implied shifting property rights from common to private. According to Allen, this did not enhance efficiency. It did not happen because 'enclosed farming was more efficient than open-field farming but because landowners could expect positive redistributive effects from this reshuffling of land rights' (Platteau 2000: 15–16).

A substantial part of formal institutional development has concerned the establishment and protection of property, especially private property.

Much of modern history – that is, the last 150 years – has also been focused on the protection of the interests of labour to counteract some of the above-mentioned asymmetries. This concerns the right to organize and to create some countervailing power to that of the capital owners, including the setting aside of common funds to finance conflicts. It also concerns safety regulations at the workplace, the length of the normal working day, laws concerning child labour, retirement schemes and so on. These changes have largely been made since society has accepted them as legitimate. According to this view, the appearance of such regulations has less to do with reduced transaction costs[14] and nothing to do with equalizing marginal costs and gains.

Following on from the above, we can also recognize that in the case of *the firm*, not only transaction costs issues are involved. Many questions concerning power and interest protection are implicated, too. First, the firm is a command structure. It is organized in a hierarchical fashion as a way to execute power. Second, as Marglin (1991) argues, the firm is a means of securing not only a higher total surplus, but also a greater part of the surplus for owners.

A short visit to the first English textile factories may illustrate the point. As argued in Section 7.2, firms or factories may reduce coordination costs. However, in the case of the first textile factories there was little, if any, coordination between the workers involved. The labourers worked their looms as they would have at home. No assembly line was established. Nor did the first woollen mills use water- or steam-based power, so this could not be the reason for bringing the workers together.

The textile factory historically followed the so-called 'outwork' system whereby the work typically took place in the workman's own cottage. The system was set up by capital owners – that is, those who had the necessary capital to finance weaving looms and support workers by providing wool. Marglin (1991) observes that the capitalists complained about the functioning of the labour market. If workers were paid more, they worked less. Once their immediate needs were satisfied, instead of increasing their effort, higher pay made them 'stretch Saint Monday into Holy Tuesday' (p. 236). Marglin also refers to 'endless squabbles over product quality as well as embezzlement and fraud' (pp. 236–7). Thus, capitalists searched for ways to increase control and thereby increase their revenues. The end result of this process was, according to Marglin, the factory.

While the history of the firm is both one of transaction costs reduction and one of power enhancement, the same is the case with *the state*. Historically, and even in many countries today, the state exhibits many predatory characteristics as a way of concentrating power and wealth in the hands of a small elite (Gustafsson 1991).

While the national state as we know it, is a relatively young structure, systems of geographical control that one may call states or 'kingdoms' have a rather long history. The pharaohs of Egypt, the kings of Mesopotamia, and the emperors of China and of Rome take us far back in time. The establishment of the structures we call western democracies has taken place mainly over the last 300 years.

Certainly, structures like the old European serfdoms and kingdoms offered from the very beginning some protection against the threat from outsiders. The story – even of rather oppressive structures like that of feudal aristocracy – is not a one-sided one of only acquisition of rents via taxation and so on. Nevertheless, the state was not established primarily to reduce transaction costs related to defence and property protection. It was, according to this view, established as a power structure to support the elites.

One issue is the process of establishing state structures. Another is the continuous evolution that took place until the appearance of what we may call the modern national state. This latter development has transformed many states into democracies of different kinds. They have developed into structures which are much more representative of the 'collective will' than the old monarchies, except for the old and new despotic regimes that still tend to evolve around the globe. Fundamentally, this implies that the power base has shifted from brute force to majority votes in elections. Certainly, making democracy work, increasing the influence of weak groups, the need to continuously control elites and 'rent seekers', and so on are still prominent issues. However, the system has changed by establishing a set of democratic rights. The history of the state is an example of how an institutional structure may evolve into something very different from what it was originally. The state is a power structure. It can as such be used to support different types of interests and values – that is, support the construction of different types of societies.

Bearing in mind our focus, the insights from this history can be captured in two important lessons:

- First, the state is not just a structure that has been used to increase efficiency. While its potential to reduce transaction costs is huge, its role can certainly also be oppressive. The leaders may be more interested in securing their own positions and wealth than in creating institutions that benefit society more at large. While the 'young North' looked at the state as a way to increase growth and efficiency (for example, North and Thomas 1973), the 'old North' acknowledges this point (for example, North 1990) – see Chapter 4.

- Second, as illustrated by the short discussions of labour rights and environmental policy, the definition about what is efficient is not primarily a technical issue about reducing transaction costs. Instead it is a question about which interests and values we want to protect by using the power of the state. The issue of efficiency and the issue of interests, values and power cannot be kept apart.

This second observation is of interest for the green revolution case previously presented. Hayami and Ruttan's explanation was that the rules concerning the distribution of the net surplus from growing rice changed as new methods were put into practice. According to them, changes, for example, withdrawal of payment for weeding, were induced by altered marginal costs or productivity created by a combination of new technology and population growth. While the former increased capital productivity, the latter reduced labour productivity.

Bromley (1989) offers another understanding. He suggests that the explanation should be based on looking at the existing institutional structure and what potential it gives the various parties to change the rules. The issue is rather about who has the power to make changes and what are their interests. The idea that population growth reduces labour productivity is in his mind a very weak argument:

> In fact, what happened is that farm operators, by virtue of more abundant labour supply, were now able to disregard the very real costs that would fall on unemployed landless workers by the implementation of a new wage institution; it is the prevailing institutional structure that allows the farm operators to define a new legal relation that will obligate the landless labourers to engage in free weeding in order to be permitted to work in the subsequent harvest, and to receive their traditional share. (Ibid.: 24)

Bromley finds it almost ridiculous that increased population should reduce labour productivity. Do people get in each other's way, he asks. Rather, the increase ('abundance') of landless labourers made it possible to reduce their wages. The explanation was based on power relations, not efficiency.

Schmid (1987: 248–9) emphasizes in a similar vein what could have happened if all had ownership in the new technology:

> What would have been the substantive performance if landless labor [instead] had been given part ownership in the benefits of technology? It is only selective perception that regards a claim on net gain to the firm as changing the marginal cost of labour. A share of net return is not a marginal cost. . . . Equilibria between marginal cost and marginal revenue is not unique to one ownership interest and thus cannot explain change in ownership. The particular equilibrium among many possible is rights dependent and cannot explain change in rights.

What this shows, is that the 'efficiency' type of explanation of institutional changes implied not only by Hayami and Ruttan's model, but also by the ideas of 'the property rights school' and at least the 'young' North, is in many ways circular. The case studied by Hayami and Ruttan actually proves the point. While landless labourers did not get access to increased yields, tenants were able to secure their access to the increased surplus flowing from the new technology. A law was passed giving them the right to pay fixed rents. This was the logic behind the sub-tenancy structure that evolved at the same time. But why was this shift in rights restoring equilibrium? Why could not giving more of the net surplus even to the landless be termed likewise? To again cite Schmid (1987: 249):

> If the original tenant can be made part owner, why not all the landless labor in the village? There is no theoretical reason that they cannot be beneficiaries of public investments in irrigation and new plant varieties as well as landlords and original tenants. . . . This ideology [that of natural equilibrium], masked as a science, is part of the power struggle used by different groups to obtain institutions favorable to them. There is no way to have welfare economics that does not require the taking of sides.

What, then – in the midst of all this – is a legitimate use of power? Certainly, solutions that are found to be unjust by some will always be challenged by those who are deprived. A system based on very visible use of power to secure privileges is especially vulnerable to critique. Indeed, it is difficult to think of a sustainable social system built on open suppression. Power can, however, take other forms. It may be built into the basic structures of society – the institutions – like access to resources and the rules defining the distribution of surplus from production. Thus, what was originally brute force is transformed into 'the way things are'. It is changed from physical power into 'systems coercion'. Nevertheless, inequalities may exist. In fact, there are good reasons to argue that the difference between old and modern societies is not only that of democratization and increased equal rights, but also that of converting visible and brute force power execution into more hidden – that is, institutionally hidden – ones.

The formal equality of the labour contract makes the differences in real power almost invisible. The institution of capital rents is of equal importance, making it possible for owners of capital to create great fortunes just on the basis of owning. The power of ownership is – through the construction of various institutions – transformed into the marginal productivity of capital.[15] In many countries, unequal rights in access to land are governed by rules or perceptions that are deeply embodied in the culture. Many examples demonstrate that even when famines strike, the landless accept their destiny despite the fact that food is exported from the area in which they live. They just do not have entitlements to that food (Sen 1981), and starvation, even death, is seen – even from within this group – as 'their lot'.

Dugger (1989) discusses four different types of institutionalizing power that transform it from visible forces to systems characteristics: subreption, contamination, emulation and mystification. Marglin (1991) focuses on the same when he suggests that it is through the construction of various constraints creating opportunity sets that may look equal in formal terms, which ultimately produces realities characterized by great inequality. To illustrate, we shall reproduce a figure from Marglin's paper.[16] In this case the emphasis is on the different opportunity sets encountered by different occupations – that is, that of lawyers and that of car workers (see Figure 7.1).

The idea is that people trade off wages against the quality of life different jobs offer. The opportunity sets for lawyers are depicted by OAB. They have a wide variety of alternatives for combining quality of life dimensions with different wage levels. They can be a judge earning less, but enjoying a high status. They can also be a corporate lawyer, well paid, but experiencing less status. Hence, their indifference curve I_L may bring them to C where they trade off some wage for a higher quality of life or social standing. The car worker is also free to choose. Nevertheless, the opportunity set ODFE actually offers them no alternative but to settle for F. The question then becomes not one about the choices that are made in the end, but what causes the differences in the opportunity sets. This is an institutional issue.

The example is also chosen to show that conflicts and opportunities should be viewed along many different dimensions. Knowledge,

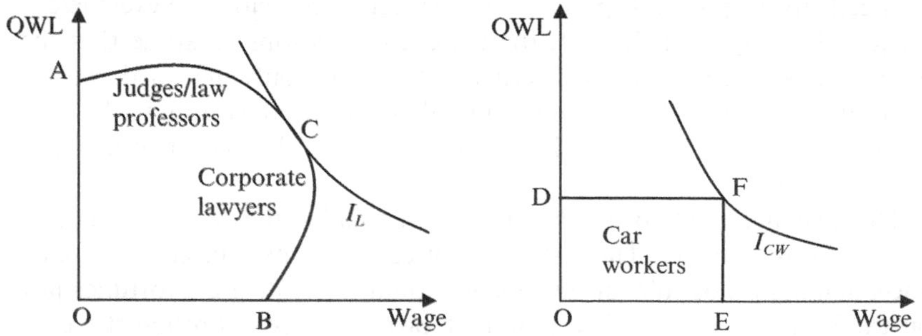

Source: After Marglin (1991).

Figure 7.1 Trade-off between 'quality of work life' (QWL) and wages for lawyers and car workers

professional status and so on may be an important source of power, not just ownership of capital/physical resources. This implies that in some cases, power has shifted from the owner of capital to the owner of a special competence. This is implicit in the above example. Even more typically we see this in the astronomic wages commanded by some movie actors and sportsmen and -women who have specific skills or 'human capital'.

Changes in the opportunity sets may occur as a result of shifts in the power of the groups involved. Workers may be able to strengthen their power of negotiation by developing organizational skills. The creation of parliamentary democracy in Western Europe was in the beginning fuelled by the will to reduce the power of the monarch and bring the government under the control of elected representatives of the 'people'.[17] This process was advanced mainly by the liberals. At a later stage this reform gave the labouring class the opportunity to become part of the electorate and form governments. The idea that every man should have the opportunity to vote – given that he owned property – was over time transferred to an equal right for all men, independent of what they owned. Finally, that right was also granted to women. It is still less than 100 years since women gained this right in western democracies.

These examples illustrate that changes in rights are a function of increased ability by different groups to define their interests and press for changes. However, it also shows that the mere logic of establishing a democracy influences which rules can later be viewed as legitimate. The values involved in the ideas behind democracy have power beyond the current interests. As the right to speech and access to democratic institutions is opened up, a process of evaluating the arguments concerning what the system should look like, is also established. This type of collective self-reflection which is here established is of great importance, an issue that will be discussed in more detail, especially in Chapter 12.

7.4 UNINTENDED EFFECTS: INSTITUTIONAL CHANGE AS REACTION TO CRISES

'Systems coercion' may not only be the effect of intended development of certain structures. It may also be the effect of dysfunctions. By this I mean that the system includes elements that tend to challenge the functioning of the system itself, even the interests of those in power. Societies are complex. Designing institutional structures such that no dysfunctions or crises appear is – I believe – impossible. The neoclassical equilibrium of sustained balance is, as an example, a fiction, as is a democratic system with continuous stability. In the previous sections of this chapter, I have elaborated on each type of explanation covered. When turning to the role of crises in institutional change, I will be very brief. This is partly due to the fact that the reaction to the crises, the institutional change itself, can largely be explained in terms of the previous types of explanations – especially those based on design. The crises – as observed – play the role of motivating institutional change.

There is a basic uncertainty involved in market economies related to the balancing of supply and demand. Firms are faced with the problem of determining future prices and quantity demands. This uncertainty not only influences the realization of the capital invested, but it may also bring the economy into phases of deep recession. Such recessions have been observed at various times, for example, in 1870 and 1930, in the 1980s, in Asia during much of the 1990s and more generally again in a period from 2001 due not least to the collapse of the information technology sector. Most EU countries have for the last 20 years experienced unemployment rates between 10 and 20 per cent. This has almost become an accepted level and due to the establishment of social security systems – not least as a reaction to the 1930s crisis – such a phenomenon is no longer really seen as a crisis.

Several models have been developed to explain these tendencies. They all relate in some way to the problems of unwanted effects of uncoordinated individual decisions. Keynesians explain it mainly by the lack of effective demand, not least the propensity by consumers to not reinvest all that they may put aside (Keynes 1936). Marxists turn to the production side and explain economic crises as a result of overaccumulation of capital resulting in too large a production capacity compared to the demand this production is meant to serve (Baran and Sweezy 1968). Others view it more as a psychological mass phenomenon whereby, during an economic upswing, investors overestimate the possibilities; when the situation reverses, they fear an ever-deepening crisis, and ultimately bring about this effect themselves by selling out and trying to save what they have. We experience what is called a 'fallacy of composition': a strategy that may be sound if only one (a few) follows the strategy, becomes a disaster when expanded to all (many).

The merit of these and other explanations will not be addressed here. My point is that the recurring crises are in themselves a major motivation for institutional changes. Therefore, over time there have been a number of reactions that have shaped the existing economic institutions:

- varying the level of public investment to counter recessions with periods of a 'heated' economy (Keynesianism);
- establishing labour market policies to balance recessions;

- creating import regulations to restrict the effects of international crises; and
- developing corporate cooperation – that is, merger strategies (monopolies) to control markets and thereby future demands.

All of these institutional changes run counter to the basic logic of the competitive market and produce their own problems in relation to that. Thus, these types of regulations have themselves been accused of being the source of crises.

While the main focus in society so far has been on the economic crises, there is another type of crisis that goes beyond that of imbalances within the economy: the imbalance between the form and capacity of the economic process and the capacity of its surrounding natural systems. This concerns both the capacity to set aside the necessary amount of natural resources to keep ecological processes intact and the ability to take care of the waste that inevitably is created by economic activity and must be dumped. We have already mentioned institutional changes that are a response to this type of crisis – for example, the sulphur protocol and the Kyoto protocol. As will be emphasized in Part IV, the basic challenge involved here is the great, but not unlimited capacity of ecosystems to absorb changes in matter flows and so on. In contrast to economic crises, where mistakes become visible rather quickly, we are here confronted with dynamics that change very gradually. However, beyond certain limits, the forces set in motion are normally so large that it is often too late to react. To the degree that we are only able to change institutions as a reaction to visible crises, this offers a rather pessimistic view of our future.

7.5 SUMMARY

In this chapter we have looked at four different explanations of institutional change: (a) spontaneous creation of institutions, (b) designed institutional change aimed at increased efficiency, (c) designed institutional change to protect certain interests, and (d) institutional change as a reaction to crises. All types offer some important insights. The tendency in the literature to favour just one is unwarranted.

So-called spontaneous creation of institutions has here been divided into 'institutional change from below' and 'pure spontaneous change'. In the latter case it is emphasized that institutions are not only developed from below, but they are also the unintended effect of several independent choices. While this model may explain some institutions, it is a mistake to view institutional change as generally unintended. Even in the case of conventions and norms – the type of institutions where spontaneity can offer insights – closer inspection reveals that intent plays an important role in many instances where changes come from below. One should be careful about confusing the (intended) change of institutions and the (unintended or automatic) reproduction of them.

As soon as conflicts are involved, the establishment of an institution to regulate these will not only be based on intent. Such acts also warrant coordination in the form of collective design of the institution. Intentional institutional change to increase efficiency is certainly a relevant type of explanation as in the case of constructing institutions like money, property rights, firms and the state. I here have in mind the ability of these institutions to reduce transaction costs. However, this type of explanation also has its limits. The very concept of efficiency, the rules and conventions by which efficiency is measured, are themselves largely defined by the actual institutional set-up, and efficiency-based explanations will easily end in circularities. The efficiency claims become embedded in the assumptions of the analysis.

We therefore have to accept that whatever institutional structure is formed, it implies the recognition and protection of some interests and the denial of others. This applies to all core areas focused on in this chapter – that is, (property) rights, for the organization of firms and the state. However, it is also the case for norms since they define and protect certain values. Certainly, the capacity of different interests to secure their protection by these institutional structures varies. Partly the relevant social groups may lack the political or other necessary power. Partly, they are not able to legitimize their interests on the grounds of arguments that are acceptable within the existing political system. The problem we face in the latter case is on what grounds can an interest be said to be legitimate. What really differentiates a justifiable interest from one that is not? While this is the subject we shall address in Chapter 8, we have here observed that some interests never need to defend themselves since institutions that are built into the basis of the system protect such interests. Thus they become 'invisible' and tend to go unchallenged.

No interest gets the ultimate protection. The dynamics, not least of market economies, create imbalances that also threaten those having the most advantageous positions. Thus, crises are important drives to institutional changes. Certainly, the negative effect of crises is still strongest for those at the weak end of the system. They have few capacities to defend themselves. One important aspect of crises in market economies is that they tend to build acceptance for more public 'intervention' or planning – whether public (state) or private (larger firms) – that is, they legitimize changes that are actually counter to the basic idea of the system itself.

NOTES

1. Sened (1997) divides between (a) spontaneous and (b) intentional emergence.
2. The distinction – as made in the literature – is often not very clear. Sened, as an example, defines spontaneous results as 'equilibria in social games without *much* intentional planning' (1997: 71, added emphasis). Thus, some (how much?) intention is allowed for even though he still contrasts 'spontaneous emergence' with 'intentional emergence' as the fundamental categories. However, it seems to be important to avoid collective decisions at various levels.
3. Remember that the fact that something is chosen because we see or believe that it functions well is not a functionalistic explanation. It is instead intentional.
4. I am somewhat sceptical of Screpanti's use of evolutionary concepts as metaphors for institutional change since it gives a stronger intimation of pure functionalist explanations than his position, in my mind, is actually based on.
5. This distinguishes it not only from the social constructivist position (see above), but also from ordinary game theory where people are individually rational maximizers and knowledge is most often viewed as common.
6. These concepts should not be normatively understood – that is, as wrong. The point is just that they are different.
7. It is reasonable to say that the rather strong fear for the state and third-party solutions more generally, may have caused adherents to the institution-as-equilibria position to opt for solutions where any institutional change is found to be spontaneous. There is thus a strong link implicit in this literature between 'what is' and 'what should be'. The descriptive and the normative go hand in hand. As emphasized in Chapter 4, this has created some very visible contradictions in this literature.
8. My presentation is based on Hodgson (1996).
9. Cited from Hodgson (1996: 110).
10. Even the latter may be questioned. At least there are both normative and formalized control mechanisms concerning the language in many countries – that is, language councils and so on. Remember also that the written language is a highly formalized endeavour, still affecting on the more dynamic oral language.
11. This example is popular as it is a fairly recent, and therefore well-documented, case of people moving into an area or a business where the law still did not hold sway and furthermore where large values were involved. Seizing assets from others was tempting and profitable. The example of the Californian gold miners is here used as an example of areas where great conflicts exist that are not yet institutionally regulated.
12. Certainly, as we know from history, even owning one's own labour cannot be taken for granted. Slavery has been a very important institution in many economies even up to the present.
13. He specifically pointed out that property enabled owners of land or stock to hold out much longer than employees in conflicts: a year or two for masters and only a week or a month for workmen (Smith [1776] 1976).
14. It should be acknowledged that organizing labour unions influences transaction costs among labourers as it influences the costs of transacting/negotiating between employers and employees. Common insurance schemes can be viewed similarly.
15. The so-called 'capital controversy' between economists of the two Cambridge universities in England and the United States – represents an important discussion in relation to this (see Harcourt 1972).
16. Among other things, I have changed the form of the opportunity sets, since Marglin's formulation seems inconsistent.
17. Men who did not own property and all women were at that time excluded from the right to vote.

8. Evaluating institutional change: the normative aspect of institutions

While Chapter 7 examined different ways of interpreting or explaining actual institutional changes, our next task is to look more directly into the normative issues involved. The issues we face here can be captured by the following two questions. How should one define what is good – that is, what is the best situation for a society? How can we next achieve such a state of affairs? In the history of economics, the answers given to these questions are many. Given the different positions within the discipline, this should come as no surprise.

The chapter will be structured as follows. First, we shall look more in principle at a set of positions concerning responses to the issue of 'what is best' (Section 8.1). We shall concentrate on four stances, that is, the utilitarian, the standpoint of modern (neoclassical) welfare economics, the Austrian position, and finally the one following from classical institutionalism as developed here. Second, we shall go more deeply into the position of modern welfare economics, both because of its dominant position in general and because it forms the basis for standard environmental economics, which will be visited several times in Part IV of this book (Section 8.2). In Section 8.3, the normative aspects of classical institutionalism will be developed in contrast to that of modern welfare theory.

8.1 DIFFERENT WAYS OF DEFINING 'WHAT IS BEST'

As indicated in Chapter 6, there was a gradual development in economic theory from an objectivist to a subjectivist account of value. We have also seen that what is good can be understood in one-dimensional or in plural terms. We shall structure our presentation of the various stands taken concerning 'what is best to do' using these two dimensions. Consider Figure 8.1.

Placing the various positions from the classical utilitarian to the Austrian within the bounds of Figure 8.1 offers just an approximate classification. To be accurate, more dimensions would be demanded. Further details concerning the various positions will be supplied in the text.

Characterizing what is good	Value dimensions of the good: Single	Value dimensions of the good: Plural
In 'objectivist' terms: weak or strong	I The utilitarian position	II The classical institutional position
In 'subjectivist' terms	III Modern welfare economics (contemporary neo-classical position)	IV The Austrian position

Figure 8.1 *Categorizing various positions concerning the normative aspects of economics*

As indicated in the figure, the concepts 'objective' and 'subjective' are not simple dichotomies when applied to our issue.[1] Concerning objectivism, two dimensions stand out. We must distinguish between whether what is good can be *defined in objective terms*, and whether it can be *objectively measured*. By 'defined in objective terms', I mean that what is good can be described, discussed and evaluated across individuals. It has *substantive* or *cognitive* content (O'Neill 1998). Statements concerning what constitutes a good life, such as fulfilment of basic needs, the development of certain skills and so on, make sense given this perspective. Moreover, 'defined in objective terms' implies that there is something specific to being human, while this interpretation does not imply that it is only one way to live a good life.[2] While living in isolation, for example, provides far fewer opportunities to develop a good life as compared to having family and/or friends, it is not in our nature that there should be only one way to live well. That is an untenable position.

The changes observed concerning how societies over time have defined what it means to be human are therefore not at odds with this perspective. Rather it supports the position that people can reason over and agree about the important elements of what constitutes a good life. These are very significant issues for a society, and the position implies that discussing what is good may produce changes concerning what people perceive as living well and what sort of institutions society should develop to support that. Following on from this, an objectivist position is not inconsistent with a pluralist interpretation of what constitutes a good life. Some may pursue a life with more emphasis on family and friendship while others may concentrate more on career and self-realization. The point is that the definition of what is good can be made in concrete or substantive terms and reasoned over. 'Family is more important than career' is an objectivist statement in this sense. However, some may argue the opposite. 'Society should support the education of its young' and 'Women should have equal opportunities to men concerning participation in the labour market and political life' are other objectives which societies have debated and many have formulated as common goals.

By objective measurements, then, I think of measurements making *interpersonal comparison* feasible. This aspect is distinct from defining what is good in objective terms. Hence, it can be claimed that what is good or a good life cannot be rationally discussed across individuals – that is, it is a purely subjective question – while at the same time one may claim that the welfare of two individuals can be compared. Person A may be classified as 'happier' than person B according to some measure, while what creates happiness for the two is incomparable.

Subjectivism then is related to the idea that a person's well-being is determined by her/his individual desires or wants. It is defined as *subjective determination*. This implies that the content of a person's well-being is determined by her/his desires or beliefs about what is good for her/him (O'Neill 1998). Subjectivism thus understood implies that what a specific person finds to be good is not open to reasoned evaluation across individuals. To cite O'Neill: 'Ends are treated as wants, and no judgement of their inferiority or superiority is allowed to enter criteria of efficiency' (ibid.: 20). What is good is a question of individual wants only, not of judgement. Furthermore, subjectivism also implies a denial of interpersonal comparisons.[3]

The stance adopted concerning these issues not only influences our understanding of a good life. It has great practical implications since it influences the way we should organize societies. The objectivist/substantive perspective of what is good puts emphasis on communication/dialogue. The *forum* – the arena for political discourse – becomes a core institutional structure. The subjectivist position supports a non-dialogical arena for human interaction – the market – an arena where we exchange, we do not communicate.

Given these arguments, we can define the following four possibilities concerning characterizing what is good:

1. It can be defined in objective (substantive/cognitive) terms and objectively measured (interpersonally compared).
2. It can be defined in objective (substantive/cognitive) terms, but only measured by subjective measures – that is, interpersonal comparison is impossible.

3. It can only be subjectively defined, but it can be objectively measured and comparisons can be made across individuals.
4. It can only be subjectively defined and measured – the latter implying that interpersonal comparison is impossible.

While position (4) is subjectivist and position (1) is objectivist in a strong sense, positions (2) and (3) are here termed weakly objectivist due to the fact that one of the two objectivist elements is accepted.

Positions III and IV in Figure 8.1 – that is, modern welfare theory and the Austrian position – are based on subjectivity concerning both content/experience and measurement – that is, class (4) above. Within positions I and II there is greater variation. The position of the utilitarians approximates that of class (3). Objectivity relates mainly to the measurement issue, even though some of these authors sometimes also discuss 'what is good' in more objectivist terms. Important writers within classical economics and classical institutional economics tend to fall more under (1) or (2).

Concerning the issue of value dimensions, the other component in Figure 8.1, the situation is simpler. Utility – understood in its classical form as 'happiness' and in its modern form as a mere ranking (see Chapter 6) – is a one-dimensional measure. The concept of plural values or preferences, on the other hand, implies that the values involved cannot be transformed to one common denominator. This is the position of both classical institutionalists and Austrians. However, according to classical institutional economics, the involved plural values can be open to reasoned discussion across individuals. One may deliver arguments over which various, yet irreducible, experiences and skills should be part of a good life. In the case of the Austrian position, with its rather radical subjectivism, it is claimed that no such evaluation is possible. Values are still considered plural since both freedom and welfare are elements of what is of value to a human, according to this stance.[4]

Given the above understanding of the character of values and their dimensions, one can produce very different positions concerning how to evaluate various states of the world – that is, making conclusions concerning 'what is best'. We shall look briefly into each of the main positions as placed in Figure 8.1. Given that these positions are fairly distinct, they also reflect a historical development in concepts and perspectives. This will be emphasized.

8.1.1 The Utilitarians: Weakly Objective and One-dimensional Values

The utilitarian position can be classified as weakly objective. As we saw in Chapter 6, it was developed among a group of philosophers in which

Bentham held a dominant position, and later taken up not least in economics. Utilitarianism consists of three elements (see also Sen 1988):

1. *welfarism*, which demands that the goodness of a state of affairs is a function of how much utility or happiness that state brings;
2. *consequentialism*, which demands that every choice is determined by the goodness of its consequences only, its ability to create utility; and
3. *sum-ranking*, which demands that utility information regarding any state should be assessed by looking only at the total sum of all the utilities in that state.

The element of welfarism implies that a state should be evaluated only according to the utility, that is, the happiness it creates. Happiness is a certain feeling – a mental association driven by urges of pleasure and pain as experienced by the individual. What is defined as good or bad is subjectively determined, but the literature is unclear on this point. One will often find references to a more objectivist understanding. This is clearly visible in Mill, the only major classical economist to take utilitarianism on board. However, neoclassical economists adhering to the utilitarian stand – that is, 'early' neoclassicals like Marshall – also use formulations that on many occasions have an objectivist flavour concerning the definition of what is good (see also Hodgson 1988 on this issue).

Consequentialism implies that only the consequences of an act matters. The intent behind the act is not important – that is, no act can be viewed as right or good in itself, and an act performed for reasons other than utility or happiness does not count in assessing 'what is best' as defined by the utilitarians.

Finally, the utilitarians claim that happiness can be objectively measured with a cardinal measure and summed across individuals. Hence, sum-ranking is a third distinct element of the position, implying that utility is viewed as comparable across individuals. The aim is to maximize a society's total welfare as measured in welfarist and consequentialist terms. The optimum – the best state – is the one that maximizes the sum of utilities for all individuals.

The utilitarian ideas were taken up in economics by Mill and later by the early neoclassical economists such as Jevons, Walras, Marshall and finally Pigou. While the idea of measuring utility in cardinal terms generally prevailed, the question of interpersonal comparison was still somewhat disputed. Jevons explicitly denied it ([1871] 1957). Marshall seems to have had some problems with it, but generally still followed the rule.

Pigou (1920) developed a welfare theory built on the utilitarian framework that went further than any other economist. Two distinct elements of his welfare theory can be identified. Following the standard conclusion of utilitarianism and sum-ranking, an optimum exists when the marginal utility of the last unit of income or money is equal across all members of society. If this were not so, the sum of utilities – that is, total welfare – could be increased by shifting income from those with the lowest marginal utility ('the rich') to those with the highest ('the poor'). Whether this implied equal income across all members of society is not clear, since the marginal utility of a certain level of income might vary between individuals. None the less, there was an egalitarian flavour attached to the position.

The second element of Pigou's welfare theory was that in optimum, private and social costs should be equal. If all goods were sold in competitive markets, this was assumed to be the case.[5] The problem was with costs that went unnoticed in markets – for example, physical externalities. Thus, while the utilitarian position basically favoured market solutions, Pigou suggested that both income transfer and public action in the form of taxes and subsidies should be used to produce the optimal outcome of the economy.

8.1.2 The Position of Modern Welfare Theory: Subjectivism and One-dimensional Values

Modern welfare theory is historically a child of utilitarianism, but has its distinctiveness from the changes within neoclassical economics taking place mainly in the 1930s. The most important change is that sum-ranking is dropped. The stance adopted is that utility cannot be compared across individuals without making value judgements, and such judgements should be avoided. Moreover, utility cannot be cardinally measured. It becomes an ordinal concept – that of a ranked order. The utilitarian perspective of welfarism and consequentialism is, however, basically supported. Nevertheless, it is important to observe that these concepts take on a somewhat different meaning since the core concept defining both – the utility concept – is changed in content.

As we saw in Chapter 6, first references to 'usefulness' disappeared gradually in mainstream economics following the shift to the utilitarian concept of 'happiness' and 'pleasure'. Next, references to mental states like 'happiness' also disappeared. Instead, utility becomes defined in terms of a preference index, as an individual ranking of goods. As such it shifts to a purely formal concept. It gives no information about what utility or welfare consists of, either in real and substantive terms or in the form of pleasure. This is the perspective that is fundamental to modern welfare theory. However, the use of the concept of utility is still somewhat confusing as employed by practitioners of modern welfare theory, and different connotations prevail.[6]

The welfare rule applied in modern welfare theory is the Paretian one. This is a logical move since sum-ranking is rejected. The Paretian concept avoids all interpersonal comparison. At the same time, the preference functions of the individuals are considered continuous, so trade-offs can be made between all goods involved. Thus utility, while not comparable across individuals, is still considered a one-dimensional concept.

While based on the idea that utility is a purely subjective notion, the move to the Pareto principle was thought to guard against subjectivism *in the social evaluation* of what is a better or a best state. Nevertheless, a value judgement is implicit in the Pareto rule. Pareto efficiency means simply that at least one person is made better off and nobody else is worse off in a move from one state to another. When applied to practical policy, however, this gives primacy to the existing distribution. A distinction between efficiency and distributional issues is therefore developed. The job for the economist is to work out what is efficient in Paretian terms, given the distribution of income. It is then up to the politicians to fix the distribution.

Within these confines, modern welfare theory has produced the fundamental theorems of welfare economics (Arrow and Hahn 1979; Boadway and Bruce 1984; Sen 1988). The first theorem states that *every perfectly competitive market equilibrium is Pareto optimal.* This demands certain conditions – that is, the core and standard application theorems of neoclassical economics must hold. There must be no externalities and markets clear all relevant transactions.[7] The second theorem states that *every Pareto optimal social state is a perfectly competitive market equilibrium.* Again, the core and application theorems must hold and furthermore there must be no economies of scale. Thus, Pareto optimality is directly linked to a set of assumptions about the world and a distinct institutional structure – that of perfectly competitive markets.

One important problem has been that many real-world situations defy the market and the Pareto principle. This is not least the case when physical interrelations exist – for example, externalities and public goods. As with Pigou, it is observed that state or some other collective action is then needed. The decision rule applied by modern welfare theory differs from that of Pigou, however, since one has moved to the logic of the Pareto principle and not that of maximizing the sum of utilities. Moreover, a specific problem is faced since in situations with physical linkages, there will normally be both gainers and losers to a change in the allocation of resources. This was not a problem for utilitarians like Pigou. The better state defined as the highest sum of utilities could still – in principle – be determined. The Pareto rule has, however, no answer to this situation. The next step in the development was therefore the establishment of the potential Pareto improvement (PPI) or the Kaldor–Hicks criterion (Kaldor 1939; Hicks 1939), which says that an improvement exists if the gainers can compensate the losers and still be better off. The improvement is potential since compensation is not thought to be undertaken. This criterion also produced some problems, however, which we shall return to in Section 8.2.

8.1.3 The Austrian Position: Subjectivism and Plural Values

The dismissal of interpersonal comparison and the introduction of the concept of utility as a mere ranking of preferences was a development in neoclassical economics strongly influenced by the Austrian tradition.[8] This tradition has its own, quite distinct, position concerning the issues discussed here, warranting separate treatment. First, the idea of subjectivism is in a way taken one step further. This is paralleled by looking at freedom as a separate goal in itself. Consequently, the Austrian stance must be termed 'pluralist' according to the dimensions of the good. It is about both individual freedom and welfare.

The Austrian tradition strongly emphasizes that beliefs about value – what is good – do not answer to rational arguments. The role of the market is to coordinate action between people with (very) different conceptions of what is good. It is especially in the evaluation of this that these authors deviate from modern welfare theory. They have no sympathy for social or collective choices whatsoever related to the allocation of goods. Concepts such as PPI – which is so fundamental to modern welfare theory – are avoided. In the case of Hayek, this is partly dependent on his belief that a state bureaucracy is unable to acquire the necessary knowledge and calculate the social optimum.[9] However, the conclusion also seems to be influenced by the supremacy given to liberty and freedom found, not least in Hayek's writings.

The Austrians, like the neoclassical economists, justify markets in welfare terms. The argument is different, though, since the former avoid equilibrium and formal model analyses as their basis and instead emphasize the discovery process. Markets are creative. They are viewed as processes of discovery and change, not as systems characterized by equilibrium. Continuous changes follow from entrepreneurial activity motivated by market opportunities. No equilibrium exists, and according to Hayek (1976: 6), 'the maintenance of a spontaneous order is the prime condition of the general welfare of its members'.[10] Parallel to this, the Austrian tradition has a strong libertarian basis for their thinking 'which allows welfare to be overridden given a conflict with liberty' (O'Neill 1998: 55). From this we can also see that there is a consistency problem in the Austrian stance. The weight given to freedom actually implies a certain objectivist perception of what is good. It accentuates a specific good – liberty. It argues for its primacy, and goes way beyond the idea that what is good is a purely subjective issue.

Here it is also important to recognize that among the Austrians, freedom takes on what is often called a negative form. This is most clearly developed by Hayek. Freedom or liberty refers to the absence of constraints on doing what one wants – the only constraint being that one shall not deliberately hurt others. This is a logical definition given the strong focus on spontaneity, individuality and subjectivity. It is still problematic, as can be illustrated by looking at the alternative – the positive account of freedom which accentuates autonomy and aspiration. Here the focus is both on the capacity to realize one's aims and on the number of options or possibilities made available to choose between. The positive definition demands a collective process to create this capacity and these possibilities. It must actually be based on some objectivist perception of what it means to be free.

8.1.4 The Position of Classical Institutional Economics: Objectivism and Pluralism

This brings us to the position of classical institutionalism. Here, the collective creation of capacities and opportunities are core issues. However, this is a positive conception of freedom, which also requires that when we develop human capacities and opportunities, we have to make choices between different conceptions of the good. We must – as a collective – make choices concerning which values and interests are to thrive. This demands an objectivist perspective of defining what is good. Consequently, the classical institutionalist position is explicitly dialogical. It is based on communicative interaction. It focuses on the process by which the members of a society can come to terms with what institutions should be established to support the development of favourable capacities and opportunities.

As we have already underlined (Chapter 5), the classical position is characterized by the view that the good is a plural entity with dimensions that cannot be collapsed into one single scale. It is not only about utility as happiness, but also about integrity, rights and commitments – that is, values that involve different forms of reasoning. Taken together, the position goes beyond that of a calculative welfare measure – be it utilitarian or based on the Paretian model.

If we return to the structure of the full utilitarian model, we can actually distinguish differences concerning all the three elements involved. Most basically, the classical institutional position perceives *welfarism* to be far too narrow, if not a completely wrong perspective. The issue is not (only) about the desires – that is, about how much utility (whatever definition) – that a certain state of affairs brings. Following Knight (1922), man is an *aspiring* rather than just a desiring being. While wants are drives for satisfaction, they are also objects for evaluation and development. We reason over our desires: 'Should I really want this?', 'Smoking is what I desire, but is it good for my health?', 'I am thirsty, but my fellow hiker needs the water more than me'.

In relation to this, Holland (2002) emphasizes that it is a problem with the welfarist model since it 'separates the cognitive and non-cognitive components of human motivation. The problem is that, shorn of any cognitive content, desires become indistinguishable from brute urges, with the result that they are unable to constitute reason for action at all' (p. 21). Human beings disintegrate into a simple machine forced by their unreflected desires.

The cognitive or objectivist account of needs does not imply – as earlier suggested – a given structure of needs common to everybody. To clarify, let us pay a short visit to the psychologist Abraham Maslow and his well-known theory about needs. His position is objectivist in a very narrow sense of the concept, since he has developed a given structure or hierarchy of needs. At the bottom of this hierarchy come basic physiological needs like air, water, food, shelter and sleep. Next come higher material needs like safety and security. When these needs are relatively satiated, the human will give priority to social needs like belonging, love, acceptance and self-esteem (Maslow 1954).

The classical institutionalist will acknowledge that there is something characteristic to being a human. Living a human life implies certain things. Many of the above points are almost self-evident given the kinds of beings we are. Therefore Maslow is correct in taking a stand against a subjectivist account of desires or wants. On the other hand, it is difficult to support a given hierarchy – especially the further away from basic physiological needs one goes. It neglects the role of the social and institutional in determining who we become and to what we aspire. But, culture does not define it all, either. We may ourselves choose between different life-plans, given the society in which we are raised. Thus, while the welfarist model is failing due to its inability to offer reason for action, the psychological model is failing because it tends to give a fixed answer based upon individual psychology only (see also Hodgson 1988 on this).

Turning to *consequentialism*, the difference between the welfare economists and the classical institutionalist is less pronounced. Also, classical institutionalists focus on the consequences of a policy. The problem encountered in utilitarianism and modern welfare theory is the narrow understanding of consequences – covering only those which produce utility or welfare. The opposite or alternative position to consequentialism is 'deontology', whereby the goodness of an action is not defined by its welfare consequences, but whether an act is right or wrong in its own sense. It is to pursue a certain value or moral standard as illustrated in the case of the unemployed chemist who did not want to take the job offered since it implied the production of biological and chemical weapons (Chapter 6). This again relates not least to the view that the individual should have certain rights. Its integrity is an aim in itself. Further, the individual may in many situations deny doing what gives the highest welfare if this is in conflict with what is perceived as morally right – see again the chemist example.

The distinction between consequentialism and deontology must not be exaggerated. Deontology does not necessarily disregard consequences. In the sense of Rawls (1971), rights may be created just to produce good consequences. Classical institutionalists would in general support such a view. Rights defend interests and values, and one reason for defending an interest or value may be that it can also produce the right consequences in the form of well-being. Sen (1988) has developed this view in a distinct way. In his plea for *rights-based consequentialism*, he suggests that there need not be a conflict between the focus on consequences and that on rights. This is the case if one accepts that consequences are measured along dimensions other than just welfare, such as the distribution of rights *per se* – that is, accepting plural value dimensions.

One important aspect in this is the responsibility we have as citizens to make it possible for members of a society to develop their skills and personalities. It is not only the well-being, but also the 'agency aspect' of a person that is of importance for creating a good life. In classical institutional thinking this goes back at least as far as John Dewey.[11] The agency aspect focuses on the individual as a doer. It is not only the consumer aspect that is important. Also essential are the creative feature – the development of individual capacities and skills – and the issue of developing one's character. Sen (1988: 59) supports this when he contrasts the well-being and the agency aspect: '[The] "agency aspect" takes a wider view of the person, including valuing the various things he or she would want to see happen, and the ability to form such objectives and to have them realized'. The interesting thing, I believe, is that even most welfarist economists would embrace such an understanding of a good life. They support the development of schools, a diverse variety of public fora and so on. As economists we might believe that the development of such institutions are not counter to our model. However, it falls outside what can be logically evaluated by the welfarist perspective.

Sum-ranking was the last element of utilitarianism. Since classical institutionalists emphasize that values are plural, sum-ranking has no meaning. Nevertheless, members of this position will favour interpersonal comparison, not on the basis of utility comparisons as with the utilitarians, but in the form of evaluating opportunities, abilities to fulfil basic needs, literacy, possibilities for education, nutrition, health care, environmental conditions and so on. They will focus on various dimensions of relevance for creating an adequate or flourishing lifestyle, and consequently make it possible to decide where it is most urgent to support changes. Neoclassical economists themselves use data on these issues in their empirical studies – for example, neoclassically based development analyses. It seems to be common sense, but to make such comparisons is contrary to the model on which modern neoclassical economics rests.

Do not misunderstand. I do not say that when theorizing, welfare economists think that the poor are poor and the illiterate are illiterate because this is what they want or desire. A distinction between efficiency and distribution is clearly made. The point here is that evaluating what should be done in concrete terms like setting up schools, launching nutrition programmes and so on, is outside the evaluation of welfare economics because it implies saying something about what is a good life in objectivist terms. To put it bluntly: a consequential welfarist would leave it up to the illiterates themselves to decide whether illiteracy is something they would prefer to avoid. This should be done only if their willingness to pay for schools is high enough to cover the costs.

One may counter and say that this is to take subjectivism outside its bounds. I do not think so. Treated consequentially, the subjectivist model is unable to treat the issue of, for example, education in any way other than superficially. The basic issue is parallel to the self-reference problem of costly information as discussed in Chapter 5. The problem in this specific case can be simply illustrated: how can an illiterate person evaluate the consequences of becoming literate? When evaluating this, one is forced to accept some objectivist notion of what a good life is like. Rather than looking at that with suspicion, one should consider how that is part of our everyday lives and look for ways of improving the processes by which decisions about the good life are made.

8.2 EFFICIENCY: LOOKING DEEPER INTO THE WELFARE ECONOMICS POSITION

In the above sections we have discussed the issue of normative evaluation in economics in rather broad terms. We shall now narrow down to look more deeply into one aspect – the understanding of efficiency as defined by the modern (neoclassical) welfare economics position. To do that, I shall first explain the reasoning behind the first and second welfare theorems. Second, I shall discuss some of the problems involved when we try to distinguish between efficiency and distribution, an essential characteristic of modern welfare economics.[12] The presentation is brief. More comprehensive treatments can be found in Boadway and Bruce (1984) and Varian (1992).

8.2.1 The Efficiency of Markets

Welfare economics focuses basically on the gains of trade/exchange. It starts off with individuals or households with given endowments. It is then shown how exchange in production and consumption can make people better off. The reasoning can be illustrated by a rather simple example. If person A owns land[13] and person B owns labour, combining the two via exchange would result in greater production than setting each factor into production alone. Besides, land and labour can be used to produce different commodities. If one producer produces grain and another produces potatoes, a higher level of utility can (normally) be obtained if these products are exchanged and both producers, who are also subsequently consumers, can consume some of both products. This is the case since the marginal utility of any good is thought to decrease with the amount consumed – certainly a reasonable assumption.

This reasoning can be enlarged to cover k agents, n inputs and m products without any changes in the basic logic, except for the important shift to a competitive market – that is, many producers and consumers. If all rights to endowments (resources) are initially distributed, if agents are (costlessly) maximizing individual utility/profit, if they never do wrong, if preferences are given, if no agents have market power and exchange itself is costless (zero transaction costs), then Pareto improvements will have to be the result of any voluntary exchange starting off from the initial endowment distribution. Exchanges will stop when no more gains from trade are achievable – that is, equilibrium is reached. Thus, the competitive market yields a Pareto optimal outcome. Given the list of assumptions and defining efficiency in Paretian terms implies that voluntary exchange must foster efficiency.

We could have stopped with this intuitive story. However, to see more fully what goes on, I shall give a brief illustration of each of the three steps involved. First we look at the production problem – that is, that of distributing the vector of given inputs or resources $\mathbf{x}$ between the various outputs or goods – vector $\mathbf{y}$. For reasons of exposition one normally simplifies and analyses a situation with two inputs – for example, x_1 and x_2 – and two

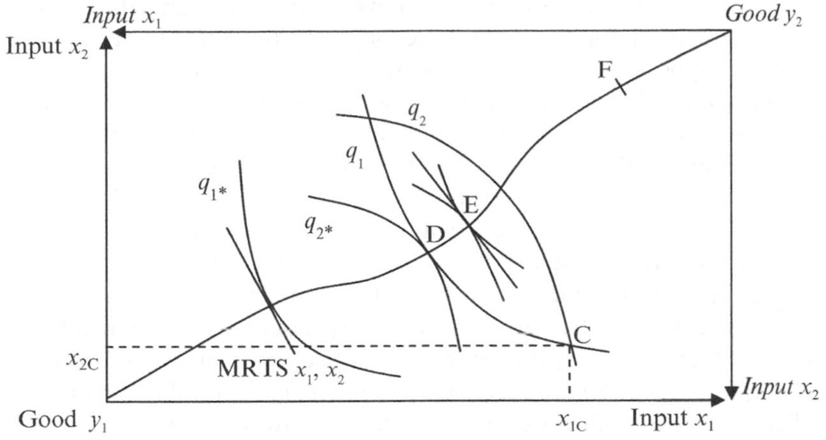

Figure 8.2 Efficiency in production

outputs – for example, y_1 and y_2. The problem is: how many of these inputs should be devoted to the production of the two outputs, respectively? The exchanges going on here can be described as in the Edgeworth box of Figure 8.2, covering how *efficiency in production* can be determined. In such a box the production problem for one good is superimposed on the other. Thus, we have the problem of combining x_1 and x_2 in producing y_1 put together with the problem of combining x_1 and x_2 in producing y_2 – see the axes that are named in italics.

Let us start in the lower left corner – that is, we consider only the production of the good y_1. If we move northeast, we see that more of the inputs x_1 and x_2 is used to produce more and more of the good y_1. Furthermore, the same amount of y_1 can be obtained by different combinations of x_1 and x_2 – for example, the same amount of grain (y_1) can be obtained by different combinations of land (x_1) and labour (x_2). All these combinations of x_1 and x_2 for a given amount of y_1 are together called an isoquant. The amount of y_{1*} can be produced with different combinations of x_1 and x_2 as illustrated by the isoquant q_{1*} in the figure. Its form is dependent on the fact that reduced amounts of one input typically must be compensated by increasing amounts of the other. Consequently, isoquants are convex to the origin. The derivative of the isoquant is called the marginal rate of technical substitution (MRTS) between x_1 and x_2 and shows how much a given reduction in the use of one input must be compensated by an increase in the other to keep output constant.

Normally x_1 and x_2 can be used to produce more goods – for example, y_2. The problem of allocating x_1 and x_2 to produce y_2 is similar to that of y_1. The allocation of the two inputs to both outputs can then be analysed by superimposing a diagram for y_2 onto the one for y_1, starting off from the opposite – that is, upper right – corner. We observe that the production of y_1 and y_2 becomes bound by the total (given) amounts of inputs x_1 and x_2. The question is now: how much of x_1 and x_2 should be allocated to the production of y_1 and y_2, respectively? The answer to this is also illustrated in Figure 8.2.

All points within the box are technically feasible, but not optimal. To illustrate this, let us assume that we are at point C. Then the amounts x_{1C} and x_{2C} are used to produce quantity q_1 of good y_1 – and the rest ($x_1 - x_{1C}$ and $x_2 - x_{2C}$) is used to produce quantity q_2 of good y_2. This is not an optimal point because the MRTS between x_1 and x_2 in producing y_1 is different from that of producing y_2. In optimum they should be equal, otherwise more of, for example, y_2 could be produced without reducing the production of y_1 at all. This we see by following the isoquant of y_1 from C – that is, q_1 – to point D. The level of y_1 produced is by definition kept constant by this move while the amount of y_2 increases. We move to higher isoquants of y_2 than the one we started off from in C – that is, we move from isoquant q_2 to q_{2*}.

The curved line that connects the two origins – 'Good y_1' and 'Good y_2' – is called the contract curve. It shows all technically efficient combinations of x_1 and x_2 in producing y_1 and y_2. They are all characterized by the fact that the MRTS of transforming x_1 and x_2 into y_1 equals the MRTS of these inputs when used to produce y_2. One such point is E, others are D and F. Discriminating between all points on the contract curve can first be done when the demand and thus the relative prices between y_1 and y_2 are determined. This issue can be studied in yet another Edgeworth box, shown in Figure 8.3. The question here is *efficiency in exchange*.

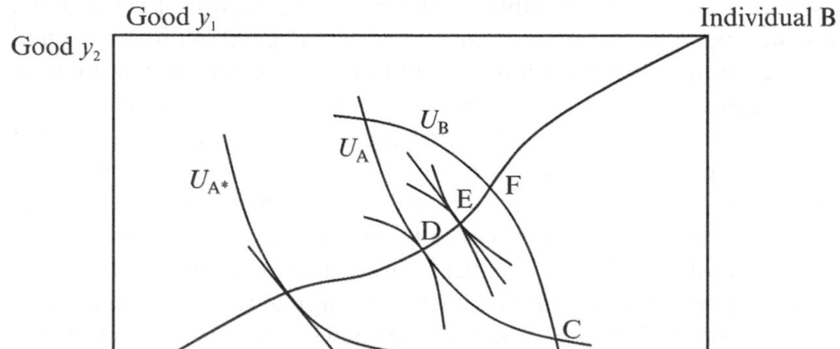

Figure 8.3 Efficiency in exchange

In this case we look at how a given amount of the goods y_1 and y_2 is distributed between individuals A and B. The basic logic is the same as in the case of production efficiency. Hence, Figure 8.3 is in principle equal to Figure 8.2, although the outputs or goods are on the axes instead of the inputs. Similarly, the isoquants are replaced by indifference curves showing combinations of y_1 and y_2 in consumption yielding the same level of utility for the consumer – for example, indifference curve U_{A*} for consumer A. The logic behind its form is also the same as in Figure 8.2. Reduced consumption of one good must be compensated by increasing amounts of the other to keep the utility level of the individual constant. The marginal rate of substitution (MRS) shows how much a given reduction in the consumption of one good must be compensated by an increase in the consumption of the other to keep utility constant.

If the two individuals A and B are endowed with amounts of y_1 and y_2 similar to that in point C, they can both gain by exchanging goods until they reach a point on the contract curve between D and F. Moving towards that curve implies that both can gain. Moving away from it implies that at least one must lose. All points on the contract curve are Pareto optimal. The MRS for individual A is equal to that of individual B, that is, $MRS^A_{y_1,y_2} = MRS^B_{y_1,y_2}$.

The essence of the two fundamental theorems of welfare economics – as defined in Section 8.1.2 – is to answer the following question: how can efficiency in both production and exchange be obtained simultaneously? To discuss this issue – that is, the issue of *social efficiency* – we must combine Figures 8.2 and 8.3. This is done by turning the contract curve of Figure 8.2 into a production possibilities frontier (PPF) in goods space – that is, with the goods y_1 and y_2 as axes. The PPF shows each efficient combination of x_1 and x_2 in producing y_1 and y_2. This frontier is then superimposed upon Figure 8.3. All this is shown in Figure 8.4.

The optimal point – that is, F – is found when two conditions are simultaneously satisfied. First the marginal rate of substitution between the goods for individual A ($MRS^A_{y_1,y_2}$) must equal that for individual B ($MRS^B_{y_1,y_2}$). This is efficiency in exchange as defined above. Second, this MRS must next equal the marginal rate of physical transformation (MRPT) between y_1 and y_2 – $MRPT_{y_1,y_2}$. This is the derivative of the PPF and establishes the link to the production problem. It is shown that given the assumptions of the core and standard application theorems of the neoclassical model, the implied market will facilitate this via price signals where the relative price of the outputs – that is, p_{y_1}/p_{y_2} – in optimum equals MRS_{y_1,y_2} and $MRPT_{y_1,y_2}$. The community indifference curve (CIC) shows the effective exchange of y_1 and y_2 between individuals A and B – the only members of our simplified society.

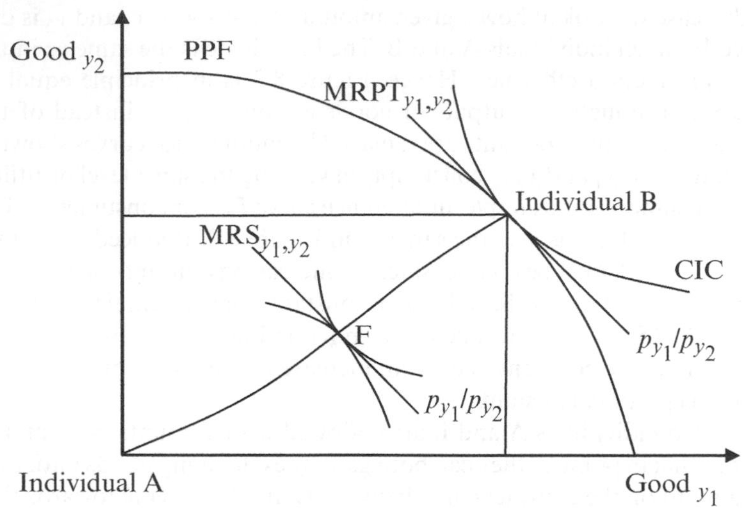

Figure 8.4 Social efficiency

Each point on the PPF relates to a specific contract curve in goods space. Furthermore, for each distribution of endowments a specific optimum will be obtained. Thus, if individual A has few endowments, implying low purchasing power, optimality will mean that individual B gets most of what is produced. Social optimality only implies that – *given the initial distribution* – nobody can increase her/his level of utility without decreasing that of others.

This is the basis for the claim that efficiency and distribution can be treated as independent issues. First, this reasoning implies that markets produce efficiency as defined. If society wants another solution to the allocation, it should redistribute income, not affect the functioning of the markets. Eaton and Eaton (1991) formulate it as follows: 'Suppose that we have identified some Pareto-optimal allocation that we would like to implement. The second theorem [of welfare economics] tells us first to redistribute the initial endowment and then to rely on competitive markets to achieve Pareto optimality' (p. 421).

Second, the reasoning supports the basic idea of modern welfare theory that economists can be safe when focusing only on the issue of efficiency. It implies – it is believed – only one (innocent) value judgement: that more is better than less. The reasoning is illustrated in Figure 8.5. Here the contract curve of Figure 8.3 – in goods space – is turned into a so-called utility possibilities frontier (UPF) in goods space. The UPF shows all the Pareto-efficient distributions of the goods y_1 and y_2 between individuals A and B, that is, all the points on the contract curve.

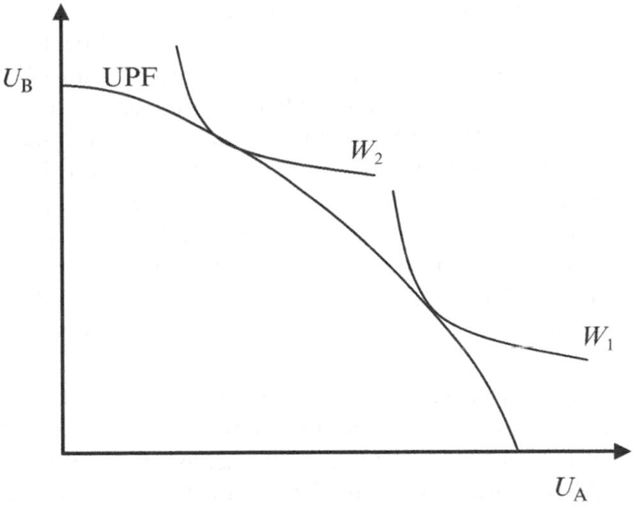

Figure 8.5 Social welfare and social optimality

When society considers distribution, it considers where on the UPF one should position the allocation. The figure shows two different social welfare functions – that is, W_1 and W_2. They imply different weightings of the utility experienced by the two individuals A and B. W_1 favours A and W_2 favours B. Thus, well-functioning markets ensure that one is on a UPF, while society, if it so wants, may redistribute the production by redistributing income.

However, if a different bundle of y_1 and y_2 had been produced, we would have had a different contract curve in goods space – Figure 8.3 – and consequently a different UPF.[14] Each point of the PPF has a specific UPF attached to it.

8.2.2 The Problem of Isolating Efficiency from Distribution

There are some problems related to making the kind of distinction between efficiency and distribution on which modern welfare theory is built. We shall focus on three here. The first is related to the language involved. The second is linked to the observations previously made on fairness and reciprocity. The third is connected to the fact that Pareto improvements are very often not possible. In most cases relevant for policy, some will have to lose, and then welfare economics advocates compensation tests (the PPI rule/Kaldor–Hicks). The problem with these is that they reveal fundamental inconsistencies in the distinction between efficiency and distribution.

Such a distinction may seem innocuous. However, saying that something is efficient implies that it must be good and should be done. This is the case, even if one starts off at point C in Figure 8.6 panel I and ends on the UPF at D. This move meets the Pareto criterion – as do all moves to the frontier between D and E. All these solutions still continue to make A much better off than B. Certainly, the latter conclusion hinges on a comparison of the utilities of A and B. One should recognize that modern welfare theory does not allow such comparisons. Thus, the neutrality of the efficiency statement is supported both by the idea that a move must be better since nobody loses and who is really best off in the end cannot be scientifically decided, even though A may end up with many more goods than B.

In Section 8.1 we discussed whether it is possible to evaluate in objective terms what constitutes a good or better life. The reasoning supported in this book is that it makes sense to make comparisons of, for example, living standards. Added to the arguments for this in the previous section, one should recognize that there is confusion of words here too – that of confounding 'untestable judgements' with 'value judgements'. The problem we may have with observing or measuring something – for example, utility – does not imply that (all) statements about it are by necessity value statements. Blaug (1992: 119) is clear on this when he says:

> [S]tatements about interpersonal comparisons of utility are not value judgements but merely untestable statements of fact: they are either true or false, but to this day we know of no method of finding out which is the case (Klappholz, 1964, p. 105; Barrett and Hausman, 1990). Value judgements may be untestable but not all untestable statements are value judgements (Ng, 1972).

I would suggest that while there are no methods for comparing utility, there are ways of comparing living standards. While these are not value neutral, the point made here is that the plea for neutrality is anyway impossible. It tends to confuse matters rather than making value issues become value neutral.

Continuing with the issue of language, we also acknowledge that while the concept of efficiency is generally positively laden, this is not the case with its counterpart, distribution or redistribution. There is an asymmetry here that in itself may affect policy. Moving from C (Figure 8.6, Panel I) to, for example, D is efficient. Moving from C to F is an inefficient move, since someone is losing. Moving from C to D or E and then to F is efficiency coupled with redistribution. The last move – that of redistribution – implies that someone loses, and it may also be viewed as if somebody 'must be helped' or is 'unable to take care of him-/herself'.

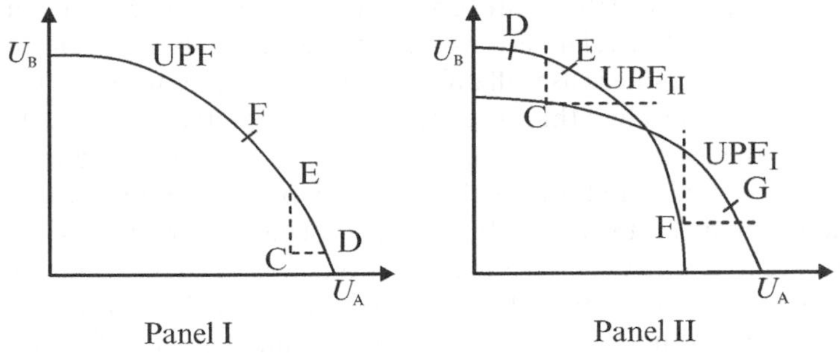

Source: Panel II is based on Varian (1992: 406).

Figure 8.6 Efficiency, redistribution and welfare theoretical consistency problems

The term 'redistribution' is used rather than 'distribution' to underline that any distributive act must occur in real time where somebody must give away what s/he has previously earned in markets (or elsewhere). Hence, there is a problem with the previous citation from Eaton and Eaton (1991). It is not about redistributing the initial endowment, it is about redistributing what came out of previous and ongoing market processes – a result that people naturally would feel is the result of their own efforts.

The distinction between efficiency (good, neutral) and (re-)distribution (politics) is therefore not placed on an equal footing. Who wants to go against what efficiency has produced? The distinction does not favour neutrality in economic reasoning. The focus on efficiency – as defined – becomes effectively a defence for the status quo distribution. Thus, efficiency is not only about more is better than less. Distributive aspects of a fundamental character worm their way back in. My position is not that still more should be done to guard economics against this problem. The position taken here is instead that it cannot be avoided. Then it is much better to develop a language that reflects that insight. This will make things much more transparent and true to its matter content.

The second issue was about social rationality – that is, fairness considerations and reciprocity. As was demonstrated in Chapter 6, such considerations played an important role concerning the way people treated distributional issues. Most proposers in the ultimatum games were willing to give more than what was required for just making both better off. Splits that were obviously unequal were systematically refused by the receivers. If people care about the overall distribution in a society, it may not be enough to ensure that a solution is Paretian. Point E, and also points near to E in Panel I of Figure 8.6, may be strictly preferred by people as compared to

D or points near D. Furthermore, going from C to D will be seen as a loss for individual B. B is relatively worse off even though the logic of the Paretian rule is that s/he is as well off as before. Even cases where both gain, but one gains (much) less than the other, will be observed as a loss by the one gaining the least (see again the results from the ultimatum games).

What is observed here may involve several explanations. One may argue that people obtain satisfaction from others being better off. This mechanism will tend to move societies in a direction of more equality than would otherwise be the case. It may also be, as is partly argued in this book, that reciprocity and fairness have been such important elements of sustaining social groupings that this has become almost a part of our genetic make-up. Finally, social constructs such as norms emphasizing equality are important in very many societies. Nevertheless, if any of these mechanisms are involved, the Paretian rule becomes much less value neutral than otherwise believed.

The third problem is that Pareto optimality or Pareto improvements – while being problematic in their own sense – are very often irrelevant. We live in the same physical world, and a gain for some will often imply a loss for others (Schmid 1987). This is certainly not least the case if we consider environmental issues, even though the problem is also related to a wide variety of other concerns. Recall the question of building a dam. It will bring profits and jobs, it may bring energy and/or better water control, but it also brings loss of habitat, replacement of people, reduced agricultural production and so on.

Should economics remain silent in these cases? This seemed to be the necessary outcome of the revolution of modern welfare economics of the 1920s and 1930s. However, that was too restrictive a conclusion for the profession, and the compensation test appeared. If the gainers can compensate the losers and still be better off, the project is termed efficient. If the sum of gains outweighs the sum of costs, it should be undertaken. This is consistent with the PPI or Kaldor–Hicks criterion.

None the less, someone would still lose. Certainly, as Sen (1988) points out, why just do the compensation test and not also support real compensations? As he argues it is really no less value laden to advise policy makers to compensate, than to refrain from saying that compensation should be done. This would make the PPI principle redundant, however, and we are back to the Pareto principle again.

Just as important, the PPI principle is shown to be inconsistent. Scitovsky (1941) showed that it was quite possible that in some situations compensated losers could actually gain by paying the winners to go back to the original allocation. The significant finding implicit in this is that efficiency and distribution cannot really be kept strictly apart even when understood in own – that is, Paretian terms. We shall look into this issue in some detail.[15]

We argued above that to every point on the PPF there is a unique contract curve in goods space with a unique UPF attached to it. Only if people's preferences are identical and homothetic,[16] will a reallocation of income yield the same bundle of outputs in goods space – that is, be allocated at the same point on the PPF.[17] If these very restrictive assumptions do not hold, there will be two different bundles of y_1 and y_2 produced which give two different points on the PPF resulting next in two distinctly different UPFs. If these intersect – and that will most probably happen when income is redistributed[18] – we may observe a potentially circular movement between different allocations. The importance of this depends upon the form of the curves and how far away from the starting point we actually move as an effect of the income redistribution (see also Samuelson 1950).

Look at Figure 8.6, panel II. Let us envisage that we start off at point C on UPF_I. To illustrate the compensation test, a move to D on UPF_{II} is potentially Pareto preferred since there is some reallocation of D – for example, E – which is strictly Pareto preferred to C. A move from C to F is similarly potentially Pareto preferred to C since here a reallocation to, for example, E is also possible. The gainer (A) can compensate the loser (B). However, as they are at F, both individuals would gain by moving to G, a move that is strictly Pareto preferred. Then, however, we are back on UPF_I including, for example, point C. Consequently, F is both better and worse than C according to the compensation principle. An inconsistency is observed.

This illustrates that there is an interconnection between efficiency and distribution. More importantly, recall that any point on the PPF is related to a specific UPF. That is, a specific bundle of goods y_1 and y_2 is implied. As soon as one moves along the UPF – that is, one is redistributing income – a new combination of the involved goods would (most probably) be optimal. Letting the market adapt to the new distribution, would 'force' a new point on the PPF. Moving along a UPF implies shifting to new UPFs, which would then probably intersect. The distinction between efficiency and distribution erodes in its basics. An envelope can be constructed around all these UPFs, called the grand utility possibilities frontier (GUPF). However, moving along that frontier implies moving between UPFs and between different allocations of inputs x_1 and x_2 resulting in different bundles of y_1 and y_2, different price vectors and different aggregate values of production.

In evaluating the discussion of these issues, Bromley (1990: 92; original emphasis) concludes:

> Only later [after Kaldor] would it be realized that one did not know – indeed, one could not know – the *value* of production independent of the distribution

> of income and the associated price vector that provided the weights to the various physical quantities being produced. That is, the new welfare economics showed the value of an unambiguous Pareto optimum, but in the absence of old-fashioned utilitarianism [that is, sum-ranking], economists were unable to say exactly what it was that had been optimized at the Pareto optimum (Blackhouse 1985). To put it more bluntly, ' . . . Pareto optimality is optimal with reference to those value judgements that are consistent with the Pareto principle' (Ng 1983: 30). Put another way, 'The Pareto criterion is not a complete preference ordering except in uninteresting societies where all individuals have identical preferences' (Hammond 1985: 424). If a preference ordering is not complete, it cannot be consistent or coherent. Samuelson (1950) soon showed that we cannot even be certain that group A is better off than group B even if A has collectively more of everything. It was beginning to seem that the very essence of economics – that more was preferred to less – was suspect.

While I believe that most economists, having thought deeply about these issues, realize the various shortcomings (see also Gowdy 2004), the messages of the above citation have not carried through to the level of practice. Despite all debates and conclusions made, cost–benefit analysis (CBA) is done as if it could give a consistent response to the question of 'what is efficient'. While the problems with CBA go wider than this (see especially Chapter 12), the ignorance of the problems is an indication that the science has a hard time accepting that value neutrality is unattainable.

The position of classical institutionalism is that efficiency cannot be defined independently of the chosen institutional structures. Actually, these structures play a crucial role in establishing what becomes efficient. We shall next look into some of the arguments for that position.

8.3 INSTITUTIONS DEFINE WHAT BECOMES EFFICIENT: THE CLASSICAL INSTITUTIONAL PERSPECTIVE

The basic point made by classical institutionalists is that efficiency – as measured in neoclassical terms – is derived from the chosen institutional configuration. It is a result – an artefact – of that structure. The basic issue, as seen from an institutional point of view, is about which institutions we want to establish and protect. This is about which values and interests society wants to support. The conclusion depends both on some fundamental ideas about the relationships between the individual and the society and on some more technical issues concerning not least the effect of transaction costs. In building the argument, I shall start with the more technical issues (transaction costs) and close with the more fundamental ones.

8.3.1 Rights, Transaction Costs and Neoclassical Efficiency

Neoclassical economics and its special brand of modern welfare theory makes the analyses based on given rights to resources. However, the initial distribution of rights is the more important one. These rights define which resource allocation becomes efficient. Let us develop the reasoning by looking at a simple example – that of safety in coal mines.[19]

We may envisage two rights structures concerning safety. In the first case – R_O – owners of mines do not have the responsibility for the safety of the miners. In the language of Wesley Hohfeld they have a *privilege* while the miners (workers) have *no rights*. In the other situation – R_W – workers have the right to their health/future labour power. They have a *right* while the owners of the mine have a *duty*. The normal assertion is that the two ways of distributing rights will yield the same level of safety in mines. Consider panel I of Figure 8.7.

MB are assessed marginal benefits of increased mine safety, while MC are the assessed marginal costs in the form of, for example, investments in safety equipment. Let us start with the situation where workers have no rights – that is, R_O. This implies that we start off in the lower left corner. Workers will here observe that the marginal benefits of safer mines well exceed the costs of some safety regulations. They may thus pay mine owners to increase safety – for example, accept lower wages against higher safety. The optimal level of safety – that is, q^* – will be reached. This will also be the solution if rights are with the workers – that is, we start off at R_W. Then owners realize that it will be cheaper to pay higher wages to workers as a compensation for accepting lower safety levels. Workers could then – if they so want – spend some of the extra income on buying insurance to cover losses if damages should occur. Anyway, the allocation

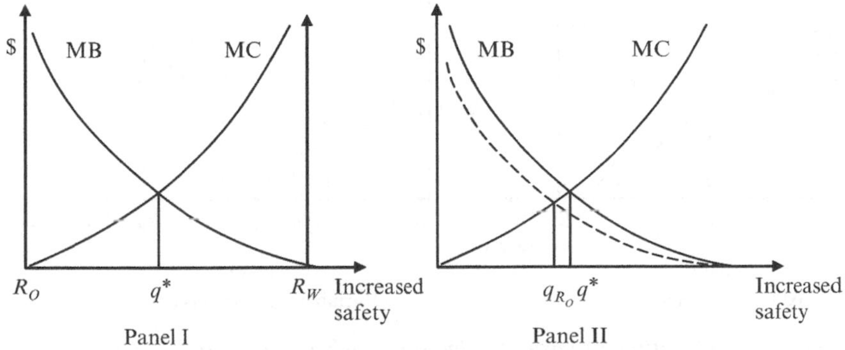

Figure 8.7 The efficient level of mine safety

of resources for safety arrangements will also be q^* in this situation. It is independent of the defined rights structure.

The above reasoning forms the basis for what is known as the Coase theorem,[20] which says that the initial distribution of endowments or rights does not influence the allocation of resources – that is, what becomes efficient. However, it demands *no income effects* and *no transaction costs*. In evaluating the theorem, let us start with the income or wealth effect.

Having or not having a right influences the opportunities that each party faces. It influences who must carry a cost and who has the right to inflict costs upon another party. Thus, in R_O workers must carry all costs of mine safety, while in the case of R_W the owners have responsibility. In the case of mine safety, the *income effect* of this may be substantial since the costs involved are often high. It simply implies that if the right is with, for example, the owners, the increased costs for workers reduces their income or their capacity to pay. This implies a lower willingness to pay for safety – that is, the MB curve will shift downwards compared to a situation where workers have the right. The higher this cost is as part of their total income, the greater the shift. This kind of shift is illustrated in panel II of Figure 8.7 – the broken MB line.

This is just part of the story. As important is the effect of *positive transaction costs*. In our case these would include costs of assessing the risks, of remediation, of making agreements on the safety level and ensuring that these are followed. Figure 8.8 illustrates the effects of positive TCs. No income effects are included to simplify the figure.

Panel I is developed to analyse the consequences of fixed TCs. Panel II fits a situation where these costs are variable. In practice, TCs will be of both kinds, and the reasoning behind the two panels must be combined.

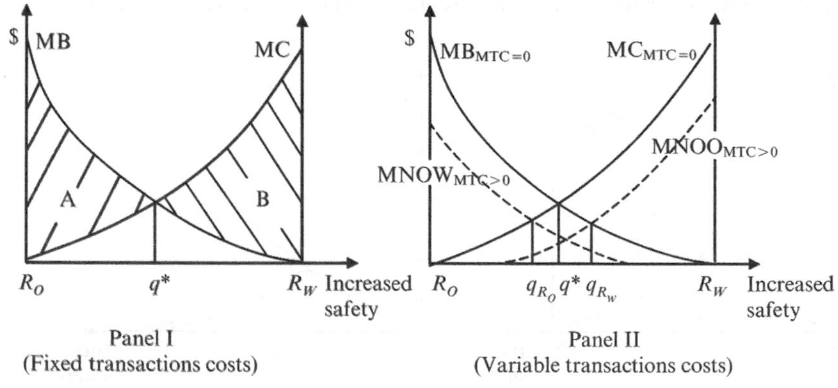

Figure 8.8 The efficient level of mine safety when transaction costs are positive

Starting with panel I, we observe two areas of gain from trading rights. Attached to R_O, gains equal the hatched area A. Starting off at R_W, the gains are area B. The simple message is that if the rights structure is R_O and fixed TCs are less than A, q^* will still be optimal. Similarly, if in the case of R_W they are less than B, the bargain will still yield q^* as the result. Gains from negotiating a new level of risk will be reduced, but still positive. In the cases where the opposite holds, where TCs outweigh gains from increased or reduced safety, the optimal level of safety will be zero (given R_O) or absolute (given R_W). The initial rights distribution defines in this case the resource use.

In the case with marginal TCs, the situation is a bit more complicated. Let us again start with R_O. Since rights are with mine owners, the workers must carry the transaction costs of bargains. This reduces their willingness to pay for safer mines. Recall that the marginal benefit (MB) curve is equal to workers' (maximum) marginal willingness to pay for safety. They are indifferent as to whether they pay that amount to obtain better safety or do not pay. If marginal TCs are positive, their offer will be reduced correspondingly. The value of a safer mine is simply reduced because it costs to transact over the higher levels of safety. Their offer curve shifts downwards to $\text{MNOW}_{\text{MTC}>0}$ – marginal net offer from workers, given positive marginal TCs.[21] The difference between the curves is the marginal TCs. The optimal level of safety shifts from q^* to q_{R_O}.

A similar reasoning can be made for R_W. In this case, positive TCs reduce owners' offers for being spared from investing in safety. We see a shift to $\text{MNOO}_{\text{MTC}>0}$ – marginal net offer from owners, given positive marginal TCs. In this case, the optimal level of safety becomes q_{R_W}. Giving the rights to workers implies safer mines than if rights are with the owners when marginal TCs are positive.

Following North and his argument that the state has the capacity to reduce TCs, letting the state also define required levels of safety could be a cost-reducing alternative to pure bargaining. Yet another institutionally dependent 'efficient' level of safety would occur if it is the state that defines what safety level should be in place.[22] In relation to this, one should note that the costs of defining what risks are involved in various mines are often very high. Therefore, knowledge may also be asymmetrically distributed between the parties. This may be an argument in itself for some kind of public engagement.

From this short story we observe two very important lessons. First, the institutional structure defines who must carry which costs, including transaction costs. It defines which costs or losses go uncompensated. Second, the distribution of rights defines what is the optimal level of involved goods. In our case this is first of all relevant to the level of safety. However, it also influences the efficient level of mining, the price of metal and so on, and finally the price vector for all goods in equilibrium.

8.3.2 Rights, Costs and Compensation

A right is a rather complex relation. One observation of great importance to us is the large differences observed in the literature between willingness to pay (WTP) and willingness to accept compensation (WTA) for a certain good or right. In the case of the miners, R_O implies that these must buy safety. Then WTP applies. If the rule is R_W, WTA is the right measure. As already mentioned, the income effect should result in some, but normally not that big a difference between the two (see also Willig 1976). After reviewing the literature, Gregory (1986) and Horowitz and McConnell (2002) conclude that the differences observed between WTP and WTA measures are much larger than what could be reasonably explained by taking the income effect into account. WTA estimates tend to be on average three times as large as those concerning WTP, even higher in some cases. This is the situation even if the values at stake are small or the goods are highly substitutable with other goods. Kahneman and Knetsch (1992) and Knetsch (2000)[23] conclude similarly. Figure 8.9 illustrates the effect on optimal safety given R_O and R_W.

The point is that the distribution of rights prior to a bargain influences the outcome far beyond that of the implied income effect. The explanation of what is going on varies across the literature. March (1994) suggested – as we saw in Chapter 5 – that bounded rationality would result in

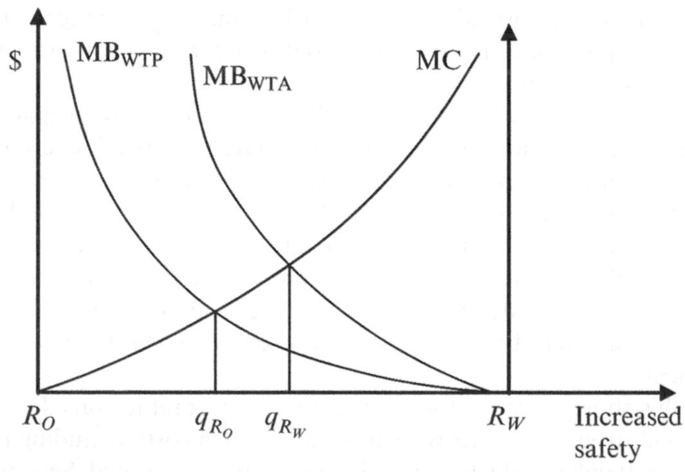

Figure 8.9 The effect of the rights structure on efficient levels of safety: the difference between WTP and WTA

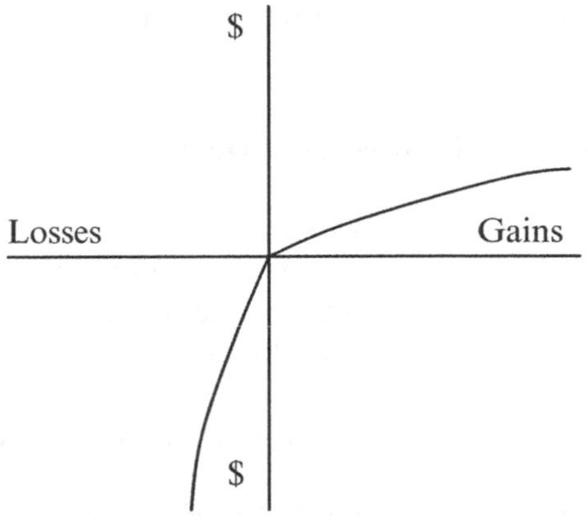

Source: After Tversky and Kahneman (1986).

Figure 8.10 Loss aversion

differences between WTP and WTA. This was an effect of the target setting implied by satisficing. Tversky and Kahneman (1986) offer a somewhat similar interpretation. They talk of 'loss aversion', emphasizing that there is a difference between gains and losses from a status quo situation. Acquiring a good or a sum of money is of less importance to the individual than losing the same good. The curve of losses and gains may be viewed as kinked at the status quo distribution – see Figure 8.10.

These results show that there are problems with the PPI rule which goes beyond those discussed in Section 8.2.2. The differences observed between WTP and WTA make it impossible to do economic analyses in the form of, for example, cost–benefit analysis, without taking a stand on rights issues. Preferences, even as measured in economic terms, depend on the institutional setting – here the rights structure.

While I believe the concept of loss aversion is of clear relevance to the analysis, one should also make some distinctions concerning what is lost. The loss of a right may imply more than the loss of income or a thing *per se*. Let me illustrate by going back to our example. In the case of R_w – that is, when rights are with the workers – the right will in itself imply that society has defined the status quo to be safe mines or workers' ownership to future health or labour power. They have an entitlement. To trade that entitlement may be viewed not only as something that relates to subjectively evaluated losses – that is, WTA becomes high due to loss aversion. It may also be found to be an act against the very intent of the rule. It is defining not only a relationship between miners and mine owners, but also the integrity of the miners. Compensation for accepting lower safety could be understood to destroy that integrity. It could be understood as accepting a bribe.

8.3.3 Institutions and Preferences Once More

The above reasoning brings us back to two issues that we have discussed previously. First, individual preferences may vary, depending on the existing institutional setting. Second, they may be plural, thus blocking trade-offs, either entirely or within some bounds. Both these issues challenge the view that efficiency evaluations can be based entirely on individual willingness to pay and that efficiency can be safely analysed without also taking rights and distributive issues into account.

Recall the point made in Section 6.4: if preferences are (also) a function of the chosen institutions, the supremacy of any order based on the preferences that this order itself produces vanishes. When preferences are a function of legal rules, these rules cannot be justified by reference to the same preferences.

To repeat, modern welfare economics has tried to circumvent this problem at two levels. First, it defines preferences as fixed or independent of institutional structures. Second, it defines distribution as a non-economic issue. It is a 'solved political problem', to cite Lerner (1972). Thus, efficiency and distribution can be kept apart. As soon as one realizes that these ideas are based on a problematic set of defences, one is confronted with the problem of defining justifiable second-order institutions. These are the institutions that structure the choices of the institutions, which then govern day-to-day resource allocations.

8.3.4 Institutions as Protecting Interests and Defining Opportunities

In our concrete case, the fundamental question is really this: on what basis should the level of safety be decided? Should it be on the basis of the parties' willingness – or ability – to pay, or should instead the collective decide some standards or goals concerning what is a reasonable level of safety? In other words: which standards concerning, for example, working conditions should society support? This goes beyond the issue of distributing tradable rights. It is about setting general standards that society commits itself to. While we have previously focused on the technical capacities of the state – for example, its ability to reduce information and transaction costs and to 'do the trade' for the workers – we observe that this is just a small part of the problem of public decision making. The basic question is that of formulating who we shall become.

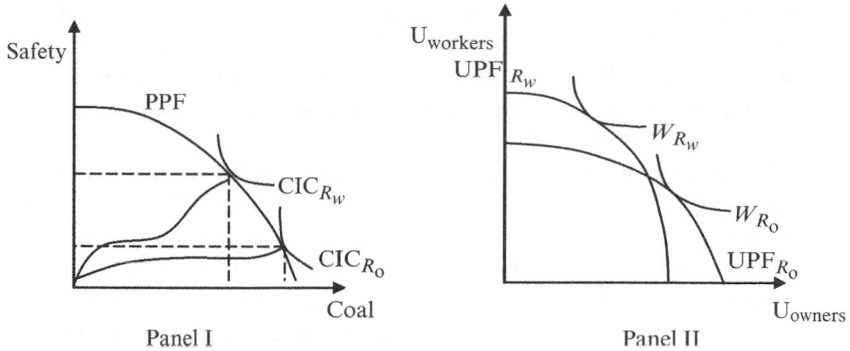

Figure 8.11 *Safety versus coal production*

An important role for institutions is therefore to define which interests or values should be defended. In our case with safe or unsafe mines, the public authorities must decide which working environment it supports – in concrete terms whether it is the interests of the mine owners or the workers that should be protected. While protecting the interests of the workers implies that, for example, less coal will be produced, greater safety, more future labour power and increased workers' integrity will be obtained – and also, more general respect for human life will be developed.

Also the issues at hand here can to some – but only some – extent be treated within the model of welfare economic theory itself. The two rights structures R_W and R_O yield two different positions on the PPF in panel I of Figure 8.11. Parallel to that, we have two different UPFs in utility space with their separate social welfare functions related to them. Basically, what society does when it defines the rights structure is to define which social welfare function – whose utility – gets protected.

With a slight rewording of Bromley (1989), we may call this an allocation[24] of economic opportunity. It is not about increasing efficiency in standard terms. A change in rights could even be viewed as reducing efficiency since moving from, for example, R_O to R_W implies reduced amounts of coal produced. There are, however, two products involved – that is, coal and safety. It is therefore not about leaving the PPF to lower levels of production. It is about moving along it. In other words, it is about changing the form of the 'pie', not about increasing its size.

Similarly, it is not about (re)distribution of income either – that is, splitting the 'pie' differently. While moving between R_O and R_W has income effects, it is basically about changing the opportunity set for the agents. It changes the possibilities for each group to secure their interests. While the reasoning behind Figure 8.11 illustrates the role institutions play in defining what 'becomes efficient', it is restricted to those issues that can be handled in pure utility terms. Although thus restricted, it still illustrates the main point, that the distinction between efficiency and distribution is not only inconsistent, but it is also unable to handle major issues of institutional development or reform.

The choice of institutional structures cannot be one based on individual preferences only. First, it is about choosing which values will get protection. We move from the level of individual choice – as in markets – to the choice made by the collective. We move to the domain of social rationality and communicative interaction. Here the distinction between the role of being a citizen and that of being a consumer is crucial. When we talk about the citizen, we talk about our role as participants in a process to choose the institutions within which subsequent consumer decisions will be made. The ideal here is that the citizen chooses between different alternative sets of opportunity – economic and otherwise. These choices are made on the basis of a judgement about not only which interests will flourish, but also which values will thrive.

The distinction between interests and values is an important one. While the concepts may overlap, the concept of interests emphasizes the interests existing as an effect of prevailing institutions. By values I mean ideas about what and who we should become. They say something about what is a preferred state that goes beyond the interests of a particular individual, or the specific interests even of a whole group in a given situation. They are at the centre of the long-run decisions to be made by the citizen. There may be no conflict between interests and values as here defined. In such a situation, existing institutions foster the values one also holds as a more detached or forward-looking citizen. The interests that these institutions produce are in line with the values that the citizens of the society want to defend.

At this institutional level, the questions are both about deciding between conflicting interests of the day and about forming the future and its interests. This issue is certainly a complex and difficult one. It is about both how different voices can develop and how they can present themselves to the general public. It is next about how these processes are developed to arrive at choices among the citizens that are binding for them when subsequently acting as 'mere' economic agents.

The reasoning involved here implies claims concerning the procedures of the second-order institutions: that is, the constitution. Procedural neutrality – that is, procedural neutrality between interests and outcomes – is often advocated. However, neutrality is not easily defined. Should weak groups be supported specifically to help them participate in the public debate? At what point is that regarded as securing neutrality rather than creating biases? Furthermore, it may be that the society even at this second-order level would like to institutionalize elements concerning what a good life should be. It may also want to bind itself to some principles concerning the outcome of the decision processes it constitutes. Typically, a constitution defines some basic values it serves to protect, not only decision rules. Remember, to establish a constitution is in itself to take a stand for a certain type of society. Its existence is counter to the very idea of neutrality between different types of societies.

None the less it is a problem that special interests – those not earning support among their fellow citizen for a certain allocation of economic opportunity – may be able to force solutions to their advantage. The concept of particular interests is used to describe solutions where some, often strong minority groups, are able to establish rules that defend or foster these interests despite the fact that their claims do not have the (prescribed level of) general support defined by the second-order institutions of the society.

In economics this is called 'rent seeking' and implies that resources are used to obtain favourable conditions for special interests and not for producing goods or services that are of value to the society. Certainly, the distinction is hard to draw. In neoclassical economics, rent seeking is often equated with producing institutions that do not follow the rules of modern welfare theory – that is, that do not result in Pareto improvements, real or potential. A move from R_O to R_W in our example could, as indicated, easily be categorized as rent seeking. Such changes will first have to be advocated, and resources must be used to argue that it is better to shift the rights to the workers. Second, resources will next have to be allocated to increase safety if miners are able to make the decision makers support their claims.

However, no interest can be articulated and gain a hearing in the public debate without resources being used for that purpose. The problem becomes how to guard against illegitimate claims – that is, claims not supported by the public. It is important to realize that the problem lies here and not with resources being spent in the political sphere, that the decisions do not follow economic theory and so on. While the market model may serve us well in many situations, it cannot on its own offer a recipe for which institutional structures to choose.

8.4 SUMMARY

The normative aspect of institutional change is about defining 'what is good', or 'what is best'. We have observed some deep cleavages in the economic literature concerning this issue. Along one dimension we see a split concerning whether 'what is good' can be characterized in objective or subjective terms only. Along another, the issue is whether the dimensions of 'what is good' are singular or plural.

The original utilitarian position has been termed weakly objectivist. It is based on the desiring individual, on the ideas of welfarism (utility), consequentialism and sum-ranking. The last implies that utility can be compared across individuals. It also implies that a single measure of 'what is good' is involved and can be measured as the sum of utilities across all individuals.

In modern welfare theory, two important changes are made in comparison to this. There is a shift to subjectivism, implying that sum-ranking is dropped and the Pareto or potential Pareto improvement rules as evaluative rods for public policy are taken up. The concept of utility is moreover shifted from measuring happiness to becoming a mere ranking of goods. All this was made in a process trying to make economics value neutral.

Moving to the Austrian/Hayekian position, we observe a strengthening of the subjectivist aspect in the sense that almost any state activity except that of granting private ownership rights is denied. Consequently, if welfare comes up against individual freedom – understood as freedom from constraints – individual freedom is given priority. Plural (two) value dimensions are actually observed.

The stand of classical institutional economics is finally both objectivist and pluralist. It starts off from the aspiring rather than the desiring individual. This implies that the normative issues are not only about satisfying given wants, but also about developing distinct human capabilities. While various dimensions of 'a good life' may be incomparable (pluralism), this position still claims that it makes sense to compare opportunities for different groups and individuals concerning their possibilities to live good lives. This grants both a certain right and a certain responsibility for society to make choices concerning its future development.

While the utilitarians believed that social choice could be made on the basis of sum-ranking – that is, the total utility summed across all individuals – modern welfare theory has retreated to the Paretian concepts, implying a denial of interpersonal comparison. The Austrian position is a more radical version of the latter, really denying the possibility of any social choice. All is individual and all is/should be market orientated. Finally, classical institutional economics is built on the perspective that social choice is both possible and very important, and has to be built on communicative processes.

The interesting discussion – as I see it – takes place between the modern welfare theoretic and the institutional positions. A main idea underlying the former is the distinction between efficiency – that is, economics – and distribution – that is, politics. It is shown that any politically chosen Pareto optimal social state can be established by a competitive market equilibrium (second welfare theorem). Politics and economics can, it is then believed, be kept apart. It is certainly a problem that all core and standard application theorems of neoclassical economics must be satisfied for the welfare theorems to hold. Another point is that even given that these theorems are satisfied, making such a division results in various inconsistencies when shifting from one market equilibrium to another – that is, the problems appearing with the PPI rule. The distinction between efficiency and distribution can therefore not be unfailingly defined.

The classical institutional position starts off from the opposite side of the problem. Not only is it emphasized that the distinction between distribution and efficiency cannot be sustained. The point is that the very essence of institutional reform is foremost about which interests or values should get protection, not about creating efficiency as the result of transaction cost-free bargains given whatever distribution is existing/defined.

This reasoning follows from two interlinked arguments. First, we have the issue of circularity. If *preferences* and *transaction costs* are themselves functions of the economic or institutional system in place, then what becomes efficient must also be a function of the system as defined. It cannot be evaluated independently of that system or on the basis of only one – that is, the market. That is if:

1. Efficiency $= f_1$ (preferences, TCs), that is, the standard neoclassical assumption if accepting positive TCs. Then if, as has been suggested throughout various chapters:
2. preferences $= f_2$ (institutional system), and
3. TC $= f_3$ (institutional system), simple substitution gives
4. efficiency $= F$ (institutional system).

Consequently, analyses like those based on the Pareto optimum and market-type institutions as the reference point, are virtually useless as criteria for institutional reform. Certainly, if efficiency is defined as what the individual chooses when acting in markets, then acting in markets must create efficiency. Problems appear if choosing in markets is costly – that is, positive TCs – and markets themselves influence preferences. Then welfare theory says nothing about efficiency. It merely says that markets are best if we assume conditions that make markets best. This does not take us very far.

The second element of the reasoning is thus the following: since 'what is good' ('efficient') is a function of the institutional systems, choices about institutional reform must lie beyond those of individual preferences and technical efficiency. It is a second-order issue concerning which interests and values society should protect. The concept of the citizen as opposed to the consumer is supported to mark the distinction between (a) choosing *among* and (b) choosing *within* sets of institutions. Subsequently, the choice of institutions must be based on arguments, on reasoning about what sort of society we want to foster, not on aggregating willingnesses to pay. The objectivist perception of 'what is best' offers the opportunity to do such evaluations.

NOTES

1. The distinctions made in the literature between an objectivist and a subjectivist stance is full of inconsistencies and confusion. For a comprehensive and clear treatment of the various issues, see O'Neill (1998), whose presentation shows that there is much more to the objectivist–subjectivist theme than is presented here. I have chosen to focus just on what I find necessary to give the reader a meaningful understanding of the main positions.
2. O'Neill (1998: 39) clarifies: 'If "objectivism" meant that what is good for a person is entirely independent of who and what that person is then it would be implausible. What is good for us depends on something about us, on what we are like' . . . 'What is of value to us cannot be independent of the kinds of being we are, and the capacities we have. That is compatible with the rejection of the subjective determination thesis that what we desire or value determines what is valuable to us. On an objectivist account we cannot choose like that. Given the kind of social creatures we are, no matter how much an individual might place a value on a life without ties of affection to others, his life cannot be led happily without them'.
3. This subjective perspective on preferences/values goes back to Hobbes, and we can follow its development, for example, via Locke, Bentham, Herbert Spencer, the neoclassical economists and the Austrians. Actually, the idea was formulated as far back as the Greek sophists such as Protagoras and Trasymachos. The sophists stood in opposition to an objectivist interpretation so typical for Plato and Aristotle. Although there have been many developments in the understanding of the natural world, of humans and of society since antiquity, this cleavage in the social sciences has thus been rather consistent.
4. Claiming that freedom is an important – indeed *the* important – element of what is of value to an individual is to take an objectivist position concerning the definition of what is good as defined here. Thus it may be argued that the Austrian position is inconsistent in its emphasis on subjectivism. We shall return to this soon.
5. Certainly also demanding that economic agents were rational in the neoclassical sense, that transaction and information costs were zero.
6. For those interested in this, the writings of Sen are of great value – for example, Sen (1988, 1991). See also Broome's comments to Sen for an account of a philosopher's understanding of the use of the concept (Broome 1991a, 1991b).
7. This is implicit in the assumptions since if all rights are distributed, transaction costs are zero and agents are individually rational, no externalities – more precisely no Pareto-relevant externalities – will appear. More on this will be found in Chapter 13.
8. Much of that influence went via Lionel Robbins who, for a period, participated in Mises' seminar at the Chamber of Commerce in Vienna.
9. Hayek's ideas have been most clearly formulated in the so-called 'socialist calculation debate' (Hayek 1948), but they also have relevance for the idea of the social optimum as defined in welfare theory.
10. See also Chapter 7 on spontaneous order. I am indebted to O'Neill (1998) for this reference.
11. Dewey was perhaps the most influential philosopher in the United States in the late nineteenth and the early decades of the twentieth centuries. He had a distinct influence on contemporary classical institutionalists.
12. Readers who are familiar with modern welfare economic theory may skip Section 8.2.1. It is included for those who have a limited background in the subject.

13. It goes without saying that person A also owns her/his own labour.
14. The envelope of all these UPFs is normally called the grand utility possibilities frontier (GUPF).
15. For those interested in going deeper into the issue, see Scitovsky (1941), Samuelson (1950), Gorman (1955), Boadway (1974) and Chipman and Moore (1978).
16. This implies that the consumers value goods in fixed proportions, implying that whether poor or rich, they buy the same *relative* amount of all goods.
17. This is a rare, if not a nonexistent situation. Normally, if redistribution implies increasing the income of the poor, more basic goods will be purchased and fewer of the 'luxury' ones. This happens even if preferences are identical between the two. However, they are non-homothetic.
18. Preferences are now assumed to be non-identical or at least non-homothetic.
19. While developed a bit differently, I am indebted to Bromley (1989) for this example.
20. The theorem is based on the reasoning found in Coase (1960), who himself did not call the proposition a theorem.
21. Marginal transaction costs are considered constant. In standard expositions of this issue – for example, Randall (1974) and Bromley (1991) – they are considered proportional to the change in gains. I find that assumption to be not particularly convincing. However, this does not influence the reasoning and conclusions.
22. There may be other reasons for letting the state or the public define safety standards. This will be discussed in Section 8.3.4.
23. Knetsch (2000) gives an example with ordinary mugs where the WTA/WTP ratio is in the magnitude of 1:3.
24. Bromley (1989) uses the concept 'reallocation of economic opportunity', emphasizing the shifting of rights. I think the idea goes wider and also includes the initial distribution within a field – therefore 'allocation of'.

18. [illegible] assumed to be [illegible].
19. [illegible] the index [illegible] sample.
20. [illegible]

23. [illegible] the WTS/WTP ratio is [illegible].

PART IV

Institutions, Environment and Policy

In Parts I–III we have focused on basic issues in institutional theory/institutional economics. We shall now turn towards an analysis of environmental policy. The more general perspectives developed in Parts I–III will form the basis for Part IV. First, we need to say something more systematically about the subject matter – that is, about what characterizes the environment and the problems involved. Part IV will therefore start with a presentation of some basic perspectives on the dynamics of the natural environment – the biosphere (Chapter 9). This is followed by five chapters that are specifically directed at the environmental policy issues.

In Chapter 10 the focus is on the issue of resource regimes, which is the basic institutional question concerning environmental management. First, we shall examine the rights issues involved – that is, who gets access to which resource, how are certain uses allowed to affect others, and how do the institutions involved treat the implied conflicts. Second, we shall look at how different regimes motivate action and influence values.

Chapter 11 focuses on the issue of valuing the environment – or more precisely, the process of defining how we should treat it, what we should preserve and so on. We shall define the concept of a value articulating institutions and look at what characterizes the process of environmental valuation. We shall show how the valuation procedure itself may influence which values become emphasized. Finally, we shall look at how well the results from monetary valuation studies themselves fit the core assumption of rational choice.

This analysis is followed by a further development in Chapter 12, looking more in depth into three different value articulating institutions used in environmental policy: cost–benefit analysis/contingent valuation, multicriteria analysis and deliberative methods. These are different ways through which people can articulate their values concerning the environment with distinct and profound effects on the process of valuation and the outcomes thereof. The strengths and weaknesses of each method are assessed on the basis of institutional theory.

We then turn to the issue of choosing policy measures. Given that a society has decided what it wants to achieve, policies for reaching these goals must be instituted. Thus, in Chapter 13, we shall combine the insights from earlier chapters on an analysis of the choice of policy instruments. Here the ideas from Chapters 5 and 6 on rationality and preferences will be combined with insights from Chapter 9 on the dynamics of the economy–environment process and the ideas developed in Chapter 10 concerning resource regimes.

Finally, Chapter 14 cross-cuts the main themes of the book. I do so by trying to look into the future, and raising a number of questions concerning what institutional reforms are needed to increase our ability to solve urgent environmental problems. Some suggestions for institutional reform that are in line with the normative message of this book are also offered.

9. The environment

The first aim of this chapter (Section 9.1) is to present a perspective on the environment as a system of interconnected processes and to give a brief introduction to some basic ecological insights that follow from this, which we shall utilize in later chapters. We shall also give a brief presentation of the laws of thermodynamics. These laws have in themselves profound implications not least for environmental economics and policy.

The second aim (Section 9.2) is to give some insights into how the environment and environmental problems have been perceived over the years. This section is especially focused on clarifying how various economist traditions have perceived the environment and how this perception has influenced the analysis of environmental problems.

The way we think about the environment and the interactions between the environment and the economy must certainly influence the way we treat it. Starting off from a model where markets and individual commodities are core concepts will result in a very different understanding of environmental issues compared to a perspective where a systems view is advocated and the emphasis is on interconnections.

9.1 THE ENVIRONMENT AS A SYSTEM OF PROCESSES

The environment can be viewed as a set of items – as trees, fish, birds, lakes, various reservoirs of metals and so on. Certainly, such a perspective is highly relevant in some cases, but too narrow to be able to treat environmental issues more generally in an appropriate manner. In this book the concept of 'the environment' will be equated with that of the biosphere, and it will include not only the involved species and the relevant ecosystems, but also the interlinked bio-geochemical processes that keep this system functioning. Throughout the coming chapters, we shall experience how important this conceptualization of the environment is and how it may yield different conclusions from that which views the environment as a set of items or commodities.

9.1.1 Life has Created its Own Conditions

A biological lexicon describes which species dominate which parts of the world, which ecosystems are typical for different climatic zones and so on. Life is characterized by an almost infinite diversity depending on variations in local climate and supplies of nutrients.

At the same time, life is dependent on a great number of biogeochemical cycles. This concept refers to the interaction among biological, geological and chemical processes. For example, the content of the atmosphere is maintained by a series of rather complicated processes, and likewise for the top-soil, the oceans and even the earth crust itself. The biological processes play a decisive role for the composition of the atmosphere and for the living soil. However, they also play an important role for some geological processes such as the formation of calcareous mountains and the development of carbon layers that ultimately produce oil and coal. Through these processes, matter is circulated and waste becomes resources again. Certainly, the geological circulation periods are quite different from those experienced for soil and air. As an example, Schlesinger (1991) suggests that on average, sodium remains in the ocean for about 75 million years, for potassium the figure is 11 million, while for calcium it is about 1 million.

The biological systems, the atmosphere and the oceans have not always looked like those of today. First of all, the composition of species has changed considerably. The general tendency has been towards greater diversity – even though extinction also occurs continuously. Wilson (2001) points to the fact that species become extinct all the time. That is part of the process of evolution whereby species that are more fit take over old resource niches or adapt to new ones. Thus, he suggests an average lifetime of a species of about 1 million years. In some periods more abrupt developments – that is, mass extinctions – have been observed. Altogether there have been five such periods over the last 450 million years.

Second, the composition not least of the atmosphere has changed dramatically throughout earth's history. This change is furthermore linked to the development of biological life. In a way, life has created its own conditions. This can be illustrated most clearly by the development of the composition of the atmosphere (see Table 9.1).

There are uncertainties involved when we try to estimate the composition of the first atmosphere. The basis for the left-hand column of Table 9.1 is the composition of gases in volcanic eruptions. There are strong reasons to believe that the original atmosphere must have had much the same composition, since it was formed by gases released from eruptions on the surface of the earth before it cooled off sufficiently to form solid rock.

Table 9.1 Composition of the atmosphere (%)

Gas	When the earth was formed	Today
Nitrogen	1	78
Oxygen	0	21
Argon	0	0.93
Carbon dioxide	12	0.035
Methane	0	0.00017
Sulphur dioxide	7	variable
Water vapour	80	0–4

Source: Graves and Reavy (1996).

Three gases dominated – especially water vapour, but also carbon dioxide (CO_2) and sulphur dioxide (SO_2). No free oxygen existed while the content of nitrogen gas was very low. This was an atmosphere that was not very friendly for life forms as we know them today. It deviates dramatically from the present composition of the atmosphere. Today nitrogen and oxygen dominate, while the percentage of water vapour and especially CO_2 is substantially reduced. We are talking about two very different worlds.

The reduction in water vapour is mainly due to physical changes – that is, reduction in the temperature that certainly followed the cooling of the earth's surface and the formation of its crust. With the fall in temperature, there was torrential rain and the oceans were formed. These were not as salty as they are today since the washing out of matter from the newly appearing land must have been fairly limited. Over time simple forms of life were created in these oceans. These micro-organisms – anaerobic life forms and later plants – were not dependent on free oxygen for their metabolism. Instead they transformed water and CO_2 into organic matter with the help of energy from the sun. Free oxygen was a waste product of this process. Over time this waste accumulated and resulted in a considerable change in the composition of the atmosphere. At the same time, the CO_2 content was substantially reduced, as the carbon was bound in the increasing mass of organic matter – not least in the form of dead organisms accumulating on the sea floor.

This development, in turn, had a dramatic effect on the potential for different life forms to evolve. The production of free oxygen created conditions for a shift to aerobic life forms dependent on free oxygen. Moreover, the oxygen in the atmosphere was the basis for the creation of the ozone layer, which reduced the amount of ultra-violet radiation reaching the earth's surface. This in turn made it possible for life to survive outside the oceans – on land and in the air. These changes did not happen abruptly, but gradually the earth was set on a development trajectory with a much larger potential for developing varied forms of life than was previously the case.

Our brief example illustrates several issues:

- Via evolution, life has in a way created its own conditions. The biosphere is a self-organized system, formed as a result of a vast amount of trial and error. The composition of the atmosphere is a product of biological activity as its structure has become a crucial element in the development of that activity itself. Similar descriptions could have been given for the development of soils and the composition of the oceans.
- Given the supply of high-quality solar energy, matter which is a waste from one species can become a resource for another. This is the mechanism whereby the balance of matter circulation on the earth is secured, thus ensuring that resources do not become non-usable waste. The bio-geochemical cycles have been crucial for the long-run survival of the system.

There is an important balance between continuity and change characterizing this system. Focusing on the importance of continuity, Ayres (1993: 203; original emphasis) suggests:

> Any disturbance to the bio-geochemical cycles is ipso facto a threat to survival. A materials cycle consists of a sequence of transformation processes and reservoirs or compartments. It can be represented schematically as *stocks* and *flows*. The condition for stability is easily stated: the stocks in each compartment, or reservoir, must remain constant (at least on the average); and, for this condition to be met the inflow into each compartment must be balanced exactly (on the average) by the outflows. If this condition is *not* met, the stock in some compartment must increase, at the expense of the stock in some other compartment. If the cycle does not re-stabilize, somehow, it will collapse.

It is important to note that while there are always changes going on – that is, there will not be full restabilization of all cycles – the system has reached a high level of stability concerning important parameters like air and ocean composition, temperatures in various climatic zones and so on. Certainly, changes in the inflow of solar energy also play a role here, and some variability will thus occur. The system is far too vast and complex to secure continuous stability. Thus, the ability of the system to maintain some macroscopic balance is dependent on the capacity of adaptation at the level of systems parts, for example, the species. Nicolis and Prigogine (1989: 218) indicate how complex systems are able to do so:

> [C]omplexity has been connected to the ability to switch between different modes of behavior as the environmental conditions are varied. The resulting flexibility and adaptability in turn introduces the notion of choice among the various possibilities offered. It has been stressed that choice is mediated by the dynamics of the fluctuations and that it requires the intervention of their two antagonistic manifestations: short scale randomness, providing the innovative element necessary to explore the state space; and long-range order, enabling the system to sustain a collective regime encompassing macroscopic spatial regions and macroscopic time intervals.[1]

The fact that the composition of the atmosphere has been relatively constant over a long period of time indicates that the system has reached a fairly stable state at the macroscopic level. None the less, species become extinct, new resource or waste niches appear giving opportunities to new life forms. Genetic mutations are the dominant source of 'short-scale randomness' continuously 'trimming' the system and increasing its resilience towards internal variation and external shocks. It has the capacity to increase what is called 'systems resilience' – that is, the ability of the system to return to its original state after a shock (Holling 1973, 1986; Perrings 1997). The problem for humans – who now have the capacity to change almost any ecosystem and certainly do so – is that it is almost impossible to describe what will happen both as an effect of single changes and as the sum of many. Through converting land into new uses, by wiping out species at a very high rate, by emitting an ever-increasing amount of substances that are unfamiliar to various environments, by rapidly increasing the amount of CO_2 in the atmosphere, we force changes at a speed not earlier observed. These changes may certainly have the capacity to alter the performance of the system in an essential way – that is, a so-called 'attractor shift' may be observed.[2]

The quality of human life – as that of any other species – is dependent on the quality of the systems that humans are part of. Human beings may have the capacity to transform the systems far beyond that of any other species. They also have the capacity to study the systems and possibly single out key species. However, they do not have the capacity to 'pick and restructure' and at the same time secure the original level of resilience. They cannot master the system in any such way. This would have implied the capacity to acquire information that has been stored in the system as a function of all trial and error processes going on for millions – indeed billions – of years. This is a vast – that is, impossible – endeavour.

9.1.2 The Laws of Thermodynamics

At the heart of the above analysis lie the two fundamental laws of thermodynamics. The first law states that in an isolated system, energy (and matter) can neither be created nor destroyed. This is also called the law of energy (matter) conservation and implies that energy (matter) can be changed in form (quality), but not in quantity (Ruth 1993). This implies that the volume of the matter cycles previously described is characterized by constancy in that what leaves one compartment of the system must appear in equal magnitude in other compartments. This is the basis for the above quotation from Ayres.

The second law is maybe less intuitive. It states that the entropy of an isolated system increases over time. Entropy is the same as disorder, implying that an *isolated system* – that is, a system where no matter or energy comes in or leaves – is characterized by increased disorder or loss of quality (Georgescu-Roegen 1971; Ruth 1993). Sun-rays, an example of free energy, comprise a high-quality, ordered structure. As they hit a body – for example, the surface of the earth – they are transformed mainly into heat, which is disordered compared to the original sun-rays. Similarly, matter in the form of ordered structures also tends towards disorder.

The second law can be interpreted in statistical terms, simply saying that the chance of order occurring in any given situation is much less than that of disorder. This is because order – as the concept implies – demands a specific structure. Order in the form of sun-rays, a tree, a bird or a specific physical structure like a house or society for that matter, cannot appear by chance. Why is it then that there is a high degree of order in the biosphere, especially in the earth's ecosystems, if the rule is increased disorder? The point is that the biosphere is not an isolated system. It is instead *closed* – implying that while matter is constant, energy may enter and leave (Barrow 1995).

The constant import of high-quality solar energy makes it possible for the self-organized processes of energy and matter transformation in the biosphere to become established, refined and enduring. There is, however, one important prerequisite for this to happen. The second law implies not only that order is possible given an external high-quality source of energy, but also that in the course of conversion of high-quality energy into ordered biological processes, the total level of disorder will still increase. This implies that to maintain order in the earth's biosphere, the created disorder must be exported. This is secured through the export of heat from the earth back into the universe,[3] and the system is stabilized at a level where the amount of (high ordered) energy coming in is (approximately) equalized by (low ordered) heat leaving.

The worries concerning global climatic change are centred on this issue. The amount of CO_2 in the atmosphere is now rising to a level where the amount of heat leaving is lower than the amount of solar energy coming in. A temperature increase will continue until a new balance is established. However, this demands that the amount of CO_2 in the atmosphere also stabilizes. A continuous growth in CO_2 emissions will result in a continuous rise in earth surface temperature. The significance of this process does not least relate to the alterations that may occur in the functioning of both natural ecosystems and humanly constructed systems. Conditions for present life forms may change dramatically.

9.2 INTEGRATING ENVIRONMENTAL AND NATURAL RESOURCE ISSUES INTO ECONOMICS

9.2.1 The Environment in Economic Thought

While this book does not offer space for any comprehensive coverage of the role the natural environment has played in economic thought, a few, core observations are warranted.[4] Basically, it seems that the role of the natural environment in economics is dependent on how the economic process in general has been perceived. Typically, in neoclassical economics the specific perception of the environment is largely the same as that of other goods – that is, it is viewed as a set of commodities or items that can be defined and replaced. Some alternative interpretations have developed over the last 30–40 years, especially within the emerging field of ecological economics. Here a more explicit systems perspective is utilized. In building our institutional analysis of environmental problems we shall not least draw on some core ideas from this latter tradition – ideas which fit neatly into the institutional model as developed here.

It is natural to start our short overview by again looking at the position of the classical economists. In the writings of Smith, Ricardo and Malthus we observe that natural resources – especially land – played a crucial role in understanding the economic process. These authors were especially interested in the role played by land for the continuation of economic growth. As mentioned in Chapter 2, they were active in the period of early industrialization – a period giving birth to an economic growth not previously observed. The question arose whether this development could be sustained. Malthus ([1803] 1992, [1836] 1968) argued that the productivity of land would become a constraint on future growth. Land was a given resource and the ability to increase its yields could not, in his mind, keep pace with population growth. Therefore over time there would be increased subsistence costs and growth would finally cease.[5] Ricardo ([1817] 1973) emphasized that new land could be cleared for agriculture and that this would counteract the Malthusian poverty trap. However, new land was less productive, so even Ricardo concluded that growth had its limits.[6]

Neither Malthus nor Ricardo included technological change in their analyses. Mill, who was active some decades later, observed this new development in technology, and the focus in his writing shifted from looking only at land and labour as the main production factors, to also including produced capital. The effect of technology or produced capital was to counteract the 'natural limits to growth' (Mill [1848] 1965). During his lifetime the use of minerals in the economy increased substantially, and he also showed an interest in exhaustible resources like coal. However, even Mill perceived a sort of stationary state in the end.

In relation to our subject area, Mill seems to have been the first economist interested in nature as something more than a production factor in standard terms – as a source of inspiration and recreation. Hence, he opposed the tendency to focus only on material goods, so dominant in the work of most other economists.

The idea of limits to growth continued to be influential for some time even into the era of neoclassical economics. Jevons at least, was interested in the issue. He took over where Mill left off. In his view, the fundamental scarcity problem was that of coal or energy supply (Jevons 1909). While we again observe changes in problem perceptions reflecting changes in contemporary resource dependencies and technological development, the basic problem was the same as for the classical economists and for Mill. Even Jevons was afraid that the resource base might not be able to sustain the growing population.

These worries soon vanished. Jevons was himself part of a movement shifting the focus from resources and production – the classical economist perspectives – to exchange processes – the neoclassical perspective. Land as a specific resource became relatively less important for the economy, and the standard production function in the neoclassical texts came dominantly to contain just capital and labour. Nature as a separate input and separate problem to worry about almost disappeared for a while. Certainly, it is easy to envisage that the tremendous changes in production capacity flowing from continuous technological change influenced this. Furthermore, the growth of markets and exchange boosted an increased interest in their operation. Soon economics came to focus almost exclusively on the rather abstract schemes of exchange that developed.

None the less, some publications within specific fields of environmental and resource economics were published in the period between 1850 and

1950. The question of optimal harvest of forests was an issue engaging German foresters from at least the mid-eighteenth century. The publication of the Faustmann formula was an important step in that process (Faustmann 1849). It solved the problem as it can also be seen as an early version of what now is called 'optimal control theory'.

The issue of mineral extraction was, as we have already seen, a topic that attracted attention fairly early. Henry Hoskold developed a formula for calculating the net present value of a mine in the 1870s. Early in the twentieth century, L.H. Gray formulated a method for calculating the optimal extraction path of an exhaustible resource (Gray 1914). Hotelling (1931) was later credited with solving this latter problem.

Pollution was also an issue that caught the attention of some. Pigou (1920) discussed the topic of negative and positive externalities – that is, physical relationships between agents not being captured by the market. He developed the idea that taxes and subsidies could be used to correct the market mechanism that was failing in these instances.[7] His work was generally accepted, but it still did not play a major role in the development of the discipline. Instead, environmental issues continued to be rather insignificant until the 1960s.

At this time an increased awareness of environmental problems – not least pollution – resulted in a growing interest in environmental issues. The problems were a function of vastly increased production volumes with a corresponding increase in waste emissions. At the same time, new, synthetic matter such as DDT (an insecticide) was being released into the environment. Rachel Carson's book *Silent Spring*, published in 1962, was both a symptom of a changing perception and a source of inspiration for a slowly growing environmental movement.

Three factors are of special importance if we try to explain why it took so long for environmental issues to feature more strongly on the agenda. First, technological change had increased rapidly throughout the twentieth century. New sources of energy were found – for example, vast resources of coal and oil, which eliminated the fear of running out of resources. Second, pollution is a gradual process. The recipients have a substantial capacity to handle changes in materials flows – see the discussion on ecosystem resilience, above. Thus, since few seemed to consider potential future harms – that is, accumulative effects over long time horizons – the problems had to become clearly visible in one way or another before it was realized that something was wrong. Third, and implicit in the above, the perception of the production–consumption–waste cycle was very undeveloped. The production system was not set up under the perspective that whatever product one produces, it will in the end become waste. Problems were rather treated sequentially. There was no culture for thinking about materials flows in more systems or holistic terms. Engineers constructed wastewater systems as if the sewage miraculously disappeared when it could no longer be seen. The typical anecdote in textbooks of environmental economics was that of the smoky factory sited next to a laundry hanging out clean linen to dry – a very visible and tractable problem. While not at all unimportant, it nevertheless illustrates the rather narrow perspective employed.

The revival of interest in environmental problems in the 1960s, spurred three different types of responses within economics. First, Coase (1960) attacked the Pigovian solution of taxing the emitter. Coase claimed that in a world of zero transaction costs, there is no need to tax externalities. Given that rights are distributed, the parties involved in an externality conflict can themselves find the optimal allocation of resources between emissions and abatement. If it is more costly for the factory to change its production process, close down or move, than it is for the laundry, then the laundry will move. Whether the factory has a right to emit or the laundry has a right to clean air, does not matter for that decision, it matters only for who has to carry the costs. Coase's paper gave birth to a long series of analyses on the consequences of this finding. While it may seem a bit odd to discuss externality problems under the assumption that transaction costs were zero – see also Mishan (1971) – the Coasean analysis made a tremendous impact not least on those economists who were against state regulations.[8]

Second, we have the revitalization – or really the start – of neoclassical environmental economics. Here Pigou's ideas were developed further, not least as a reaction to Coase. The concept of an externality was refined. Different policy responses were analysed – for example, taxes and tradable quotas. Finally, issues concerning the effect of different market structures and uncertainty on optimal policies were analysed. The establishment of the *Journal of Environmental Economics and Management* in 1968 was part of this process. Baumol and Oates (1975) summarized much of the development in this period. In the years to come the issues were further developed in a vast literature in the form of both journal articles and textbooks. The literature tended to divide into those covering environmental economics and those covering the economics of natural resources.

Third, we observe the beginning of a tradition later called 'ecological economics'. Here Boulding's article on 'Spaceship Earth' was an early representative (Boulding 1966). Later came the work of Georgescu-Roegen (1971), with its strong focus on thermodynamics. Finally, we have the report from the Club of Rome, '*Limits to Growth*' (Meadows et al. 1972). The last was a model study developing scenarios for resource stocks and environmental quality up until 2100. Meadows et al.'s forecasts indicated great problems of sustaining economic growth beyond 2020 due to resource exhaustion and pollution. The Malthusian message had revived in a modernized form.

The report spurred a long list of reactions. It became an important document for the growing environmental movement. However, it also provoked strong counterattacks. The model used was vulnerable to a set of technical criticisms – in particular concerning the structure of the model (for example, see Cole and Curnow 1973; see also Kula 1998 for a review of the debate). But, most important were the fundamental arguments against resource scarcity as a problem for long-run economic growth – perspectives developed by respected neoclassical economists such as Partha Dasgupta, Geoffrey Heal, Robert Solow and Joseph Stiglitz. Their argument was principally that the capability to replace natural resources with man-made capital could sustain growth over time given that technological progress was rapid enough (Dasgupta and Heal 1974; Solow 1974; Stiglitz 1974). If natural resource rents are reinvested (in human-made capital), consumption could be held constant over time (Hartwick 1977). With technological change it could be increased.

The above analysis depends on assumptions about the possibility of replacing natural with human-made capital. If the so-called elasticity of substitution (σ) between natural and human-made capital is larger or equal to 1, this implies in principle that no natural resources are essential to the economic process. It can finally run on no (if $\sigma > 1$) or very low (if $\sigma = 1$) amounts of such resources (see also Toman et al. 1995).[9] As the limits to growth model by Meadows et al. can be criticized for some of their assumptions, presuming $\sigma \geq 1$ is principally very problematic – not least since it goes against the second law of thermodynamics.

The debate on the issue of technological change, growth and resource scarcity has continued since the 1960s. The debate has shifted somewhat from focusing on growth to focusing more on the issue of long-run sustainability. The arguments of the neoclassical economists have nevertheless been fairly stable, basing the argument on the potential for substitution, implying that no natural resources are really essential.[10] It was observed, however, that for the path to be sustainable, agents must also meet prices covering the full social costs of their activities – that is, internalizing all externalities now and in the future. Thus, two conditions must be simultaneously fulfilled – both the substitutability between natural and human-made resources, and the existence of right prices on all resources, goods and services (Asheim 1994). Concerning the latter, Common and Perrings (1992: 15) comment that 'either . . . there exists a complete set of markets including a complete set of contingent markets from the present day to infinity; or all agents in the system contract in current markets on the basis of "rational expectations", about the future course of resource prices'. That is, the idea demands that humans have perfect foresight. This was also acknowledged by, for example, Dasgupta and Heal (1979). However, it does not seem to have encouraged any strong interest among neoclassical economists towards exploring other, more realistic assumptions about human capabilities.

In opposition to this stance, there was a development within the confines of ecological economics that looks more directly at the specific role of different natural resources in sustaining economic processes. The focus here has been not least on the restricted possibilities for substitution. I shall close this section with a brief overview of the arguments developed here.

In his book *Steady-State Economics*, Herman Daly (1973) developed the systems perspective introduced by Boulding. Daly emphasized that the *total scale of the economy* compared with the environment is a core issue – indeed, more important than the question of allocations within the economy. He therefore focused on the issue of 'throughput' of resources – that is, the flow of natural resources through an economy – necessary to maintain a constant stock of physical wealth (artefacts) and a constant population in a steady-state economy:

> [T]he economy, like an organism, lives on a continual throughput of matter and energy taken from the environment in the form of low-entropy raw materials (depletion) and returned to the environment in the form of high-entropy waste (pollution). The biomass of an organism, or a population of an organism, grows to some mature or equilibrium size. The throughput then functions to maintain the size and structure of the organism. . . . The throughput flow (depletion→ pollution) is the *cost* of maintaining the population of people and artefacts and is not to be maximized, but rather minimized, subject to the requirement that the equilibrium stock of people and artefacts be maintained. (1973: 140; original emphasis)

This is a very radical or different idea compared to that of neoclassical sustained growth and it boosted debates about the involved resource 'pessimism', how to measure throughput, how to organize such a society and so on. Richard Norgaard, among others, developed the idea further by introducing the concept of 'coevolution', capturing the various types of relations between humans and nature as well as the size of throughputs (Norgaard 1984, 1994).[11] His idea was that humankind is part of nature and has co-evolved with it. Humans have had the special capacity to develop their own niches, not only to exploit given ones. When co-evolutionary, this process does not challenge the basic functioning of the ecosystems. It rather develops within these bounds. Parallel to this, changes in the use of nature also involve changes in the social system. Using the paddy rice culture of Asia as an example, he continues:

> The land intensive practice of slash and burn agriculture was gradually abandoned over many centuries as investments were made in dikes, terraces, and water

> delivery systems for increasingly intensive paddy agriculture. This ecological transformation provided superior weed control and greater nutrient retention. The environmental system modification process, however, was not unilateral. In order to maintain the ecological system in its modified form and to acquire the benefits of modification, individuals changed their behavior and the social system adapted to assist and reinforce appropriate individual behavior. In the case of paddy rice, the benefits from ecological transformation could only be acquired through complex social changes that facilitated property ownership, water management, and labor exchanges. (Norgaard 1994: 41–2)

The paddy rice system was able to sustain a much larger population than the old system, not only for a brief period, but for a long time. The alterations, which were substantial, did not challenge the long-run stability of the system. On the basis of his analyses, Norgaard emphasizes that it is very difficult to define which developments are coevolutionary – that is, sustainable – and which are not; which developments keep the system within its fundamental bounds, and which ones force it into more abrupt and irreversible shifts. He therefore concludes that multiple small experiments – that is, changes – are better than a few big ones since the stakes then become smaller and the chance of forcing the system beyond its bounds is less probable. He also suggests that diversity in coevolving systems is an inherently good thing due basically to the same kind of reason. If societies do different things, future 'collapse' is less likely. Finally, he concludes that experiments entailing very long time commitments – for example, nuclear waste – should be avoided.

Norgaard argues that prior to the significant use of hydrocarbons, the economies/cultures seem to have basically coevolved with ecosystems: 'With the exploitation of fossil hydrocarbons, cultures coevolved around hydrocarbons, apparently becoming increasingly free of ecosystems for the last century. . . . The transition to sustainable development will not be easy because of the extent to which hydrocarbons have driven a wedge between cultural evolution and the biosphere' (1994: 47).

What differentiates this analysis from the previous neoclassical studies is the perspective of nature as a dynamic system – rather than just a stack of physical resources or 'things'. The concepts of critical natural capital and resilience were next introduced into the debate, developing the idea of coevolution further. As the notion of sustainability took hold – especially after the presentation of the UN report *Our Common Future* (UN 1987) – these concepts got a core role in clarifying what sustainability could mean. In the UN report, sustainable development was defined as 'development that meets the needs of the present without compromising the ability of future generations to meet their own needs' (ibid., Section 2.1, point 1). What this could mean prompted new debates.

Following the neoclassical idea of substitution, it could be argued that sustainability would be obtained if:

$$dC/dt + dR/dt \geq 0,$$

where C = human-made capital and R = natural resources. This is the so-called 'weak sustainability' definition, that is, the development path is sustainable if total capital is maintained or increased. This definition is consistent with a development where R is significantly reduced (even to zero) as long as the growth in C compensates for this loss. In contrast to this, 'strong sustainability' was defined as:

$$dR/dt \geq 0.$$

This implies that the amount of natural capital should not be reduced. Leaving aside the vast problem of measuring C and R, both definitions are probably equally extreme or irrelevant. Rather, it is so that certain natural resources are substitutable while others are not. While some natural resources or biosphere processes can be changed or exchanged, others are not thus substitutable. On the basis of such reasoning the concept of critical natural capital has evolved – for example, Nöel and O'Connor (1998). By such capital, we mean natural resources that are essential to the functioning of the ecosystem – that is, cannot be substituted for.[12] One example is phosphorous, which is physically scarce and fundamental to the photosynthesis of plants. One cannot envisage replacing that element by any other.

Recognizing this is crucial. However, defining what is critical leaves us with a new set of challenges. Common and Perrings (1992) and Perrings (1997) have used the ecological concepts of stability and resilience – as defined by Holling (1973, 1986) – in a response to that problem. Holling defines stability as the propensity of populations in an ecosystem to return to equilibrium after a perturbation (external shock) has taken place. Resilience, on the other hand, is the propensity of a system to preserve its organizational structure intact after a perturbation. This ensures that its future functioning is in principle undisturbed. The distinction can be simply illustrated as in Figure 9.1.

A and B may be viewed as two different attractors ('basins') of a system. A may describe the existing climate system of the globe or characterize an oligotrophic lake. Its stable state is found in x. External influences, such as increased CO_2 emissions (climate) or nutrient inflows (the lake), may imply changes in the state variables of the system (that is, the level of CO_2 in the atmosphere or the amount of nutrients in the lake) moving it to position y. In the case of the climate, temperatures may increase, precipitation may

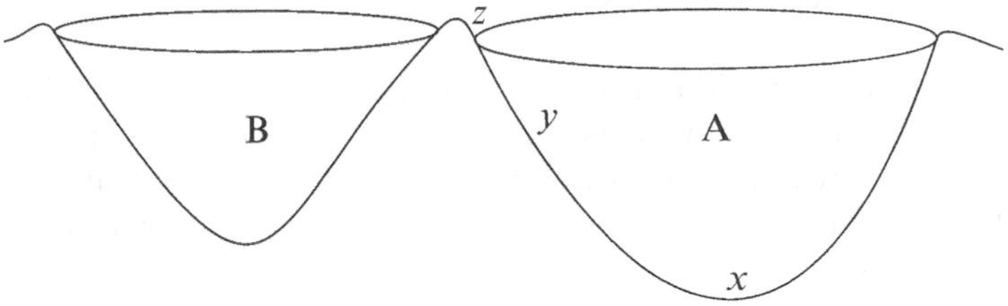

Figure 9.1 *Systems stability and resilience*

change. However, the structure of the system is not changed. This implies that if the CO_2 concentration falls back to its reference levels, the system will move back to state *x*. Similarly, in the case of the lake, the volume of each species will shift due to the increased amount of nutrients, but the basic structure of the lake remains and reduced nutrient input will over time cause the lake to return to *x*.

The system is more resilient at *x* than at *y* – that is, it can handle a greater shock before it leaves the basin/attractor A. If external influences such as continuous CO_2 emissions force the system up to *z*, it may easily 'flip' and enter the attractor B. The distance from *x* to *z* or *y* to *z* can be interpreted as the resilience of the system A given these two states. Shifting to attractor B implies a shift to a system with a different, often unforeseeable structure and dynamics. In our cases, the shift induces irreversible changes in ecosystem structures and functioning. The climate is irreversibly changed, for example, by the potential to divert the Gulf Stream, melting the polar ice caps and so on. In the other case, the lake has shifted from being oligotrophic to becoming eutrophic.

From this analysis we see that what is 'critical' about natural capital is not least the structure of the system, its flows and its ability to resist changes in these. Identifying essential systems functions that keep the system running relatively intact becomes crucial. While this gives meaning to 'what is critical', knowing what keeps us within an attractor, that is, knowing what may bring us from *y* to *z* and so on, is still a very difficult task. Indeed, it is impossible to measure since we cannot know the boundaries of an attractor. In practice this has encouraged the development of the so-called 'precautionary principle' and the idea of the 'safe minimum standard' – see also the previous points made by Norgaard. While different in some respects, both the precautionary principle and the safe minimum standard focus on making decisions so that we do not risk passing beyond *z* in Figure 9.1. Certainly, not knowing where this point lies, forces us to be extra careful; an issue we will return to later.

9.2.2 Three Different Concepts of Risk

Parallel to the above variations in perspectives, we also observe differences in the understanding of concepts such as risk and uncertainty. The standard application area of neoclassical economics contains zero information costs. However, the notion of positive information costs has been accepted in many studies and has stimulated a comprehensive number of studies concerning the treatment of risk.

Risk and uncertainty can be subdivided into (i) ordinary risk; (ii) uncertainty; and (iii) radical uncertainty. *Ordinary risk* implies that outcomes are not certain, but the agent knows both which outcomes may occur and the chance or probability for each of them. Hence, a throw of the dice may result in any number from one to six. Nevertheless, the agent knows that the probability is equal for each possible outcome. Gambles have the same structure and agents can then calculate the expected value as the sum of the value of each outcome times its probability.

In the case of *uncertainty*, the various outcomes are again known, but their probability is unknown, and the expected value cannot be calculated. However, the literature often assumes that the agent can calculate subjective probabilities about outcomes[13] and in this way the problem is reformulated into risk. In the third case, *radical uncertainty*, not all possible outcomes are known. Calculating probabilities has no meaning.

The concept of ignorance relates to the above. Ignorance means that we simply do not know. In cases characterized by ordinary risk – that is, knowledge about all outcomes and their probability can be established, previous ignorance can be reduced to a complete set of outcomes with probabilities. Certainly, in some cases it is possible to increase knowledge further by establishing certainty. Then it is not the problem as such, but our lack of knowledge that characterizes an issue in risk terms. If the problem is characterized by uncertainty or radical uncertainty, some level of ignorance cannot be avoided. Certainly, the practical implications are greater in the case of radical uncertainty.

The risk concept dominates in neoclassical analyses. This ensures that the standard rationality assumptions can be kept intact, while one has to shift from maximizing utility to maximizing *expected* utility. As strongly emphasized in Chapter 5, the inherent problem in determining even expected utility in the case of positive information costs cannot principally be avoided. Normally, it is not raised or it is simply assumed away as in the case of rational expectations.

Following from the description of the natural environment given in Section 9.1, uncertainty and even radical uncertainty are fundamental characteristics of these systems. While effects of different changes in various ecosystems and so on are potentially determinable as long as the system is not forced beyond the thresholds of the actual attractor, these thresholds cannot themselves be fully defined. This is a strong argument for using precaution as a rule for environmental policy.

9.2.3 The Two Models

In the above descriptions we have identified a substantial difference between the way neoclassical environmental economics and how more ecologically inspired perspectives consider the functioning of the environment and thus the interrelations between it and the economy. Figure 9.2 offers an illustration of some elements in this.

The environmental economics tradition tends to look at the economy and the environment as two clearly disparate spheres where interactions occur mainly along defined points such as mines, fishing grounds and so on. It furthermore distinguishes clearly between the input of resources to the economy – that is, mainly treated under the heading of resource economics – and the emission side – that is, mainly treated as pollution/environmental economics. There are no explicit links made between the two. Finally, the internal dynamics of the economy is emphasized much more than both the interrelationships with the environment and the dynamics within the environment.

The ecological economics perspective is different. Concerning the issue of matter and energy, the focus is first of all on the economy as an open subsystem of the environment or the biosphere, with only partially defined demarcations between the two. The focus is on the flow of matter and

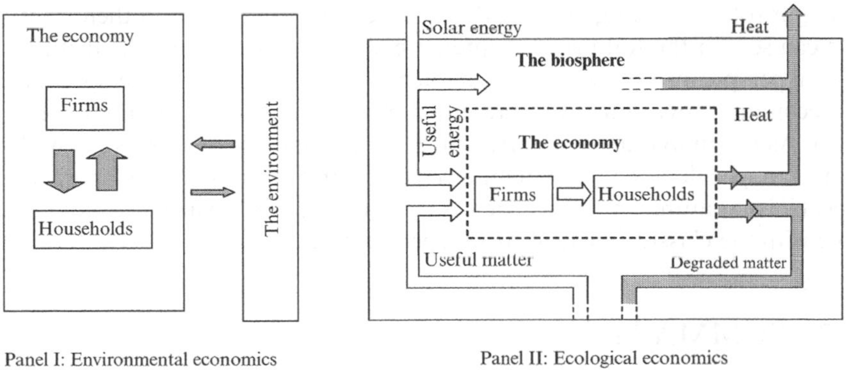

Figure 9.2 Two ways of interpreting the economy–environment interactions

energy through the system – the throughput – and the thermodynamic laws governing these processes. Since any input will become an emission, the focus concerning the regulation of pollution, for example, may as well be on the input as on the emission side of the economy.

Figure 9.2 is in no way complete. For example, it does not cover the different perspectives on the dynamics of the environment, including the different implied understanding of risk and uncertainty. On the other hand, it may also exaggerate the differences. There are some interesting examples where environmental economic textbooks have captured important elements of the ecological economics perspective – for example, Pearce and Turner (1990) – or where the two perspectives are to some extent combined – for example, Perman et al. (1999). These tendencies are interesting, illustrating that while various positions may be defined and distinguished, there are also developments where one perspective may prompt another.

However, there are limits to such cross-fertilization. At some point, adopting new perspectives may threaten the whole paradigm, or in guarding against that, the authors may choose to let the integration be only partial. Pearce and Turner, and also in part Perman et al., are good examples. In the case of Pearce and Turner, a good exposition of issues such as materials flow, thermodynamics and so on is given early on, while these insights seem not to influence at all the treatment of the various policy issues dealt with later on in the book. They are formulated within standard neoclassical confines. It seems like the understanding of the natural environment as captured by ecology/ecological economics cannot be reconciled with the neoclassical model of the economy.

It is not an aim of this book to go into the various philosophy of science issues involved in this, but theories – as social constructs – certainly do play a major role in guiding us towards what we are looking for and then what we can see. In the following chapters we shall try to combine the institutional perspectives presented in Chapters 1–8 with the distinct perspective on economy–environment as presented by the ecological economics position. As the individualistic perspective of neoclassical economics fits well with – maybe demands – an 'itemized' perspective of nature, a systems perspective on nature demands a view of societal processes which are more in line with the classical institutionalist perspective.

9.3 SUMMARY

The environment is here understood as a system of bio-geochemical processes which is shaped by various life forms as it also sustains these forms. The highly ordered structure of the biosphere is dependent on a continuous inflow of high-quality solar energy and a similar export of waste heat. The dynamics of the system must follow the laws of thermodynamics. The first law says that in an isolated system, matter and energy is constant. The second law tells us that in such a system disorder (entropy) is increasing. The earth is isolated concerning matter but open concerning energy. In such a situation – that is, in a closed system – complex order can be established and sustained, while this state is dependent on a balance between the import and export of energy.

The understanding of the role of the environment for the economy has varied both over time and between positions. While classical economists like Malthus focused on production and looked upon the finite stock of land as a strong restriction to growth, their neoclassical successors shifted to looking at exchange and ignored the issue of resource constraints almost entirely. Between these positions lies the observation that technological change may remove resource constraints. The belief was that this development could be sustained.

In the 1960s there was increasing criticism of this idea. The observation of growing pollution combined with insights from ecology and thermodynamics indicated that resource constraints had not been removed; rather, they had been pushed 'outwards' in time and space. Following this we also see a renewed interest in environmental issues among neoclassical economists. The need for correcting price signals when externalities were involved was emphasized. However, the main message was that continuous growth could be ensured through technological development and substitution.

The emerging tradition of ecological economics opposed this conclusion. Here the argument is that certain elements and processes of nature cannot be substituted for – that is, critical natural capital exists. Furthermore, it is critical to sustain the resilience of the biosphere in order to counteract the serious shifts in its dynamics that a growing economy threatens to provoke. Securing resilience demands a perspective on resource dynamics and policy very different from that of the neoclassical position.

While the various views concerning the economy–environment interactions reflect a very different understanding of the dynamics of the biosphere, this difference in perspective nevertheless seems to be defined by the economic model itself. Primarily, the perspective of nature is a reflex of which perceptions the structure of the economic model is able to handle. The neoclassical model focuses on exchange and substitution. Understanding nature as consisting of (easily) demarcatable objects that can be traded (wide sense) fits this vision of such an economic process very well. A perspective emphasizing interdependencies and processes instead of items does less well. A reform of economic theory taking these latter characteristics seriously seems to demand that we relinquish the idea of economics being about maximizing individual utility. It implies that we must focus explicitly on how we intervene in each other's lives and conditions, systematically and extensively. Our lives and resources are interconnected. This will be the subject for the following chapters.

NOTES

1. Nicolis and Prigogine are natural scientists. The way they use the concept of behaviour and choice is somewhat different from the way it is used in the present book as it may indicate that nature has 'intentions'. The authors should not be interpreted in this way.
2. Visually, an attractor can be thought of as a basin. All movements in the basin tend towards its stable 'bottom'. This does not imply that movements cannot be directed away from this stable state. Nevertheless, the system or the basin will tend to stabilize towards that bottom. The idea is that (a series of) events may have the force to make the system shift the attractor. If the system has stayed essentially within the same attractor and after some shock, we know that it is able to retreat from such a change in state variables back to its stable position. As an example: adding nutrients to a forest will change the growth of various species and their internal competition will be altered. The species composition will change. If withdrawing the fertilizers – that is, stopping the 'external shock' – returns the forest to its original species composition, the attractor is intact. The system is resilient towards such a shock. If the forest does not thus return to its original state, an attractor shift is observed. More on this follows in Section 9.2.
3. If we look at the total (isolated) system, such as the position of the sun and the earth in the universe, the entropy of this system will increase over time, while it is still possible to increase order in one part of the system such as in the biosphere of the earth.
4. Barbier (1989), Spash (1999) and Röpke (2004) offer more comprehensive expositions.
5. Malthus claimed that while yields could increase only linearly, population growth would follow an exponential path. Adam Smith, who was the 'founder' of the classical position, on the other hand did not focus much on the problem of scarce land – see Commons's remark that Smith's theory was that of abundance (Box 6.1).
6. See also Barbier (1989) on this.
7. Therefore it is standard to talk about Pigovian taxes in environmental economics.
8. See also the section on the property rights school in Chapter 4.
9. If σ between natural and human-made capital > 1, no natural resources are essential. Natural resources can be fully replaced by other – for example, human-made resources. The question of natural resource renewability is irrelevant. If $\sigma = 1$ – that is, so-called Cobb–Douglas technology – natural resources are necessary, but their productivity has no upper bound. In reality this implies no real restriction on resource use since it is assumed that technological change can make it possible to run the economy on small amounts of natural resources.
10. The following quotation from Solow is instructive: 'history tells us an important fact, namely, that goods and services can be substituted for one another. If you don't eat one species of fish, you can eat another species of fish. Resources are, to use a favourite word of economists, fungible in a certain sense. They can take the place of each other. That is extremely important because it suggests that we do not owe to the future any particular thing. There is no specific object that the goal of sustainability, the obligation of sustainability, requires us to leave untouched. . . . Sustainability doesn't require that any *particular* species of fish or any *particular* tract of forest be preserved' (Solow 1993: 181, italics in the original).
11. Other important contributions in the process of developing an ecological economist perspective are found in Martinez-Alier (1987) and Costanza (1991).

12. They define critical natural capital as 'that set of natural environmental resources which, at a prescribed geographical scale, performs important environmental functions and for which no substitute in terms of manufactured, human or other natural capital currently exist' (Nöel and O'Connor 1998: 78).
13. In 'plain English', guessing about outcomes – whether 'qualified' or not.

10. Resource regimes

Concerning resource use, the institutional issues can basically be divided in three. First, we have the question of who gets access to which resources – that is, the issue of *resource distribution*. Second, we have the costs of setting up and running institutions for the individual or common use of a resource – that is, the *transaction costs* involved. Third, we have the effect a regime may have on how problems are *perceived*, which *interests* it defends and which *values* it fosters. The aim of this chapter is to cover these questions, and thus offer insights into the role of regimes in the protection and sustainable use of the environment.

As we have just seen, natural resources are characterized by a series of interconnected processes. This implies that our use of the environment influences other possible uses. The utilization of one resource will necessarily influence the quality of other resources. In this way, independent choices may accumulate into changes of whole ecosystems and their services. The problem we face as humans is thus foremost about coordinating our activities and use of the various interconnected resources. What is rational for each individual to do, may become very negative when the effects of all individual choices accumulate. The institutional structures established to regulate resource use will here be called 'resource regimes' or just 'regimes'. They can take a variety of forms and include (a) a property rights structure (private property, common property and so on) which governs the access to the resource, and (b) a set of rules concerning transactions over the results from the use of the resource. Given the variations in such structures, resource regimes may function very differently.

The literature on resource regimes is split into several traditions. Again, the main divide is between an individualist perspective and a perspective based on social interaction and construction. Those adhering to the individualist model treat the coordination problem mainly within the game-theoretic framework of strategic action. This is a logical consequence of the individualist model. People with given preferences and endowments act in isolation – that is, interact in 'games' with each other – where they adopt the strategy offering the highest individual utility (profit) given the existing regime, the 'rules of the game'.

The social constructivist tradition, on the other hand, also emphasizes the effect of regimes on the perspectives and interests of the participating individuals, on their possibilities to communicate and willingness to cooperate. The position is taken that the regime not only influences the formal rights structure. It also influences the understanding of the problems involved, and the kind of norms and routines applied.

While the literature is generally split into these two traditions,[1] the individualist model and its game structures should be viewed as a special kind of social construct. Hence, regimes may be instituted to foster individually calculative choices. Alternatively, regimes may advance the role of common discourses and common responsibilities.

This chapter is divided as follows. First, we shall define the basic concepts of property rights, property regimes and resource regimes (Section 10.1). Then we shall focus on the relationships between resource characteristics, regime characteristics and transaction costs (Section 10.2). Third, we shall look at the relationships between regimes, interests and value (Section 10.3). The fourth concern is to present a simple model for analysing resource regimes (Section 10.4). Section 10.5 will finally focus on the most important field of resource regime construction today, that of international agreements and conventions.

10.1 PROPERTY RIGHTS, PROPERTY REGIMES AND RESOURCE REGIMES

Property rights define who has access to which resources or benefit streams and under what conditions. They distribute access to resources between the members of a society and regulate conflicting uses. The issues involved are both about distribution and about order. The latter refers to creating conditions for uses that do not result in unwanted outcomes.

Resource regimes normally consist of both rules defining access to resources, transfer/inheritance rules and so on – that is, property rights – and rules concerning how the products of using the resource may be transferred between interested parties – that is, market transactions, public allotment and so on. Furthermore, we may distinguish between formal and informal aspects, as defined in previous chapters. While in many situations informal institutions like norms and conventions are crucial to the functioning of a resource regime, we shall still start our inquiry by looking at the issue of sanctioned rights.

10.1.1 Rights and Property Rights

A right is a socially defined relation. It is an institution offering individuals or collectives an assurance that other people will behave in a specific way towards them. A typical example is that the children of a society may have, for example, the right to nine years of schooling, implying that the community has the duty to offer such an opportunity. Rights may be defined in relation to a very wide set of issues such as 'human rights' (the UN declaration), rights to medical care, rights to retirement funds and rights to use of natural resources.

Property rights are a specific type of right of fundamental importance to resource allocation issues. A *property* is often thought of as a thing – that is, a house, a piece of land, a chair. This is wrong. Even *a property right is a social relation.* It is a relationship between the *rights holder* and *the rights regarders* under *a specific authority structure* like the state granting legitimacy and security to a specific resource or benefit stream. Hence, rights run from the collective to the individual level. They have to be defined and defended through socio-political processes.

Two distinctions are important. First, someone may be able to protect his or her interests in a resource or benefit stream through the use of physical force. However, this is not the same as having a property right. It is mere physical possession. Second, one cannot derive rights from characteristics inherent in the quality of being a human. This is clearly the case both for human rights and for property rights. They are all social constructs.

Ownership is a right to specific resource or benefit streams. This is not least important to acknowledge in the case of natural resources where the same 'object' – that is, a piece of land – may offer a range of benefit streams: timber, berries, pastures, wildlife, the capacity to transform waste and so on. While some may own the right to the timber, others may have the right to pick berries or use the pastures.

Ownership to a benefit stream may also mean different things in different situations. Honoré (1961) distinguishes between 11 elements of ownership necessary to talk about full ownership:

1. The right to possess: the right to exclusive physical control. This right cannot be arbitrary, else there is no ownership.
2. The right to use: to harvest some resource for own use and so on.
3. The right to manage: the right to decide how and by whom the thing owned shall be used.
4. The right to income: to capture the surplus or the yield from the resource.
5. The right to capital: the right to consume, destroy or sell the resource to others.
6. The right to security: immunity from arbitrary appropriation.
7. Transmissibility: the right to transfer to successors, inheritance rights.
8. The absence of term: ownership that runs into perpetuity.

9. The prohibition of harmful use: ownership does not include a right to harm others.
10. Liability to execution: the liability of the owner to use the property to settle debt.
11. Residuary rights: rules concerning what to do if existing property rights are no longer relevant.

Some of Honoré's points seem to overlap.[2] One should also acknowledge that ownership may not imply full ownership as defined above. Society, when granting the right, may have formulated terms that restrict some of the above rights' elements. Specifically concerning natural resources, the right to capital may be restricted. The owner may have the right to cut down trees, but not destroy the forest by turning it into a residential area, remove the soil and so on. The owner may not be free to sell to whom s/he wants.[3] Terms may be formulated concerning the ownership period and so on.[4]

It is also important to stress that, according to Honoré, even full ownership does not grant the right to harmful use – that is, point 9. This is an important issue in the case of natural resource use, but is not easy to define. Moreover, as soon as interconnections exist, prohibition of harmful use may imply restrictions on some of the other points – see the examples above concerning the right to cut down trees, remove soil and so on. Hence, there may seem to be a conflict between point 9 and the rest of the list. In the case of interconnected resources, full ownership of some benefit stream will have to imply some harm to others. We shall return to this.

From these brief comments, it is obvious that to state that something is privately owned is actually saying very little. One must also list the more specific content of the rights and duties involved. Whether the owner, manager or worker are different persons or not will influence the dynamics of the rights involved too.

10.1.2 Property Regimes

Types of property regimes

A *property regime* is the structure of rights and duties characterizing the relationships between individuals[5] with respect to a specific good or benefit stream. In the literature it is common to differentiate between four property regime types:

- private property;
- common property;
- state (public) property; and
- open access.

While the rights holder in the case of *private property* is normally thought of as an individual, *common property* is likewise private property for a group of co-owners (Bromley 1991). In the case of a *state property* regime, the ownership is in the hands of the state. Ownership at lower public levels, like the county or the formalized municipality level, is largely of the same form and could, by changing the label from state to public property, be explicitly covered by this category.[6] Finally, *open access* is a situation where there is no property.

If we look at a specific piece of forest, it can be privately owned – for example, owned by a private person such as a farmer or a private corporation as in the case of a forest company. This grants the individual or the firm certain rights and obligations concerning the use of the forest. It can also be owned by all inhabitants in a certain municipality – a village – or by a specified subgroup of these inhabitants. It is common property. In this case there are two kinds of rules: (a) those defining who are members of the commons and who are not, and (b) those defining the rights to use various involved benefit streams – that is, which benefit streams can be utilized, by which members, to what degree and maybe also by what means.

The forest may also be owned by the state. This implies that while in principle the resource is owned by all persons having state membership, state-authorized representatives make decisions concerning resource use. Finally, open access implies that whoever wants to use a benefit stream from the forest may do so. A privilege exists for everybody. Certainly, open access is 'what was there in the beginning'. Transforming it into one of the other three property regimes could be motivated in two different ways. First, it could be motivated by the wish to get exclusive access to a benefit stream for some individuals or collectives. This is the distributional aspect. Second, it is important as a means to regulate the use of the resources – to avoid overuse or to regulate external effects of different uses – to avoid the 'tragedy of open access'.

In addition to the formal rights of a property regime, often there are also norms and other informal institutional elements that supplement its functioning. As in the case of private property of land, unformalized rules such as locally defined management practices may play an important role concerning its use. Systems of informal redistribution of the crop to, for example, landless people as members of the wider community are other examples. This implies that the formal structure does not define all aspects of use.

In relation to this, while the core of private and state property is a set of highly formalized rules, informal regulations seem to play a more fundamental role in the case of common property. This is especially the case in countries with a weak state where common property rights may even be unsanctioned by the state. This seems to be related not least to how this type of regime has often evolved. Historically, land dominantly belonged to groups of people. It was where they lived. Norms developed concerning their uses, and we may talk of 'common property' even though it was not property in the sense of any state protection. Perhaps a 'group of elders' within the community defined and controlled the core norms and rules concerning individual use. If members of a neighbouring commons decided to try to seize some of the land, no external power could be called upon. This possibility was established first when the single communities, after long and complex processes, were able to establish a state type of 'third-party' authority.

However, even the 'group of elders' in the village is a 'third-party authority'. Thus, one should be careful about making too strict distinctions here. Certainly, a modern state has many more opportunities to issue rights and control its execution than a village council. None the less, there are also important similarities. The 'war' between two local communities – common properties – is in principle no different from conflicts or wars between states and the subsequent lack of an international authority to decide on these matters. Both a village council and a state parliament have a limited geographical jurisdiction. While the delegated powers are certainly different, a third-party system that can handle conflicting claims within the relevant territory exists in both cases. However, since common property tends to be the property regime that is the least formalized with important rules embedded in the whole social structure, it is easy to make the mistake of equating a commons with open access. One cannot just look into the 'books of codified rights' to find the rules. One must study the culture and establish insights on the norms and the conventions. So, when Hardin (1968) wrote his famous article on 'the tragedy of the commons' – that is, of what he believed was a (unregulated) commons – he confused the terms as he was actually writing about 'the tragedy of open access'.

Distinguishing between the regimes

As already indicated, the defined property regimes are rather broad categories encompassing a variety of forms. To draw a strict line of demarcation between them is thus not easy. We certainly consider something owned by a family or corporate firm as private property. However, we also observe that in both cases it is reasonable to talk about a group of co-owners. At the other extreme, state or public property is also a form of co-ownership, where politically elected representatives/agencies act on behalf of all members of the nation state, county and so on.

Since all three categories of ownership may imply some kind of co-ownership, the main justification for distinguishing between them really concerns *the type of relationship among the co-owners*. This may be reflected as differences along all 11 of Honoré's components. Since we cannot go through all of them, I have chosen two – the right to capital (including transmissibility)[7] and the right to manage – to illustrate what distinguishes different types of co-ownership.[8]

Concerning the *right to capital*, the archetypal case of 'co-owned' private property is a situation where each co-owner owns a specific part or share[9] that s/he can freely dispose of. This is what characterizes, for example, a joint-stock company. Everybody is in principle free to buy and sell individual parts, even though rules may exist defining restrictions on this freedom.

In the case of common property such as a common pasture, the situation is quite different. Here items, shares or benefit streams are not individually owned. The pasture is common. What the common ownership grants is a right to every member of the common property regime to use the pasture as long as the internal rules for use are followed. The co-owners do not, however, own any piece that can be individually sold. The only way to get access to the resource is to qualify for membership of the commons. Access to or disposal of the resource/benefit stream is governed by mechanisms other than trade.

Concerning the right to capital, modern forms of state property may be viewed as a special type of common property since, as in the standard case of common property, no individual share can be distinguished. However, some distinctions to common property can be made. First, any member of a state is automatically granted the right of ownership to the state's properties. Furthermore, while common property is *private* property for a group of co-owners, state property is not. Finally, state or public property will normally be multi-objective – that is, concern a wide variety of benefit streams/capital. The public may run schools and hospitals, manage forests and infrastructure and so on. Common property may often be narrower in focus as it often concerns only specific sets of benefit streams and a subgroup of the inhabitants in an area – for example, access to a common pasture in the mountains only by farmers of the nearby valley. However, this again varies between societies, since in many developing countries local communities and common property regimes may cover many of the benefits that states offer in, for example, more western types of societies.

An interesting analogous case is the status of subnations such as indigenous peoples executing common property rights to specific resources like fishing and hunting grounds or pastures situated within a state that is dominated by another nationality. There may be cases where all the inhabitants of a region are members of this subnation or ethnic group, and they may even be co-owners in a common property regime. This ethnic group may historically have its own authority structures to grant, regulate and control the rights of the commons. The distinction with state property becomes rather subtle in a situation like this. In such cases we typically observe conflicts over who owns the basic resources such as the land (the capital) – that is, whether the property right is a usufruct right only and so on (Hahn 2000; Riseth 2000). In addition, the understanding and role of property concepts may vary across cultures. This can be seen in the confrontations between European settlers and Native Americans. The formal institution of property, as discussed here, is largely a construct of 'western' culture.

With regard to the *right to manage*, we also observe differences and similarities across the groups. In the case of private property, management is concerned with obtaining the goal(s) of a single enterprise – its management objectives. Whether we talk of a one-person firm or a large corporation, we still talk about one set of goals. This set may be narrow, for example, maximizing profits, or it may be wider, to also encompass some of the interests of the employees. Nevertheless, there is one defined set. There may be specific principal–agent problems, especially in large corporations where owners and managers may have competing ends (for example, Galbraith 1971), making the above conclusion somewhat simplified. However, this does not influence the basic characteristics of the purposes of management and is not only a problem for large, private businesses.

In the case of common property, management is about coordinating the individual uses of the co-owners. Actually, it is about regulating the effect of one co-owner's use on the possibilities for the others. Each co-owner has his/her own goals. They may utilize the common resource to a different degree or in a different form as long as use falls within the rules of the property regime. In the case of a common pasture each member may have a right to let a certain number of animals graze. The animals themselves may be privately owned, and the commons normally do not engage in what each member does with the products made from these animals. The responsibility of the commons is to manage the common resource, rather than to involve in the result each co-owner may obtain from participating. Certainly, close contact through membership in a commons and its management operations, may build trust and engagement across the group of co-owners, fostering reciprocity, different types of collateral security nets, common values and so on. This may be an important effect of common property regimes, and may even influence the participants' perspective on how private their engagement in the commons is. However, the rules of the commons are normally focused on what each member is allowed to do.

Concerning management, state property may vary substantially across fields, partly resembling a firm and partly acting more like that of common property. This depends not least on the type of activity. If the state property is a hydroelectric power plant, the management situation may tend to look like that of a firm. The difference may be found in the goal function that may encompass other or wider elements than that of a private enterprise. Often this is the reason for keeping some productions, which otherwise could be easily privatized, under public governance. What otherwise may become a 'negative externality' of a private enterprise, may be 'internalized' just by making the production a public responsibility and, in this way, influence management purposes directly.[10] In relation to this, the question of the efficiency of private versus public bureaucracies is also an important issue.

Typically, the state is also engaged in facilitating individual resource uses, like building and maintaining infrastructure, and delivering public goods such as establishing national parks, running public schools and public health-care systems. In these cases, managing state property is more like that of a common property regime. However, there is a difference concerning who undertakes various tasks. In the case of common property, the co-owners themselves normally execute these as in the case of maintaining irrigation systems or fertilizing the common pasture. In the case of state property, agents take care of the functions.

10.1.3 Resource Regimes

Finally, regarding the concept of a resource regime – or simply regime – we observe that different property holders may want to conduct transactions with one another over the products they make when utilizing the property they hold. Thus a resource regime consists of two elements: (a) the property regime that governs the use and transfers of the right to a resource, and (b) the rules that govern the transactions concerning the results from the use of the resource. Firms may sell their products in markets. That is certainly the most typical, but private firms may also be involved in producing some public good, like health care on a state licence, implying that it is, for example, social criteria and not purchasing power that governs the distribution of the product. States/public owners may dominantly allot their produce to the citizen on the basis of social or community-based principles such as giving priority to certain age groups, health status and so on, or goods may be offered free to all. We generally think of access to schools, hospitals, roads, police protection and so on. The state/public authority may also engage in market transactions. Co-owners of a commons are also often engaged in market transactions over their produce. In this sense, co-owners often act like ordinary firms, even though they may also be involved in community-based distribution. Again this is defined by the wider institutional context.

Thus, private owners, co-owners of a common property regime and state agents all may operate in markets. However, in the literature we often find an implied relationship between the type of property regime and the type of economy (resource regime). Cornes and Sandler (1996) talk about market solutions as opposed to state property, common property or open access. This may be a bit confusing. Apart from open access, all regimes should be viewed *internally* as systems of direct coordination, as *command and control* systems. Furthermore, goods produced within any of the four systems may, as we have seen, be traded externally. This relationship depends on which broader set of institutional arrangements the actual property regime is embedded in. The principal difference is again concerning the focus of the management and what can be traded, not the existence of markets *per se*. While members of a common property regime are not allowed to trade their right to utilize the benefits of the commons, they certainly may sell the products they individually produce utilizing these benefits. They cannot sell their right to graze animals on the common pasture, but they can sell the meat that their privately owned animals produce. So while markets may be considered much more important in the case of private property, it is by no means exclusive for that kind of system.

10.2 PROPERTY REGIMES, RESOURCE CHARACTERISTICS AND TRANSACTION COSTS

It is unnecessary to conclude that undefined or unclear property rights may yield both large conflicts and great losses (see Chapter 7). Following the simple model of individual rationality, one may also be tempted to conclude that private property is the only efficient solution – see the arguments delivered by the property rights school (Chapter 4). This will yield the best incentive structure, as everybody will secure the fruits of their individual efforts. Concerning common property, on the other hand, there may be a problem since the harvests reaped by one will also influence the opportunities for other members of the commons – that is the alleged 'tragedy of the commons'.

This reasoning misses the fact that any property regime except open access – be it private, common or state/public property – may have very precise rules or norms establishing the necessary incentives for resource use and maintenance. It further overlooks the fact that private property regimes also have incentive problems when externalities appear, and following the perspective developed in Chapter 9, externalities are neither minor nor accidental issues. They are pervasive phenomena. As emphasized, this follows from the fact that resources and natural processes are interconnected – linking various resource uses necessarily to waste production. In economic terms this can be translated into *high costs of keeping different agents and their uses apart*. If it were possible to costlessly demarcate all streams of benefits, all processes, there would be no external effects. Each agent would own and consume only their own parts. Given the existing interrelations in natural resource systems, this is impossible to obtain. And even if it were possible, it would ruin the quality of the resources, since their very functioning depends on their working together.

Therefore, while demarcation will always be only partial, the potential gains of demarcating must be compared to the costs, which are a kind of transaction costs. Moreover, the level of such costs is an effect both of the property regime and the character of the good. This is important for understanding which property regimes are used for different resources.

The standard reasoning about the positive relationship between private ownership and 'efficient' resource use holds only if transaction costs are zero. However, as discussed in Chapter 4, any regime/property regime has the same technical efficiency characteristics under such an assumption. Simply formulated, since information is cost free, all individuals will have perfect knowledge. Further, since communication is cost free, these individuals can freely coordinate their actions with others. Finally, all objects, even all complex natural processes, can then be costlessly demarcated. This ensures that any property or resource regime will yield the same result as long as we assume that the regimes have no direct effect on preferences and the distribution of income or endowments.[11]

In neoclassical resource economics there is a tendency to relate property regimes directly to types of resources. Concepts like 'common property resources' and 'open access resources' are standard.[12] This wording is based on *confusing regimes with resources*. One should not deny that there are relationships between resource characteristics and property regimes. There is, however, no direct relationship as the above concepts imply. There is no 'natural' force working here. Regimes and resources are to be handled as distinct entities, and resource characteristics are only one among many factors that influence the choice of a regime.

When discussing this issue, we may start out from the simple categorization given in Figure 10.1, which is based on the standard distinction between exclusion costs (transaction costs/TCs) and rivalry in use or in consumption.

The exposition is simplified not least because one would normally think of exclusion costs and rivalry as a matter of degree and not distinct categories. Costs of exclusion concern the costs of demarcating the good or benefit stream and formulating the necessary rights so that the owner can

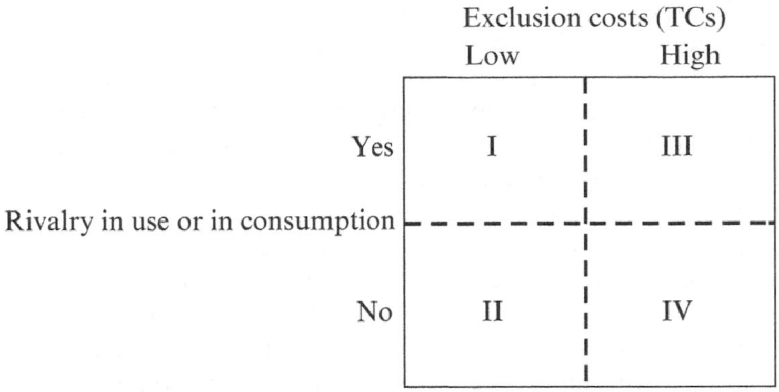

Source: Simplified version based on Randall (1983).

Figure 10.1 Characterization of resources or goods according to costs of exclusion and rivalry in use or in consumption

control it only. This group of costs will also include the necessary costs of policing. High exclusion costs imply that dividing the good into individually owned parts is expensive. Rivalry in use or in consumption implies that when someone uses a good, others cannot use it as well. When a piece of bread is eaten, it is not available for others. On the other hand, a scenic view can be 'consumed' without influencing the view for others. It is a non-rival good.

While goods of type I are typically labelled 'private goods', types II and IV are called 'club goods' and 'public goods', respectively. With regard to type III, 'common-pool resource' is an often-used concept (see Ostrom 1990). In the tradition of leaping from resource characteristics to regime implications, type I is often equated with private property and markets, II with clubs (a special type of common property), III with standard common property or open access and IV with state/public property. This is a problematic practice. But before looking more into that, it is necessary to make some clarifications concerning the dimensions of Figure 10.1.

First, the allocation problems concerning rival and non-rival goods are very different. In the case of rival goods, the core problem is that of external effects. This is typically the case for many environmental resources where use – at least beyond a certain level – will reduce its quantity and/or quality. In the case of non-rival goods this is by definition not an issue. If it is a naturally produced asset, no allocation problem will then exist – for example, the case of the air before it became a scarce resource. The good is there 'by itself', and one use will not influence other uses. Open access will suffice.

Hence, in the case of non-rivalry, problems appear first when we turn to humanly produced goods since there is no incentive to pay for them. Public goods such as military defence is a typical case where all individuals will benefit from its existence independently of whether they have paid for it. The cultural landscape is another case involving environmental assets more directly. Consequently, some mechanism other than the market must be created for financing the good – for example, establishing 'clubs' with membership fees (type II) or using public provisioning (IV).

In cases characterized by rivalry, the costs of exclusion – that is, transaction costs – come to the fore. For the so-called private goods (I), costs of excluding others from use are low, according to Figure 10.1. Demarcation is easy, and if demarcation is complete (that is, no externalities), private ownership will only exhaust resources that it would be preferable to exhaust individually. Furthermore, as long as there are no externalities, other individuals will not be hurt thereby.

If demarcation or exclusion costs are high, splitting resources into individual parts (properties) may be too costly. Such demarcation may cost more than it pays, which may result in the 'tragedy of open access'. Because of the high exclusion costs, everybody is free to use the resource. It is then rational to use the resource to the point where total costs equals total gains or revenue, not taking into account the external costs borne by other users of the resource. The person who allows an extra animal onto the pasture does not have to take into account that this animal reduces the quality of the pasture for the animals owned by others and that are already there. Given this dynamic, resource rents will be driven to zero. The common resource will be exploited in a way detrimental to all users. However, it may not be destroyed, as is so often implied in the literature. This may occur only if there is a threshold below which the resource is not able to regenerate (see also Box 10.1, below). The chance of this happening depends not least on the capacity of the technology involved. Many resources under open access have historically been protected by low harvest capacities. It is the increase in such capacities that has created many resource conservation problems.

If regulation via private property is too costly, the potential destruction may be thought of as a rational result. The cost of splitting up the resource into individual lots, fencing it and so on is too high. It is not matched by a similar increase in resource rents. This might be the case for a fish stock, a pasture or for the air. The productivity of the resource plays a role here. The more productive it is, the higher are the demarcation or transaction costs that can be tolerated. However, even in a case of such a productive resource as the air, the resource characteristics are such that demarcation costs obstruct individual demarcation.

However, *transaction costs vary between property regimes*. To avoid the tragedy of open access, one can establish common property. Here a balance is made between enlarging the group of co-owners to a level where exclusion of 'others' is possible, but internal coordination between the co-owners is still not too costly or difficult. Instead of splitting up and fencing individual parts of a pasture for each owner, the use is common and instead governed by access rules and so on. From this perspective, common property is about striking a balance between the cost of exclusion and that of dividing (Oakersson 1992).

This trade-off problem is basic for the technical aspects of managing environmental goods. As emphasized earlier, natural systems are characterized by a large number of functional interrelationships. Demarcation of specific resource entities in such a situation is mostly deemed to be only partial and often very costly. Plots of land may be demarcated by fencing. This may keep some out, but it cannot efficiently govern the vast biogeochemical cycles that the land is continuously involved in. As already mentioned, if demarcation is technically successful or complete, the resource is likely to be destroyed due to this isolation from its interrelationships with the rest of the system. In such a situation, private property may have little to offer. Demarcation may be only formal; in practice it may be complete along only a small subset of dimensions. In the end, few natural resources would fit the description of type I in Figure 10.1.

Summing up, we see that two problems surface here. First, demarcation is costly and the cost of different types of demarcation should be compared with the potential gains involved. Second, demarcation will normally only be partial. So, while it may be possible to demarcate the right to some resources – that is, the right to harvest a field – the emissions from the land in the form of losses of nitrates, ammonia, nitrous dioxides, carbon dioxide and so on cannot be equally demarcated. While these flows in natural systems have been adjusted to the needs of the biota – that is, as an effect of its self-organization – these flows are expanded beyond such limits in many modern humanly developed systems, such as industrialized agriculture, with heavily increased volumes of inputs.[13]

In some cases individual demarcation may be technically feasible, but the resource may still not be productive enough to carry the exclusion costs involved. Hence, Bromley (1991) argues that private property only appears where the productivity of the resource is high enough to carry the extra demarcation/transaction costs involved. It is not so that economic yield is a function of the property regime, which is the standard argument of the property rights school (see Chapter 4, Section 4.1.1) and often is also mirrored in neoclassical economics texts. Instead the causation may be the reverse: the chosen property regime is a function of (potential) economic yield.

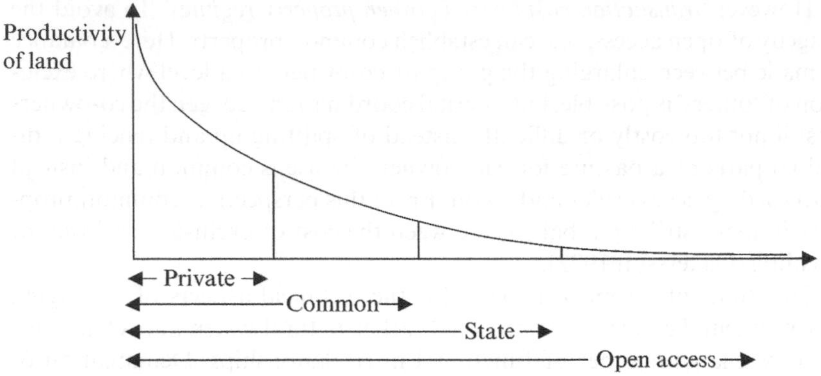

Source: An adjusted version of Bromley (1991).

Figure 10.2 Productivity of land and possible types of property regime

As illustrated in Figure 10.2, we tend to find private property on the most fertile land, common property (pastures and so on) on less productive ground, then state property and finally open access on what is left. Thus, we cannot use the productivity observed for certain property regimes as a test for which regimes are best. They may just fit different situations.

In fact, as is implicit in Figure 10.2, we can turn the efficiency argument on its head. While all regimes may be established for highly productive resources – that is, being able to carry the transaction costs involved – private property will not function for resources where the ratio between productivity and transaction costs is low. Transaction costs are a function of the property regime chosen. Thus, when analysing these issues, transaction costs become a core technical question. The problem is that these costs are often very hard to observe since they are primarily preventing certain solutions. Thus, they simply do not materialize.

In relation to Figure 10.2, one should also be aware that even in cases where the resource is highly productive, establishing private property will for some resources still be irrelevant due to the high costs of exclusion. This is the case with air, for example.

A large proportion of transaction costs is often fixed. Furthermore, as soon as a system is set up, it can be used for several purposes. It is therefore hard to make *ex ante* evaluations of which consecutive gains can be earned by developing a certain structure, say a state or a common property regime. Once such structures are established, many externalities may over time become efficient to regulate – in economic language they become Pareto relevant.

Consequently, when intentionally constructing a property regime, one must also evaluate or make qualified guesses about the future development of external costs – costs due to technological change, new products and so on, implying changes in matter cycles and habitat interactions prompted by the regime. Without this, there is not much left of the efficiency argument.

10.3 REGIMES, INTERESTS AND VALUES

As we have seen, the technical efficiency arguments so often emphasized when discussing regimes, boil down to the question of transaction costs and resource productivity, and may merely have the effect of making some property regimes too costly to establish and run. The more basic issue remains, that of which interests and values are to be defended – that is, what kind of society we want to develop through the regimes we set up. One issue is what is possible (TCs), another what is reasonable (values).

As we have seen, neoclassical economics has aspired to be value neutral. However, interest defence is at the core of the problem of resource allocation, and the fundamental aspect of any property regime. Regimes are interest neutral only if transaction costs are zero, and preferences go unchanged, as do the distribution of costs, wealth and power. Certainly, these assumptions have little relevance for real-world situations.

While we have so far focused on a rather general description of the four main types of property regime, in this section we shall discuss in more detail the elements of each of them and to some extent also the wider regimes of which they may be part. Any regime must have rules about who can shift costs to whom and under what conditions, about distribution of resources and redistribution of wealth and so on. Such issues may be handled differently within the regimes. In some property regimes, formalized rights and punishment structures dominate, while in other situations, culturally embedded rules and practices have the upper hand. Even in highly formalized regime structures, the wider sphere of social embeddedness is important. Thus, when we talk about property and resource regimes in practice, we cannot only talk about the main categories. We also need to look at specific structures of each.

10.3.1 Regimes, Wealth Creation and Wealth Distribution

The distribution of resources is basic to how individuals and groups can develop their lives. The main argument in favour of private property is its capability of dynamic wealth creation through its specific incentive structures. The individual who owns a resource gets the profits from uses that the market supports. This is a drive towards changing resource use so that it is orientated towards the strongest demand. This is a powerful feature. It is, however, vulnerable to some caveats implying that societies normally regulate private property rights to counteract pertinent problems.

First, initial distribution of access to resources strongly influences the possibilities for various individuals to sustain their lives. Hence, saying that there is private property says little if anything about social conditions and a society's ability to secure the satisfaction of basic needs. Second, while private property is instrumental in supporting economic growth, it also tends to create or increase inequalities. The internal logic will be that of continuous accumulation of wealth in the hands of those winning the competition for resource use. The rule is as simple as it is obvious: returns on ownership or capital go to the capital owner only. If 'left on its own' this dynamic not only secures growth,[14] but also an accumulation of surplus in the hands of those capable of investing – that is, those owning more resources than what is necessary for sustaining their daily needs. Over time this may foster large inequalities, as observed both within and between countries.

Inequality may, however, also encourage counter effects. It may stimulate the establishment of programmes for redistribution to avoid reduced legitimacy of the overall system. It may also motivate the creation of various organizations or collectives on behalf of those losing out in this process, giving them the power to counteract more directly the concentration of wealth. This often leads to strong state involvement in societies whose basic property systems are still private. Thus the fruits of a growing economy may be redistributed so that all groups of the society get a larger share of this growth. However, the programmes instituted for redistribution will always face the problem that they create 'disincentives' for those wanting to invest. Redistribution interrupts the core mechanism of the system – the motivation established by securing for the investor the fruits of his/her own investments. Striking the balance here has become increasingly difficult in a globalizing world where capital can be easily moved (Martin and Schumann 1996; International Forum on Globalization 2002; Stiglitz 2003).

It is no surprise that great inequalities are most visible in many developing countries where redistribution programmes are often weak as the level of organization is also generally low. As an example, land is accumulated by the few through processes whereby farmers who do not have enough land to support themselves – for example, through periods of drought – are forced into debt to wealthier farmers or others and in the long run drift into the landless classes via mechanisms of ever-increasing obligations (see, for example, Baland and Platteau 1996). As Sen (1981) shows, entitlement structures are critical not least when famine strikes.

More specifically, the breakdown of assurance structures, which often follows the privatization of natural resources, strongly influences the long-run position of the rural poor (Lane and Moorehead 1994). As illustrated by the work of Jodha (1987), the privatization of local commons contributes especially to this process. There are reasons to believe that the larger capacity of a commons to resist specific natural crises also plays an important role in explaining their assurance capacities. This follows from the mere statistical fact that a commons will normally be more diversified in its resource base than smaller, individual plots. So if, for example, a drought damages one resource or part of a commons, other resources/areas may still offer yields that can support its members (McKean 2000).

While in a system of private property, change and inequality is both a prerequisite and an effect, the situation is somewhat different if we turn to the common property regimes (CPRs). First, these regimes seem less dynamic than those based on private property. Resources under common property are not tradable in the same way as those owned individually. This clearly restricts the potential for economic growth and changes in resource use, but it may not be altogether negative. The slower development may actually be important in fostering learning and the capacity to maintain the overall ecological functions involved – the resilience of the system. This depends on the quality of the property regime, though. A poorly working CPR may in practice deteriorate into open access, which may then spoil the resource.

With regard to distribution, CPRs seems to be built around mechanisms that counteract inequalities (Runge 1986; Oakersson 1992). CPRs are based on cooperation and, as Runge shows, increasing differences within a community make cooperation more difficult. There are therefore strong reasons to believe that a well-functioning commons depends on the maintenance of fairly equal distribution. Different drives towards inequality will have to be counterbalanced, otherwise it will create a potential for breaking down the CPR. Counterbalances may take the form of redistribution or various types of collateral safety nets. If these mechanisms fail, the status as a commons may be challenged, moving the regime from being cooperative to drifting into open access.

State ownership may materialize into very different types of development, depending on the wider political system of which it is a part. Some of the most unequal societies observed have been characterized by state ownership. This is not least the case for nepotism, which is similar to individual property for a single ruler. The experiences in Eastern Europe until 1990 also illustrate the problems concerning creativity and change in systems where state property dominates almost every sector of society.

On the other hand, state-owned firms might compete well in markets. Even more importantly, many societies prefer schools, health care, care for the elderly and other services to be publicly provisioned, mainly to secure equality with regard to both access and quality. Certainly, public goods in the sense of Figure 10.1 must be publicly provisioned in some sense, while still their production may be licensed to private firms.

10.3.2 Regimes and Opportunities for Cost Shifting

A property right is a social relation defining an exclusive right to a benefit stream. The focus is on access to resources. The effect on others by using the resource – that is, external effects or cost shifting – is a secondary issue. Property systems have different capabilities to handle such effects. The issue of cost shifting is a crucial one when we consider natural resources – that is, in a system of interrelated matter and energy flows. Open access is in a way 'institutionalized' cost shifting.[15] The effect of what one does – for example, emitting waste to a body of water or grazing the pasture – must just be accepted by the others. No rights exist, so everybody is free to do what they like and they must tolerate the consequences of what all others do.

However, the opportunity to shift costs onto others also exists under the other types of property regimes. Recall that in the case of environmental or natural resources, demarcations will never be complete. Rather, they will be grossly incomplete. Hence, costs will be shifted – unconsciously or consciously. Concerning the latter, Kapp (1971) suggested that if costs can be shifted, and done so without violating any previously established and enforceable rights, then this will occur under assumptions about individual rationality. The individual will gain from it.

The property regime not only defines what kind of cost shifting is allowed, but it also influences the costs of instituting and enforcing rules concerning which costs can be shifted and which cannot. Under private property, prohibition of harmful use may be formally defined more or less strongly. Even if formally prohibited, it will still not be in the interest of the individually rational property owner to make potential costs to others become transparent. In the case of environmental resources, there is in addition a long time span between, for example, when emissions take place and when damage to others can be observed. This makes cost shifting an easy and therefore very tempting strategy.[16]

Furthermore, who really harms whom? The debate within the EU about the right to use various additives in food and regulating the field of genetically modified organisms (GMOs) are both good illustrations of the complex relationships between competing interests here. On the one side we have the interests involved in setting up a regime advancing unconstrained competition and on the other we have the interests involved in avoiding (future) harm. In general there seems to be great conflict between the interests of giving priority to competition on the one hand and to installing rules like the precautionary principle on the other.

Again, in the case of common property, similar dynamics may appear between different commons or between the commons and other types of properties. The internal logic of such a property regime is still very different. Also in this case, each individual may be tempted to shift costs on to others. It is, however, a basic logic of the whole property regime to guard against such practices and an important reason for this type of institutional structure to be developed in the first place. While a private property regime assumes no physical interrelations between agents and will treat these as unforeseen surprises – 'externalities' – a well-functioning CPR is characterized by a set of rules for explicitly acknowledging and handling such interrelations. While a private property regime will primarily give emphasis to the ability to compete and not to the secondary effects of competition in the form of externalities, CPRs are constructed to regulate competition over resource use and therefore the external effect or cost shifting. These logical differences may set the two systems on very different trajectories. However, a CPR may also fall short due to, for example, a lack of awareness of long-run effects of certain accepted practices or a failure to establish and enforce rules. A CPR is by no means a simple panacea (Ostrom 1990; Agrawal and Gibson 2001).

Given the complexity and interrelationships of natural systems and the various types of time lags involved, a crucial issue for any regime is who has the burden of proof concerning external effects. One way of formulating this burden is to ask the person who is about to release some matter into the environment to prove that his/her act will not imply shifting costs onto others – either now or in the future. The alternative would be to make those who may have to carry future costs prove that damage will occur. Due to the difficulties involved in proving anything concerning biosphere dynamics – that is, the existence of radical uncertainty – the distribution of responsibility is crucial (Lemons 1998). The reasoning can be illustrated with a reference to the classical type I and type II errors in research.

A type I error is to claim that a relationship exists where there is none – that is, one accepts a false statement to be true. A type II error implies that an existing relationship is denied – that is, one accepts a false negative result. In a situation of great uncertainty it may be very difficult to prove anything with the necessary degree of confidence – that is, at the standard 95 or 99 per cent levels. The introduction of alien species may serve as an

BOX 10.1 PROPERTY REGIMES AND COMMON-POOL RESOURCES

A common-pool resource is characterized by rivalry in consumption and costly exclusion. In a situation with open access, the effort of individually rational agents will result in harvest levels where the resource rent is driven to zero. Consider the following figure for a biological resource such as a fish or a forest stock:

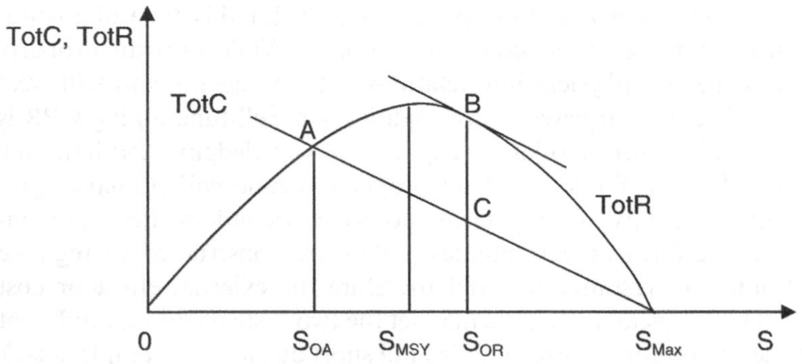

where TotC = total costs, TotR = total revenue (the bell-shaped function), S = stock, S_{Max} = maximum stock, S_{MSY} = stock at maximum sustainable yield, S_{OA} = stock given open access and S_{OR} = optimal stock given one of the other property regimes.

In a situation with open access (OA), the optimal strategy for each harvester is to adapt so that the marginal costs equal marginal revenue, without taking into account that the action influences the costs of others. This implies that in equilibrium TotC = TotR – that is, rents are driven to zero. This is point A in the figure. If a property regime is installed where the individual agent faces the effect of his/her actions on total costs, the equilibrium (new optimum) will be B and total rents will be BC. This can be obtained by ensuring that the resource is owned by one owner – that is private monopoly or state property – or by establishing a CPR regulating resource withdrawal to a level which keeps the stock at S_{OR}. (For those not familiar with the model, see any standard textbook in resource economics, for example, Hartwick and Olwiler 1998 or Perman et al. 1999).

As the figure is drawn, there is no problem related to securing the survival of the species involved, however. All points on the revenue curve are sustainable. This is very important, since the dissipation of resource rents is often equalized with the potential destruction of the resource. For this to happen, a threshold for the stock must exist below which the actual species is unable to regenerate and it may become extinct. This is illustrated by the figure below where S_t and S_{t+1} are the stock at time t and $t+1$, respectively. Below S_C it will deteriorate and finally become extinct since $S_{t+1} < S_t$.

In the case of fish stocks this kind of problem may be of particular importance, making the issue of choosing a property regime not only about not driving resource rents to zero, but also about avoiding extinction. The likelihood that extinction will occur also depends on the technology, the cost function. Going back to the first figure in this box, the level of the stock will be smaller, the lower the total cost function, increasing the chance of falling bellow S_C.

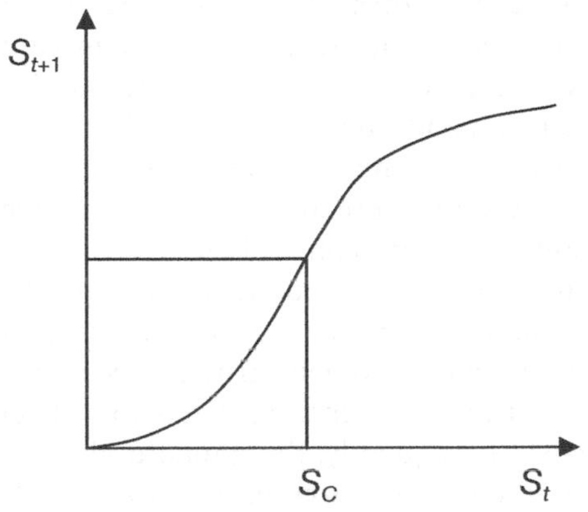

illustration. We have already experienced the negative effect of such introductions – for example, problems have been caused by the establishment of the zebra mussel in the Great Lakes of the United States and Canada. However, distinguishing *ex ante* between when it may and when it may not happen is almost impossible. This was clearly stated, for example, by the US delegation to the UN's International Maritime Organization (IMO), regarding the process of defining international rules for ballast water (Færøy 2003). Determining in advance whether a species carried in ballast water will be able to establish itself where it is released and subsequently take over important niches is practically impossible. Similarly, in the case of releasing insect-resistant genetically modified plants, it is practically impossible to say anything certain about the specific long-term effects on species or ecosystems dynamics (Strand 2001). Areas such as climate change, use or release of matter that is unfamiliar to a system and so on are all riddled with the same type of uncertainties.

There are two crucial observations to be made in relation to this. First, it is very often the case that it is equally difficult, often impossible, to both prove and disprove that there is a relationship between some act and a specific future status of environmental qualities. In economic language, information costs may be prohibitively high. We face radical uncertainty. This implies that whoever has the burden of proof – for example, the emitter or the potential victim – really determines the use of the resource and the final (but unknown) shifting of costs. If the emitter has the burden, s/he will often be unable to prove 'innocence'. Thus, no emissions will be termed legal. Similarly, if the burden of proof is with the victim, many releases cannot be prevented. The proof can be delivered only when the effects finally become visible. Even then it may be difficult to say what has really happened. Furthermore, at that time substantial costs are already carried by the victims and in many cases the process cannot be reversed. It is even possible that those responsible (firms or individuals) may no longer exist and cannot be made liable.

Second, while there is a symmetry concerning the problem of proof, there is still an imbalance concerning who in practice carries the burden. As there is a strong tendency in research to focus on avoiding type I errors, there is also a strong tendency in resource and trade regimes to put the burden of proof on the potential victim. There may be several reasons for this. There is first of all an asymmetry in the problem structure: economic activity comes first, the possible negative environmental effects come later. Moreover, the effects are potentials. Consequently, the standard idea that restricting someone should be based only on proven harm may easily be invoked, as is basic in, for example, criminal law. Finally, turning the burden of proof the other way around might be thought to restrict economic activity unduly. It is much easier to argue for the immediately observable losses of economic opportunity than to argue for uncertain or unspecified losses concerning the functioning of the environment and so on, even though these in the end may be vast, even critical.

One may only speculate, but behind the latter point might also lie a deep cultural reason. At least the 'western' perspective has long been that of controlling nature. Perhaps this is the very essence of this culture. Expansion is the aim, as built into the fundamental institutions of property rights and markets. Two issues then become central. First, given this perspective, ignorance is just something to overcome. Irreducible ignorance does not fit the basic perspective and thus to focus on type II errors is to restrain progress unjustifiably. Second, the idea of expansion and controlling nature tends to win just because of its character. It is the very essence of progress. Again we observe an asymmetry between the act of change and controlling nature coming first and the act of protection and observed inability to control coming later.

The burden of proof is crucial for environmental questions. While important, it should also be noted that none of the standard property regimes is adequate to address the problem sketched here. CPRs may have some specific advantages concerning the treatment of external effects as these simply get a stronger focus. However, the members of such regimes may also be as ignorant as others concerning the long-term consequences of their actions. In addition, CPRs are normally best at managing rather local common-pool resources. Concerning the regional and global common-pool resources, more complex regimes are needed, such as agreements between national states. Whether we call these agreements CPRs with states as members may be considered an issue of convention. Nevertheless, it is a regime type that has some distinct features that are very different from that of governing a local pasture.

International environmental treaties or regimes have been developing rapidly over the last years, reflecting the increasing interest in globalization of environmental issues. In parallel, there has also been a strong development of international trade regimes. The above-mentioned conflict concerning the burden of proof has become a core struggle between these types of regimes. We see it surfacing in discussions in the World Trade Organization (WTO) and the development of rules for trade on the one hand, and the UN Convention on Biological Diversity (CBD) with its precautionary principle on the other. It also permeates the climate debate both within and outside the International Panel on Climate Change (IPCC) (Spash 2002).

10.3.3 Regimes Shape Interests and Values

Institutions protect interests and values, and also reinforce and influence them. Since the basic ideas concerning these relationships have already been presented in Chapter 6, I shall only illustrate the main issues in our context by offering some examples. The values and norms a regime fosters depend not least on the relationship it establishes between the involved individuals. Following Bowles (1998), we can make the observation that while markets create conditions for anonymity and relations that are normally short-lived,[17] community allocations are more personal and durable. Bureaucracies are – in their ideal form – anonymous but durable. Figure 10.3 summarizes this in four ideal or principle types.

Bowles's distinctions do not fit fully with the regime definitions as used here. We have previously made a distinction between internal dynamics of a property regime and the wider institutional structures that property owners operate within. Bowles's distinctions cut somewhat across these levels. In view of the focus of this subsection, I shall use his structure since

	Anonymous	Personal
Short-lived	Ideal markets	Ascriptive markets
Durable	Bureaucracies	Communities

Source: Based on Bowles (1998).

Figure 10.3 Different structures of allocation rules

it is well suited to discussing the dynamics of interests and values development. However, there are three important qualifications. First, I shall simplify by assuming markets to be based on transactions between private property holders only. Second, concerning bureaucracies, I shall focus only on state bureaucracies. Certainly, private firms are bureaucracies too, but that aspect is of less importance here. Finally, I shall make CPRs a core example of community allocation. Bowles cites a fourth 'ideal' type – that of ascriptive markets – and gives racially segmented spot labour markets as an example. However, we shall ignore this type here.

We would expect that the relations that develop between individuals operating within these structures would normally be different. Markets simplify transactions by the making of commodities and establishing an anonymous setting. Through this, individual calculative rationality is fostered (Kagel and Roth 1995; Ostrom 2000). Communities – such as those established by CPRs – are different, giving more room for fostering group identity and reciprocity. State regulation may have a less uniform effect since the governance structures here are quite different across societies. The relationships between the state and the civil society – the community of communities more at large – seems to be of special importance.

In modern societies we observe a mix of these institutional structures striking a balance between simplifying transactions concerning 'mere commodities', securing rights and securing the maintenance of community values via a functioning civil society. When choosing between (resource) regimes, there are thus two important issues concerning the value aspects. First, we have the question of which regime is best for which resource or resource-use problem. In some cases, the market transactions regime with its commodity perspective is favourable since 'not much else' than individual utility is at stake. In other cases, such a regime is not found acceptable because what is at stake is considered a common good. Second, we have the issue of balancing the interests related to growth and expansion more in general against the need to build and maintain social coherence, against the need to secure environmental values.

While the first is fostered through competition and individuality, the second is crucially linked to a vibrant community. Regimes may have the capacity to develop fast, but they may undermine themselves by damaging the social fabric or the environment on which any regime in the end must rest. As Levi (1990) argues, a moral foundation is important not least for the cost of operating a regime – the transaction costs. If social and moral foundations are eroded, these costs may become insurmountable. If people take advantage of every opportunity to cheat others, the need for control will increase substantially. Thus, in the case of private property solutions, the role of the social fabric and the civil society is also fundamental. Here we actually face a situation where a system depends on something that is partly counter to its own logic. Markets foster individuality, but also demand that individuals restrain themselves and do not utilize all opportunities to break the rules. Similar reasoning could be made not least concerning state property. In this case the relationship to the civil society is as crucial, often determining whether state property functions well or the resources are appropriated by an elite.

In the following we shall look at three main relations or distinctions flowing from Figure 10.3: the distinctions between market and community allocation; the differences between state and community allocation; and the distinction between market and state allocation. Note the point made earlier that any kind of property regime might also be involved in market transactions. Here we shall restrict ourselves to market allocations under private property.

Market versus community allocation

Over a long period, the dominant trend in resource regimes has been a transformation of community allocations – like CPRs – into private property and market allocations. This has followed a reduction in the transaction costs of establishing and maintaining private property. However, it also seems to be an effect of cultural changes, of changes in values and power as emphasized in Chapter 7. Which of the two mechanisms has been most important is difficult to ascertain. Whatever the reason, a potential for rapidly increasing growth, new products and technologies has been created. The twentieth century developments illustrate this with great force.

However, the story is also one of problems or pitfalls following from underestimating the role of the existing institutions and the detrimental effects of destroying the existing values or 'social capital'. While better market access may be of great value to societies, a rapid transformation not taking into account the importance of existing institutions and the problems arising when these erode, have over and over again resulted in failures. The following quotation from Bowles (1998) is instructive concerning shifts in values and local coherence. Bowles refers to the work of Mallon (1983) on the growth of markets – especially labour markets – and the erosion of community institutions in central highland Peru:

> Central to the institutions of local solidarity among residents was the practice of contributing labour to road building, irrigation, and other communal projects: 'Community membership itself, and access to village resources was defined in terms of a quota of labour time that the household owed to the community as a whole.' With the extension of labour markets, many found employment in distant mines for extended periods of time, eventually converting the labour dues they owed to the community to cash payments. ... But 'migration, by commodifying relationships and separating them out from intricately woven fabric of local life, was changing the very context within which community could be defined' (Mallon 1983, pp. 264–5).
>
> Traditional institutions were further undermined by the sale of common lands (or charging fees for the use of the common lands) and the use of the proceeds to build schools and roads. Increased access of the richer peasants to distant markets for their produce freed them of dependence on the locality. The obligation to provide communal labour – or even money payments in their stead – thus became unenforceable, and the practice declined. (Bowles: 1998: 96)

Thus, the whole society with its values shifted as an effect of new institutions. Mainly, the role of reciprocity was reduced. As mentioned by Bowles, the ethnographic literature is full of examples of the environmental degradation of local commons due to similar processes. Certainly, the increased role of markets reduces the immediate dependence on local natural resources and local interaction over their uses diminishes. Consequently, the interests in and capacity to maintain well-functioning interactions is reduced (Baland and Platteau 1996).

However, one must be careful about equating common property with sustainable resource use. As Baland and Platteau show, traditional rural communities cannot, as an example, be viewed as 'inherently conservationist'. Population growth may create demands that go beyond the capacity of the resource, leading to degradation. Furthermore, people participating in such regimes will also face the problem of understanding the long-term consequences of their behaviour, and in situations with rapid ecological changes they may need external support to be able to take counteracting measures.

On the other hand, one should also acknowledge that local institutions tend to respond much faster to negative environmental feedbacks than, for example, centralized or socially detached ones (Folke et al. 1998). From this literature it is quite clear that it is the lack of understanding and not the lack of incentives to preserve – as is so often believed among 'collective choice theorists' – that is the fundamental problem in CPR management.

Baland and Platteau (1996) give numerous examples of negative effects of eroded CPRs and community structures for resource preservation. It is the erosion of the regimes, not the regime *per se*, that causes the problems.

CPRs may be especially vulnerable to rapid market integration and technological change. The case of reindeer pastoralism in northern Scandinavia is quite typical. The resource allocations of these societies were based mainly on conventions and norms. The lack, in this case, of a manifest 'third-party structure' implied low capacity to transform to stricter management rules as the process of market integration accelerated. Combined with access to new technologies, market integration caused the traditional rules concerning herd stabilization to erode. The result in this case was devastating overgrazing and strongly increased internal distrust (Riseth 2000).

In recent years it has been argued that privatization of natural resources is a necessity to establish better resource management. This has been the view of the World Bank and the International Monetary Fund, where a shift from common property to private ownership of resources has often been demanded as a condition for supporting developing countries with credits. This concerns agricultural land, forests, water resources and so on. The idea has partly been to 'mobilize' resources for development – for example, making it easier to use the resources in other ways and/or obtain financial credit. However, it has also been couched in the language of resource conservation, effectively turning what was believed to be open access into private property. The existence of vibrant CPRs has simply been overlooked due to a failure to understand their existence and dynamics. This process has deprived many rural poor of access to important resources (Goldman 1998; Platteau 2000).

Platteau made a comprehensive study of the privatization of land in Sub-Saharan Africa, on which I shall make two observations. First, Platteau accentuates that formalizing land rights has not had the expected effect of making the economy more dynamic. Use rights accompanied by a few fundamental transfer rights seem to grant sufficient security to induce farmers to invest. These are typically well guaranteed already by village communities and existing CPR structures. Second, land titling may itself produce increased insecurity. Community allocation offers a flexibility that is of great importance, given the overall conditions of the area. As an example, when people wish to return to their village due to the exhaustion of external job opportunities, this is much easier under a community system. Privatization reduces that possibility substantially. Thus, as Bowles (1998) and Agrawal (2002) also emphasize, the wider context is of great importance for how various allocation mechanisms work. This refers both to the institutional context more generally and to the kind of resources involved.

Some of the conservation problems that have been observed do not follow from the type of regime *per se*, but from how specific elements of the rights structure are formulated. In the case of private property, the durability of ownership rights plays a role. Short-term private resource-use contracts – as in the case of forest logging in, for example, Indonesia – seem to have resulted in substantial conservation problems. The motivation structure will certainly be different in the case of long-term private ownership, where maintaining the future ability to produce timber also becomes an issue for the owner to consider. None the less, if demands on returns are high, or other uses like agriculture or housing give larger profits, long-term ownership will not in itself secure forest protection. The question of who the private owner is may also influence uses. Local ownership seems to be more positive for valuing long-term investments in natural resources than distant ownership. Furthermore, it seems as if community membership and inheritance rules are also important.

Shifting from open access to private property, as in the case of catch quotas in fisheries, may seem to be a rational response to the dangers of overfishing. However, even in this case one must be cautious. Jentoft (2004), who has studied the development of fishing communities over many years, suggests that 'healthy communities and vibrant civil institutions are essential to ensure the norms, values and knowledge that promote the ethical and moral consciousness required by sustainable fishing practices' (p. 146). Moreover, the change in the regulations, moving to private rights in fishing stocks – for example, tradable quotas – is changing the relationships between fishermen and altering the interactions within the societies to which they belong (Pálsson and Pétursdóttir 1997). Certainly, the overfishing becoming increasingly visible in the 1960s and 1970s demanded changes in existing institutions. However, moving from open access in high-seas fisheries, via the establishment of exclusive economic zones (EEZs) in the 1970s, to introducing private rights (licences) in catch quotas, has not solved the problems. Partly this may be due to the control problems involved (unreported catches, the 'dumping' of small fish and so on).[18] Partly it may be due to continuous fights for larger quotas to cover the costs of investments in an excessively large fishing fleet. These problems, however, are dependent on the legitimacy of the regulations within the fishing communities themselves. Jentoft (2004) emphasizes that compliance with the rules depends on viable societies and reciprocity. This may be breaking down in a globalized fisheries business, threatening to erode the very basis for any effective regulation.

State versus community allocation

Moving to the state–community distinction, we can observe similar problems. A transition to state property may also induce a breakdown of local institutions and values. Typically such transitions imply distant management of resources, often lacking the local knowledge that is so important to good management. For example, in many African societies, it is observed that when areas become protected under state governance, local people lose rights of access to these resources. Not only does this alienation lead to 'continued but now illegal use', but also that the previous long-term local management approaches are turned into a 'grab what you can get strategy' (see Vedeld 2002; Jankulovska et al. 2003). Consequently, the protection efforts that motivated establishment of state national parks may even result in less protection.

Ostrom (2000) similarly suggests that internal rules or norms of a commons seem to function better than (formalized) external rules. The latter do not carry the same legitimacy and may not fit the local perception of what is fair. We thus have many examples of functioning CPRs breaking down because of the imposition of external rules, but also due to the failure of state authorities to recognize the right of the members of a commons to organize themselves.

As a result of these problems, there is a trend towards establishing/re-establishing community management for many common-pool resources (Platteau 2000). The experiences from this process have also been rather mixed. Establishing local CPRs from above – that is, by state action – is a challenging endeavour (Agrawal and Gibson 2001). The crucial point is the way in which the local community gets involved. Ostrom (2001: ix) comments:

> Even if legislation or policy boasts a 'participatory' or 'community' label, it is rare that individuals from the community have had any say at all in the policy. Further, many of these centrally imposed 'community' programs are based on a naïve view of community. It is unlikely that any policy based on such views has a chance to produce more than a few minor successes.

When discussing the issue of state property versus CPRs, one should therefore be aware of the precise relationship between the state and the civil society. The lack of integration here may be a serious problem in many societies, while the distinction between what is state and what is civil society almost vanishes in other cases. While the integration of the state in the civil society seems fairly comprehensive in Scandinavia, for example, it is less so in Southern Europe and very weak in many developing countries. This has a variety of historical reasons – not least including colonial rule – but should not be overlooked when understanding the differences and similarities between state/public property regimes and community-based allocations.

State versus market

Moving finally to the state–market or state–private distinction, the lively debate around the effects of neo-liberalism and the shift from public to private provision of, for example, education, transport and health care is a core issue in understanding the differences. Since the 1980s there has been a strong move towards the privatization of such 'public goods', which still continues. We shall look briefly at the experiences to see what they may say about where the line should be drawn between public and private provision of goods and services.

The official aims of privatization have been to reduce costs and produce better quality – to make the production of these goods more efficient in economic terms. The recent debate over these issues concerns on the one hand whether the provision has become cheaper, the quality has become better, and on the other, the distributional effects. The debate illustrates, however, that the change is also about what values are involved or should be supported.

A recent Canadian study (Romanow 2002) concludes that Canadians regard health care to be a moral and not a business issue. While Romanow concludes that in this case ongoing privatization has neither reduced costs, nor increased quality, the main point is the reaction from the population concerning what health care is. It is not viewed as a commodity. Similar reactions have been observed in other countries.

In New Zealand, the privatization movement has been among the strongest. However, the experiences have been somewhat mixed, because the social dimensions involved in public provisioning, and the shifts in motivation that the system change itself induces, have been underestimated. Because of the negative experiences, the Swedish minister of social affairs and the New Zealand minister of health hosted a conference in Sweden in the winter of 2003 to discuss their experiences. They delivered some strong messages concerning the failures involved in abolishing public financing and making health a commodity (Enquist and King 2003). They pointed to the fact that New Zealand had therefore reversed the process, establishing an 'intense reform to reinstate *common property principles* within health care' (ibid.: 5, author's translation, added emphasis).

The reason for the fact that private solutions are often neither cheaper nor better than public bodies in the supply of 'public goods', is not least related to the costs of contracting and controlling. While public supply is based on internal command, private supply of public goods such as welfare services for old people, waste treatment and primary education is based on contracts between firms and public bodies. This latter structure may in many situations be very resource demanding as it also creates the need to control the quality and quantity of what is finally delivered. Here one should acknowledge the difference in motivation structures. While private businesses are predominantly geared towards maximizing profits, the functioning of public bodies is much more dependent on direct command and on professional working codes. The more complex the good, the more difficult it is to do the contracting, and the less favourable this solution is compared to public delivery where professional codes (institutions) define what should be produced.

Certainly, there is always a danger that public systems may deteriorate with respect to both quality and efficiency. While the profit-seeking private firm – as long as it believes it will not be caught – is motivated to deliver less than what is contracted, the problem with public delivery is the danger that lack of competitions will make it stagnate or become overly involved in internal processes. The chance of this happening depends not least on the internal culture of the public body and on the relationships with the surrounding civil society. If the latter is functioning well, the very existence of this contact exerts control pressures on the public suppliers. The more vibrant the civil society is, the more complex and genuine the good is, and finally the more important the good is in social terms, the lower is the chance that private will be better than public provisioning.

10.4 ANALYSING RESOURCE REGIMES

Studying the effect of human action on the natural resource base involves not only studying the direct relationships between the two. From the above we can conclude that it also involves studying the institutional structures under which choices are made. If resources are (unintendedly) depleted, the problem is foremost that the regime does not fit well to the characteristics of the resource involved and/or the values of the societies involved. Figure 10.4 is useful in the analysis of resource management issues or problems.

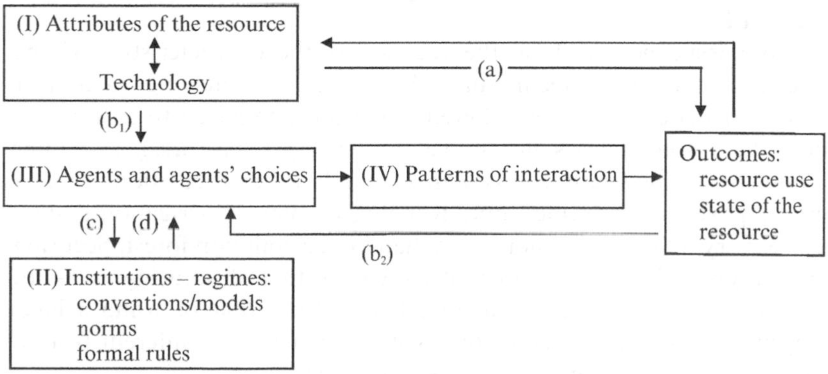

Figure 10.4 Framework for analysing resource-use problems

The model is inspired by the work of Ostrom (1990), Oakersson (1992) and Ostrom et al. (1994).[19] Resource use and the state of a resource depend – according to the model – on four factors. First we have the attributes of the resource and the technology available for its utilization (I). These characteristics define whether a coordination problem exists. If the resource is vast compared to the capacity people have to utilize it, there may be no need to regulate access. The situation is characterized by non-rivalry. In other situations, such regulations are crucial for a good result. As emphasized in the figure, resource characteristics influence outcomes both directly (a), and via the way they influence agents' choices (b_1 and b_2). As an example, the state of a pasture will depend both on the capacity of the pasture and how many animals are brought there. Finally, these aspects may influence which institutions are chosen (c) and the motivations they create (d).

The institutions or the regime (II), consist (normally) of all levels previously defined – that is, conventions, norms and formal rules. However, some regimes may lack a system involving third-party control, implying that regulations are built into the conventions and norms of the regime. Typically, many 'traditional' societies have lacked a type of formal power, making them vulnerable to changes in resource use following from (rapid) market integration and technological change (Baland and Platteau 1996; Riseth 2000).

With regard to the institutions, we may also distinguish between operational rules, rules concerning the defining of operational rules and finally external arrangements/rules. The operational rules define the everyday regulations of the regime. The rules defining how to decide upon operational rules can be thought of as the constitutional rules of the regime. Finally, the external arrangements may concern international 'law' in the case of states and national laws and regulations in the case of, for example, a local CPR.

A core issue here is how the regime fits the characteristics of the resource and involved technology. A regime for a moving stock (for example, fisheries) demands different solutions compared to a situation where the resource does not move (for example, pastures). A winter pasture-constrained reindeer system demands other regulations than a summer pasture-constrained one. Running a common irrigation system involves very different decisions from handling a common forest. Securing access to groundwater is a different issue from that of securing access to water from a river, which is again different from that of using a lake. A regime for global CO_2 emissions will face issues very different from a regime for local emissions of nitrous oxides (NO_x).

To clarify the point, a system for rights in river water must take into account the fact that some are situated upstream and others are downstream. This is not an issue in the case of a lake, although reduced water levels may have different consequences for the various residents or property holders who live along its shore. The issues involved here are not only about private versus common versus state property. They are also about the more detailed elements of each type of regime and the way these may be linked.

Agents and agents' choices (III) are the next factor to consider. Here the motivations of the agents play a crucial role. As emphasized, these are heavily influenced by the institutional structures themselves. Certainly, the opportunities given by technology and the characteristics of the resource also influence choice. In relation to this, the previous discussions about incomplete demarcations (the difference between formal and real property boundaries), options for cost shifting, costs of control and the various types of motivations – interests and values – instituted by the regime in place are all elements that play a significant role. A system may fit the dynamics of the resource well, but may still be undermined if the involved agents are motivated to break the rules and so on. This is a question about the general legitimacy of the regime and social coherence of the group involved.

Finally, problems appear as a consequence of the interaction of choices made by several agents (IV). They do so because of difficulties concerning the combined dynamics of the other three factors. Typically, the regime is not able to motivate coordinated action in accordance with what is demanded given resource characteristics, technology and the number of agents. This may be because a coherent regime has not yet been established. Regimes may, however, also fail because the dynamics of the resource and/or the understanding of agents' motivations are wrongly interpreted. If agents tend to act in a reciprocal fashion, rules based on strategic action will fail and vice versa.

If outcomes are not in accordance with what is expected or wanted, agents can change institutional structures. In principle this may occur at all levels – with regard to conventions, norms and formal rules. However, changing conventions and norms normally demands much time. Changing formal rules such as those concerning access, maintenance and withdrawal is easier. However, the capacity to make such changes depends not least on existing norms and social coherence. This is exactly why there has been so much concern about the dangers involved in destroying the social capital of any regime or society. It also destroys the capacity to change the institutions when necessary.

Box 10.2 offers some observations concerning the process of transforming fisheries from an originally open access regime to a regime based on state

BOX 10.2 REGIMES IN OPEN-SEA FISHERIES

Fish stocks have a complex ecology involving migration over large distances of open sea for many species. Historically, sea fishing was done near the coasts using simple technology and small boats. While the resource was dominantly under open access, still influenced by local community norms and values, it was nevertheless vast compared to the fishing capacity. As the technology changed, problems of overfishing and conflicts between different fishermen and different countries engaged in fishing surfaced.

The history of open-sea fisheries regulation dates back over 100 years. As far back as 1893, the lowering of total catches was discussed in England. The first international fishing conference was held in Stockholm in 1899. Already at that time local fishing communities had developed local regimes to regulate the fishing. However, this development was little known outside these groups of people (McGoodwin 1990). It is only after 1950 that more regional and international regimes appear. New technology was developed, and this increased the pressure on the stocks, resulting in increased conflicts within the industry/between countries (Stokke 2001). The 'cod war' between the EU/Great Britain and Iceland in the early 1970s, and a similar conflict between the EU and Canada in the late 1980s are typical examples.

The first step in the process was a unilateral declaration by states to enlarge their exclusive jurisdiction over resources in their nearby coastal areas. The United States did this in 1945 (Nawaz 1980). In 1952, Chile, Peru and Ecuador signed the 'Santiago Declaration' in which these countries defined exclusive sovereignty in the nearby oceans as 200 sea miles from the coast (ibid.). Other coastal states such as Iceland, Norway and Russia, later did likewise. This in itself gave rise to conflicts. In 1958 the Geneva Convention on Fishing and Conservation of Living Resources of the High Seas was established to balance the interests of the 'long-distance fishers' and local coastal nations. According to Boyle (2001) this was the least successful of all the Geneva conventions.

The demand among coastal states for establishing internationally accepted EEZs gained increased force. In 1982, after a long process of UN conferences, the UN Convention on the Law of the Sea (UNCLOS) was finally drafted (UN 1982). It was ultimately signed by 119 nations, but first in 1994 it was signed by a sufficient number of countries to become functioning (UN 2003).

It granted EEZs to coastal nations extending 200 sea miles from the coast – that is, it established state property. The establishment of EEZs did not solve the problems of regulating open-sea fisheries or stocks straddling between zones. The management of these continued to be unresolved, while UNCLOS set out responsibilities for countries that fished for stocks appearing both within and beyond the EEZs (Stokke 2001). Nevertheless, some agreements have been signed, such as the Icelandic–Norwegian–Russian agreement concerning the so-called 'Loophole' – an area outside the EEZs of the three countries southeast of Svalbard (ibid.).

Despite these regulations, there are considerable problems regarding the capacity to regulate the fisheries. A clear example of mismanagement of sea fisheries was the collapse of the Canadian cod fisheries. The conflict over harvesting and regulations was very typical in that fishing fleets from different countries – not least Canada and EU – were competing and there were also conflicts between the local fleet and offshore industrial trawlers with large catching capacity. A maximum catch was registered in 1968, but the numbers subsequently declined. Organizations were established to try to regulate the conflict of the North Atlantic fisheries – first the International Committee for the Northwest Atlantic Fisheries (ICNAF), later the Northwest Atlantic Fisheries Organization (NAFO) – but the agreements set up were too weak to avoid continuous overfishing. NAFO demanded a halt to fishing in 1986, but the EU did not follow this up (Gezelius 1996). The development also revealed a conflict between local fishermen along the coast of Newfoundland and the Canadian Atlantic Fisheries Scientific Advisory Committee (CAFSAC). The fishermen, observing drastic reductions in catches, argued that recommended quotas by CAFSAC were far too large. A debate ensued on the models used to estimate the stocks (Finlayson and McCay 2000). As a result of the collapse in the industry, the cod fishery was closed in 1992 and has not been reopened. The breakdown occurred just a few years after CAFSAC gave a positive evaluation of the development. According to Gezelius (1996), the Canadian authorities attempted to focus on the responsibility of the EU fishing industry to avoid focusing on its own role. He explains the collapse as due to problems of cooperation and overestimation of the stocks, rather than a lack of political will. The case is quite typical for a situation with weak institutions, increasing catch capacity and complex conflicts of interest.

property to areas of the sea (EEZs) combined with private permits or licences to catch quotas. This sector has been through a comprehensive institutional change over the last 50 years. Still, there are many unresolved issues. Partly, these relate to the fact that not all open-sea fisheries are yet regulated. Partly, they relate to the problems of agreeing on total allowable catches and the ways these should be distributed between various interests. While the box focuses on open-sea fisheries, inland fisheries and fisheries of local stocks even at sea have long been subject to various types of mainly common property regimes.

10.5 RESOURCE REGIMES AT THE INTERNATIONAL LEVEL

We live in an era where international agreements, conventions and protocols are being developed for many resources. This is a natural and necessary response to the level of globalization reached with its subsequent effects on the quality of regional and global common-pool resources. Regimes thus now exist at all levels from governing the use of individual private plots of land, small local commons – for example, the rules governing access to a fish stock in a lake – to regimes at the international level – for example, EEZs.

However, there are some distinct differences between local/national and international regimes. Not only are the size and the number of agents involved different. The most prominent distinction concerns the basic authority structure. While local resources are dominantly placed within the confines of one state offering the opportunity to anchor the regime in that authority, the international regimes are by definition without such a superstructure. International regimes are agreements between states and depend on their individual approval. This creates two problems. First, there is no common normative authority to either make a solid foundation for establishing the rules or ensure that any such rules are followed. We cannot talk of international 'law' or internationally sanctioned property rights in this sense. Second, there is always a great danger that too few countries approve of a certain convention, making it weak or challenging its very existence.

The first problem is fundamental, implying that the process of regime building is in danger of being dominated by purely strategic reasoning on behalf of the involved parties/countries. Countries will not sign if the convention or treaty does not offer them net gains. This implies a systematic tendency by those dominantly causing a global environmental problem not to ratify an agreement, since they will often be worse off after a regulation than before.

However, Young (2002) emphasizes that even under these circumstances, non-strategic action is observed and there appears to be a common understanding and a common set of basic norms, which many states feel obligated to follow. He also points to the fact that states are complex agents, and that conflicts within a state concerning what is best to do, may influence its acceptance of international regulations. It is rare for citizens of a state to either all lose or all gain from such rules.

Furthermore, when a treaty is signed, the parties may treat it strategically – that is, not comply if the control system is weak. And again, due to the lack of an international policing and court system, both the possibilities and the temptations are great. Young (ibid.) again comments that the situation is not that simple. Signing a treaty may in itself evoke the view that it should then be followed up. Total anarchy does not ensue. He instead reminds us:

> [A]ctors commonly adhere to the rules of regimes as a matter of habit or because such behaviour is taken for granted as a result of socialization or routinization. Even more fundamental in these terms are consequences of discourses and role definitions. States that respect each other's EEZs generally do so because they tend to comply with international law as a matter of course and because they accept the proposition that coastal states possess jurisdiction over adjacent marine areas. (Ibid., pp. 34–5)

Over the last 30 years, many treaties that regulate resource uses have been established. We have already mentioned the EEZs in relation to the fisheries, which were created during the 1970s and formalized in the UNCLOS agreement (UN 1982; see also Box 10.2). UNCLOS not only covers resources in the sea, but also grants the relevant states rights to the resources on and underneath the sea floor.

Other important examples of international regimes consider biodiversity, ozone layer depletion, acid rain and the emissions of greenhouse gases. Regulations concerning acid rain have the longest history of these, as it is regulated mainly via regional treaties – for example, the UN Geneva Convention on Long-range Transboundary Air Pollution, which was the first internationally binding instrument to deal with air pollution on a broad regional basis.

With regard to biodiversity protection, the basic treaty – the UN Convention on Biological Diversity – was established in 1992. It is a so-called 'frame work convention', to be followed up by more specified and operative conventions in specific fields. The Cartagena Protocol on Biosafety was established as such in 2000 and came into force in 2003 after it had been ratified by 50 countries. More information on these agreements is given in Box 10.3.

BOX 10.3 THE INTERNATIONAL REGIME FOR PROTECTION AND USE OF BIODIVERSITY

The development of multilateral agreements focused on the protection of biodiversity has a history dating back to the 1940s – that is, the International Convention on the Regulation of Whaling from 1946. From the 1970s and onwards under the lead not least of the UN, there have been more comprehensive developments. The Ramsar Convention on Wetlands of International Importance was established in 1971. In 1973, after the UN Conference on the Human Environment in Stockholm in 1972, the UN Convention on International Trade in Endangered Species (CITES) was established, and the Bonn Convention on the Conservation of Migratory Species was agreed upon in 1979 (Brack 2002).

Under the Rio summit in 1992, the more general UN Convention on Biological Diversity was established (UN 1992a), and by late 1993 enough countries had signed to make it operative. The CBD is based on the precautionary principle. The underlying aim is to balance the opposing interests of economic utilization of biodiversity/genes and the need for protection. In many ways this was a North–South conflict. In accordance with this, the CBD focuses on conserving biodiversity, sustainable use of its components, and sharing the benefits from, for example, commercial use of genetic resources. Furthermore, it is a typical 'frame work convention', implying that more specific treaties or conventions will be developed under the CBD – for example, the International Tropical Timber Agreement (UN 1994) and the Cartagena Protocol on Biosafety (UN 2000).

The development has demonstrated a rather deep rift between the ideas of the CBD and those governing international trade – that is, the WTO agreement. This became evident not least in the process behind the advancement of, on the one hand, the WTO Agreement on Trade-Related Aspects of Intellectual Property Rights (TRIPS), and on the other, the Cartagena Protocol under the CBD. Three issues have been addressed: the right to patent biological material, the right to trade GMOs, and the distribution of the income from the use of local species/genes in the GMO industry. These issues are interlinked. The TRIPS agreement emphasizes the importance of securing free trade and the right to use patents in the case of intellectual property – including modified genes and organisms. Certain compromises have already been made in the TRIPS agreement, implying that while micro-organisms are patentable, plants and animals of a higher order are not (WTO 1994). In 1999, a process of revising the agreement was started. In this process, the US claimed that the exceptions from patentability in the current agreement should be removed. They were unnecessary and the US already treated plants and animals in the same way as micro-organisms. Many developing countries – such as India and many African countries – favoured a more restrictive policy (WTO 1999). As a result of these conflicts, the TRIPS revision is still not completed (WTO 2003).

In the development of the Cartagena Protocol the same disagreements became apparent. Although the US has not signed the CBD, it was still allowed to participate in the process. The main conflicts appeared between the so-called 'Miami countries' (Argentina, Australia, Canada, Chile, Uruguay and the United States) and 'The Like-minded Group' consisting of many developing countries. The EU took a middle position. The Miami countries wanted a solution that gave preference to the WTO rules. The developing countries wanted to give the Cartagena Protocol precedence over the WTO agreement. The result was a protocol that granted more rights to regulate imports than the WTO rules did in the specific field of biological material (Melchior 2001), and the precautionary principle is granted a basic status in the protocol (UN 2000). However, the protocol cannot be interpreted as changing a party's rights and duties in other areas of international law. The field is still characterized by competing rules and agreements.

The emission of ozone-depleting matter was regulated on the basis of the Montreal Protocol from 1987. It was mainly directed towards phasing out CFCs which were used in cooling systems. These emissions threatened to destroy the ozone layer which would lead to a strong increase in the amount of ultra-violet radiation reaching the earth's surface.

With regard to greenhouse gas emissions, the Convention on Climate Change was established in 1992. This convention instituted a process that led to the more operative Kyoto Protocol (UN 1997), which formulates emission reduction requirements for each party to the protocol and a time limit – 2012 – for reaching the goals. While the treaty is modest in its requirements, it was only in the autumn of 2004 that enough countries ratified it. According to the treaty, countries that together account for 55 per cent of total emissions must sign before it becomes operative. This threshold was passed when Russia finally ratified it in November 2004. See Box 10.4 for further details.

BOX 10.4 THE KYOTO PROTOCOL ON CLIMATE GAS EMISSIONS

The greenhouse effect of atmospheric CO_2 was identified as early as 1859 by John Tyndall. Almost 40 years later, Svante Arrhenius argued that the ice ages of the past could be explained by variations in the amount of CO_2 in the atmosphere. He also noted that the growing use of fossil fuels could lead to a doubling of CO_2 in the atmosphere and would cause an increase in temperature of about 5–6°C. He estimated that such a doubling would occur in the 4th or 5th millennium.

It was some time before the issue surfaced on the political agenda, but after the Second World War the use of fossil fuels increased considerably and it was feared that a doubling might not be that far away. Throughout the 1970s and 1980s, interest in the issue increased further and the IPCC was established in 1988 under the World Meteorological Organization (WMO) and the United Nations Environment Programme (UNEP). It has since delivered three main reports based on comprehensive model analysis, concluding that a doubling of the amount of CO_2 in the atmosphere is likely by the end of the twenty-first century. These assessments have encouraged a political process under the UN to try to find ways of avoiding such a rise. At the Rio conference in 1992, the UN's Framework Convention on Climate Change was signed. It emphasized that the industrialized countries had a particular obligation to reduce emissions. It was suggested that these should be stabilized at the 1990 level by the year 2000 (Cicero 2003).

At the Rio summit the so-called Conference of the Parties (COP) was instituted. The first COP was held in Berlin in 1995. At COP3 – in Kyoto in 1997 – the Kyoto Protocol was agreed, establishing a plan for reducing the emissions of greenhouse gases to a level about 5 per cent lower than that of 1990 by 2008–12. To be operative, the protocol had to be ratified by at least 55 countries, covering at least 55 per cent of the total emissions of the so-called 'Annex B' countries – that is, industrial countries plus Eastern Europe. The reductions were not equally distributed among the Annex B countries. While the EU and many Eastern European countries were asked to reduce their emissions by 8 per cent, the US quota implied a reduction of 7 per cent. The level for Russia was zero and Australia was allowed to increase its emissions by 8 per cent (UN 1997). There was an intense debate over the principles for allocating the reductions (Torvanger 1999). The protocol also included three so-called 'flexibility mechanisms' to reduce costs; it opened up for trading emission quotas between the Annex B countries; it instituted joint implementation (JI) between the same countries – that is, one Annex B country can pay for measures in another; and, the clean development mechanism (CDM) was instituted, enabling Annex B countries to invest in measures in developing countries.

The idea of JI was first promulgated as a way to invest in cheaper measures – not least by industrial countries paying for measures in developing countries. According to Grubb et al. (1999), the developing countries reacted negatively: 'Why should African governments let their land be used as a toilet for absorbing emissions from Americans' second car' (p. 99). The final solution was that the JI became an option between Annex B countries only, while the idea of the CDM was developed as a different way of bringing in the developed countries.

The Kyoto process was difficult, but that which followed has been even more so. At COP4 (Buenos Aires, 1998) the US demanded meaningful participation also by core developing countries. There was also little progress at COP5 in Bonn. At COP6 in The Hague in 2000 an intense debate developed between the US and the EU on the use of natural 'sinks' such as forest sequestration. In 2001 the US finally withdrew after the Bush administration came to office. Then, at COP7 in Marrakesh in 2001, a detailed set of rules was agreed on by the states attending. This agreement was weaker than the initial Kyoto Protocol in two ways: since the US did not join, a substantial part of the original reductions was unobtainable; and the rules themselves were made less stringent, a compromise to ensure that countries such as Japan, Russia and Canada were willing to ratify the agreement (Cicero 2003).

In late 2004 Russia finally ratified the agreement, which means that the threshold of 55 per cent of emissions has been reached, and the protocol is thus operative from February 2005. Following the collapse of the Soviet Union, the emissions from Russia lessened and are already much lower now than they were in 1990. Thus Russia has 'fulfilled' its obligations. There may be many reasons why Russia has hesitated to ratify. For example, ratifying may reduce the opportunities for economic growth. Many observers have, however, interpreted Russia's behaviour as part of a strategic positioning (Tjernshaugen 2003), reflecting its attitude towards countries that want to buy emission quotas from Russia.

The effect of the protocol is uncertain. A 5 per cent cut is very low compared to what seems needed to avoid substantially increased temperatures and changed weather patterns throughout the twenty-first century. The Russian ratification of the Kyoto agreement brought some new optimism into the process. Still, to be able to achieve significant reductions the US must agree to share responsibility, and solutions must be found that encourage developing countries to participate more actively.

The United States has stated that it will not ratify the Kyoto Protocol. It has not signed the CBD, either, although it is pressing the hardest to develop a strict international system for regulating the use of ballast water to reduce the introduction of alien species – under the IMO (Færøy 2003). This apparent ambiguity seems to stem from the fact that the United States has different interests in different sectors. On the one hand it is an oil-dependent economy with a very expansive biotechnology industry. On the other hand it has experienced severe costs in connection with the establishment of alien species in its waters following the release of ballast water – for example, the zebra mussel. Hence, different national interests seem to dominate the field, creating positions that seem rather inconsistent if only the environmental aspects are considered.

This seems to be an example of strategic positioning, but several countries that have signed the Kyoto Protocol, for example, will experience losses as a result – at least in a situation where only the minimum signing level (55 per cent) is obtained. However, they have accepted the rulings of the protocol. It is evident that the will to find solutions varies both between countries and within subject areas.

Except for the acid rain regulations, all the above-mentioned treaties and protocols are instituted under the UN system. The process has been running through a set of major meetings or 'summits' – for example, the Stockholm conference of 1972, the Rio summit of 1992 and finally the Johannesburg summit ten years later. Typically, international treaties grant the national states the right to define proper action within the requirements made by the parties to the signed treaty. Thus, while the Kyoto Protocol defines reduction requirements, it is up to the different countries to choose their own measures. However, the protocol defines certain options concerning measures involving more than one country, for example, emission quota trading and joint implementation.

The different protocols and conventions are placed differently within the UN system. The IPCC is put directly under the General Assembly, while the CBD is under UNEP. Young (2002) argues that this reflects how the UN system wants to link the specific issue to other questions. The IPCC seems to be placed under the General Assembly to accentuate the North–South dimension.

Preceding, but also parallel to the development of international resource regimes is the development of international trade regimes. The first institutional structure developed here was the General Agreement on Tariffs and Trade (GATT), established in 1947. The aim was to create rules for international trade and reduce trade barriers. In 1994 the eighth round of GATT talks led to the establishment of the World Trade Organization. This development has strengthened the process for establishing increased free trade.

Over the years, several conflicts between the GATT/WTO goals and the aims of the various UN conventions have emerged. These have surfaced in two different ways. First, we observe conflicts over the formulation of environmental protection conventions, since their content may be perceived to restrict institutional developments fostering free trade. The process to establish the Cartagena Protocol (see Box 10.3) is instructive in that respect. Second, we have cases where countries have been taken through the WTO court system because some other countries claim that import restrictions have been established based on arguments concerning environmental protection that are not in accordance with those accepted by the WTO/SPS agreement on sanitary and phytosanitary measures (Shaffer 2001). The ongoing EU–US conflict on the import of GMO/GMO products is a typical example.

The first type of cases is about the standing of different international treaties. If there is a conflict, which should take precedence – for example, the UN treaties for environmental protection or the WTO rules regulating international trade. The second is about the right of individual countries to define rules concerning the protection of their environment. These issues are certainly linked. The arguments made by the claimants are dominantly that countries use the environment as an excuse to support national industries. The counterargument is that the problem is real and that the specific trade threatens national resources. The dispute between the EU and the United States over import rules for GMOs is again quite typical.

The conflicts we see here are among the most fundamental issues of our time. They link back to the burden of proof issue, and relate fundamentally to the rights of individual countries to protect their environment. They relate equally vitally to the international institutions developed to counter rent seeking in world commodity markets. The latter implies avoiding trade restrictions merely to favour national industries – that is, restrictions that are not based on the need to defend environmental qualities or other types of public goods. The dilemmas raised here have no simple solution. The problem is to define what is rent seeking and what is not, given that effects on the environment are often difficult to prove with certainty. This is why the UN has so strongly advocated the precautionary principle, potentially challenging the establishment of free trade. If the rule that 'no harm' is to be proven before a certain trade is accepted, then the expansion of markets and the quality of the environment would be very different from applying the rule that 'harm' has to be proven before regulations are allowed. The difference between the precautionary principle/avoiding type II errors and liberal trade rules/avoidance of type I errors is substantial, not least in our age of fast technological development.

10.6 SUMMARY

A right is a socially defined relation. It defines what people may or may not do to each other. A property right is more specifically a right for the property holder to claim support by the collective if non-owners do not comply with their duties. A triadic relationship exists – that between the owner, the non-owners and the collective authority granting the right. While owners have different rights to benefit streams, it often also follows that harmful use – that is, shifting costs to others – is prohibited.

We have distinguished between property and resource regimes, where the latter covers both the property structures and the rules governing transactions of the products resulting from using the property. Property regimes are divided into four types: private property, common property, state property and open access. We have observed that a clear distinction is lacking in much of the literature between common property and open access – between property in common with defined rights and duties and no property. The often alleged 'tragedy of the commons' is really a 'tragedy of open access'.

While this distinction is frequently lacking, the difference between the other three types may on the other hand be exaggerated. It is demonstrated here that all of them cover co-ownership structures. Private property is normally thought of as property for single owners. However, in modern economies, the growth of corporate property in different forms has changed this dramatically. What distinguishes the three is more the kind of co-ownership that is involved.

While the co-owners of a private firm own a part or share which they can sell, they are as owners still engaged in one collective activity – the running of the firm. Common property is also private property for the co-owners. It is, however, instituted to govern a common resource, and the selling of parts is thus not an option. Furthermore, the CPR is not directly involved in the outcome of individual participation. The regime focuses on the use and maintenance of the common resource, not the individual results flowing from that use. Rules concerning equal access among owners seem important. State property falls in between these categories: in some situations, such as the production of public goods, it resembles common property; in the form of a public enterprise it is more like a private firm.

We have emphasized two main sets of reasons behind the choice of regimes for the governing of a resource: the costs of running the regime – the transaction costs; and the issue of which interests and values it supports. With regard to transaction costs, we have observed that the character of the resource – basically how easy it is to exclude others – plays a crucial role in defining which property regime is feasible. Hence, private property is a possibility only for a subset of resources. CPRs have a much wider potential, but are especially important when the cost of demarcating a resource into individual pieces is too high or privatization tends to destroy the functioning of the resource.

With regard to the interest and value aspect, first, we pointed to the role of the regime in defining who gets access to the benefits from a resource. Second, we accentuated its role in creating incentives for economic development and the distribution of the fruits of this development. Third, we focused on the way the regime influences the possibility for cost shifting. Finally, we looked at differences concerning which values the regimes foster and protect. While the situation is rather complex, we have emphasized that private property is the most dynamic with regard to economic growth and change. However, it is weak concerning the capacity to handle external effects or cost shifting. This may be important in an environment of interrelated processes. Common property is, on the other hand, specifically instituted to handle such effects, while it is less dynamic. Finally, state or public ownership can in many instances be considered much like that of common property. However, the principal–agent structure involved is vulnerable to the great distance that is often apparent between the 'owners' – that is, the members of a state – and their representatives.

In the last 20 years there have been substantial changes in property regimes in that many resources or goods that were under common or state governance have become privatized. Similarly, state management agencies have become state enterprises. Some commons have also been transformed into state property. The experiences with these changes are mixed. So, while these processes are still ongoing, we observe a counter trend, also implying (re-)establishment of CPRs. The lessons to be learned from this are straightforward: it is far simpler to tear down common property regimes than to (re-)create them.

In the globalized world where we live, even the environmental issues have become global. Given that the risks we face due to both the increasing size of the economy and the many new forms of economy–environment interactions that appear, we are faced with many challenges that existing regime types are not well adapted to handle. Thus, over the last part of the twentieth century there has been increasing engagement in the creation of international environmental agreements. One may view these as a kind of global CPR, with states as the involved agents. However, this observation says little about their dynamics and capabilities. Certainly, this development is just at its beginning. Moreover, it runs parallel to a similar, and in many ways much stronger trend directed at fostering the creation of a global market for goods and services. We have observed great tensions between these two processes and sets of institutions. The future of our common environment is not least dependent on our ability to handle this conflict in a way that gives ample recognition to the characteristics of the environment. Our ability to form institutions has enabled us to create an unprecedented economic development. The issue is whether we also have the capacity and the will to treat the ensuing environmental effects in a sustainable manner.

NOTES

1. See Young (2002) for a good exposition of this split. Young uses the concepts of 'collective-action' and 'social-practice' models for describing what is here called 'individualist' and 'social constructivist' perspectives. While the concept of 'collective-action' models is used in much of the literature, I avoid it. This is because it really deals with situations where no true collective is involved. Thus, to call it collective action tends to confuse the issues.
2. For example, 1 and 7 may be considered subgroups of 5, and 8 may be considered an aspect of almost all other elements. Points 2 and 4 cover much of the same.
3. Polanyi ([1944] 1957) offers a series of examples on this. See also Chapter 1.
4. Typically, Norwegian authorities instituted a reversal after 70 years when they sold the rights to utilize Norway's hydro-power resources to private companies in the early twentieth century.
5. 'Individuals', as referred to here, do not necessarily imply individual people, since various types of groups such as a family, a firm, an organization and so on, certainly qualify holders.
6. There is still one distinction to be made. While the state also has the power to be both the property holder and the organization guaranteeing the property right, public property at lower levels depends on the state authority in that respect.
7. See note 2. I view transmissibility as supplementary to the right to capital.
8. What follows here is based on Vatn (2001).
9. In the case of family property, how specifically these 'shares' are defined may certainly vary. However, when someone leaves, there will normally exist rules defining what amount of resources is to be withdrawn or compensated for.
10. Instead of internalization through regulation with taxes or other remedies manipulating the choice set as in the case of a private firm, the state authorities require the state agency to perform in a certain way, to include certain wider considerations in their management. See also Chapter 13.

11. According to the argument of this book, they will have such effects. However, this is besides the point here. We cover this issue in Section 10.3.
12. See as examples: Dasgupta and Heal (1979); Field (1994b); Cornes and Sandler (1996); Hanley et al. (1997); Perman et al. (1999).
13. As an example, the amount of biologically accessible nitrogen to the biosphere has doubled over the last 50–60 years due to increased use of fertilizers.
14. However, as emphasized in Chapter 7, crises or recessions may appear due to various dynamics.
15. Certainly, if uses are non-rival there are no costs to shift.
16. If agents' beliefs are that a practice is and will be unharmful, one cannot talk of cost shifting as far as this demands a conscious act.
17. One should be aware that this may be quite different in local markets, where relations are more personal and often enduring.
18. An intense debate is going on in Norwegian fisheries concerning the practice of dumping small fish from trawler catches. Fishermen who work on the trawlers are opposed to this themselves, and have reported the practice to state authorities.
19. It must be remembered, though, that Ostrom in particular built her analyses mainly on game theory and strategic behaviour, while this exposition puts more stress on the aspects of social cohesion, reciprocity and communication.

11. Valuing the environment

In this chapter we shall go into more detail about the process of deciding how we should treat the natural environment – that is, which environmental resources we can consume, which resources we want to modify and which ones must be preserved. This implies that some form of assessments or evaluations must be undertaken to prioritize between actions.

Such an assessment includes two core components. On the one hand, we must try to figure out the physical consequences of an act, of a project or a change in some institutional structures. Someone may propose that a new road is needed, and a decision about whether the road should be built will depend on which consequences it implies. Thus, building the road will mean larger transport capacity, but also that habitats will be lost, maybe more pollution will occur and so on. Someone may propose that the regime governing the use of a fish stock needs to be changed. Again assessments of effects that may emerge are important.

On the other hand, one is faced with the issue of evaluating which of the physical consequences are the most important or valuable and whether there are rights or moral commitments involved that should be respected. This implies articulations of preferences or values, and an evaluation of rights. The assessment may be undertaken in economic terms, using contingent valuation (CV) and cost–benefit analysis (CBA). It may take the form of a multicriteria analysis (MCA), or a citizens' jury or some other kind of deliberative institution may be set up. These methods differ not only concerning how consequences are assessed, but also concerning the status they grant to, for example, monetary bids, arguments and moral claims.

We shall start this chapter by introducing the concept of a value articulating institution (VAI) linking back to the issue of resource regimes (Section 11.1). The perspective advanced here is that various methods for assessing environmental goods influence which values can be expressed, how they can be expressed, and consequently which choices are found favourable. The rest of the chapter focuses on how well the institutions of monetary valuation fit environmental decision making. A thorough inspection is warranted not least because of the strong position this practice has acquired. In Section 11.2 we shall therefore give a short overview of the main neoclassical VAIs. In Section 11.3 we shall take a step back and discuss what characterizes the process of individual valuation of an environmental good in monetary terms. In Section 11.4 we shift from the level of the individual and look at the principal challenges involved when moving from individual evaluations through some aggregation to social choice. Through these steps the basic challenges involved when making monetary evaluations of environmental goods are framed and inspected. In Section 11.5 we turn around and ask how well the results from monetary valuation studies fit the model of rationality as maximization on which the practice is founded. This is a core issue when evaluating the practice. Section 11.6 makes a link back to the questions of invisibility, uncertainty and precaution – that is, areas where monetary assessments are especially challenging.

While the present chapter will focus more on the fundamental issues involved in monetary evaluations, Chapter 12 will focus more on the alternatives to market appraisals and how they are able to handle the challenges involved when making environmental decisions.

11.1 VALUE ARTICULATING INSTITUTIONS

As suggested in Chapter 6, institutional structures influence the preferences we hold, which preferences become 'activated', and in which way we find it right to express them. In Chapters 8 and 10 we took these ideas a bit further, emphasizing that the choice of institutional structures affects what then becomes efficient. Following this view, Sunstein (1993: 229) suggests:

> When this is so, there is no acontextual 'preference' with which to do legal and political work. A government deciding on environmental issues cannot be neutral among preferences when – and this is the key point – it does not know what preferences are until it has acted.

This has important implications not least for environmental goods. The type of evaluation – the institutional structures in which it is embedded – influences the outcome. The different valuation methods that exist – for example, contingent valuation, multicriteria analysis and different deliberative methods – are all *value articulating institutions* (Jacobs 1997).

In general a VAI defines a set of rules concerning the valuing process:

1. Participation:
   - who participates;
   - on what premises (position/role); and
   - how are they supposed to participate (in writing, orally, individually, via meetings and so on).

2. What counts as data and what form it should take (prices, weights, arguments and so on).
3. The kind of data handling procedures involved:
   - how data is produced; and
   - how data are weighed or aggregated.

The problem with environmental goods is, from a neoclassical perspective, that they do not have a price since they are not traded. To correct for this 'market failure', one may establish simulated market assessments like CV, where people are asked about their willingness to pay for a certain good. In principle everybody who is in some way affected or has an interest in an issue should be given the opportunity to offer her/his bid.[1] The role offered to the respondents is that of the consumer and respondents participate individually by questionnaire or interview responses. Data take the form of prices. Data concerning the characteristics of the good are offered to the respondents in verbal or written form, photographs and so on. It is assumed that respondents, on the basis of that information, can define which issues are involved and which value dimensions are relevant. From the offered bids, WTP functions (demand curves) and total value are calculated by summing individual price bids.

Alternatively, one could use VAIs which accentuate the information problem more explicitly,[2] ask the individual to act as a citizen,[3] and demand data in the form of judgements or arguments. It could be a process evaluating arguments instead of collecting, for example, price bids. This general idea of the forum contrasts the idea of a simulated market (CV) on almost all the above points. Different VAIs of the forum type exist – like citizens' juries and consensus conferences. We shall give a presentation of these in Chapter 12. Here the point is to illustrate the differences that may exist between various institutional structures and highlight some of the consequences of using them.

From the material presented in the preceding chapters it is easy to see that the choice of VAI may have profound effects on the results of an evaluation process. Market- and forum-type institutions will – through defining different roles, what is data, and how to interpret data – influence which preferences or values become articulated. In the case of CV, one would expect stronger emphasis on individual utility, while in the case of citizens' juries or consensus conferences, one would expect more focus on what the common interests and values are – which are the better arguments in the sense of a common assessment. Furthermore, the weights of various individuals are quite different in the two cases. Summing individual bids is something very dissimilar from evaluating arguments and judging which of these should count the most.

Finally, offering bids versus participating in a communicative process like that of a citizens' jury or a consensus conference puts different emphasis on learning and developing insights about the issues raised. Economic valuation techniques like CV are built on the assumption that the respondents have the necessary understanding of what is at stake given that the good is defined. Typically, in the case of preserving a landscape, various descriptions of its characteristics are given. Photographs may be attached to the written material. In addition to this, forum-type institutional structures offer opportunities for interactive learning about different ways of interpreting what is at stake through communication between participants, with experts and so on. This should lead to the development of deeper insights into often complex and unfamiliar issues. It also facilitates a discussion about which interests and values may be involved and should get protection.

There is a clear parallel or link between the perspective of VAIs and the issue of resource regimes. More precisely, any such regime has implicitly or explicitly defined mechanisms for value articulation. In the case of private ownership and markets, it is the individual willingness and capacity to pay – the exchange value – that forms the basis. The CV is trying to mirror that institutional setting. In the case of a common property regime, the allocation of the common resources follows a system based on communication, deliberation and maybe voting. These systems are in many ways fostering the same kind of processes that one tries to accomplish by using a citizens' jury or a consensus conference.

Choosing between regimes, or choosing between different VAIs, implies choosing between rationalities and values. To emphasize what has previously been concluded: efficiency evaluations built on the assumptions that values are independent of the institutional context are deemed to end in circularities. More specifically, it becomes inconsistent to use the evaluative logic of one institutional structure – such as that of private property and markets – to evaluate outcomes from another institutional system. The performance of a publicly owned arrangement such as a school will most probably not be well understood or captured by a market-like evaluation tool like CBA. Likewise, such an assessment comparing tourist and communal Sámi use of pastures, forests and lakes will most probably be incompatible with the values and conflicts involved (Johansson and Lundgren 1998; Hahn 2000).

11.2 ENVIRONMENTAL VALUATION VIA THE MARKET INSTITUTION

The neoclassical position is to view environmental problems as 'market failures' – that is, the failure of markets to appear for some goods. It is argued that since certain goods have a price and others have not, resource allocations will not be optimal. The priceless goods will be overutilized. To correct for this market failure, the use of indirect market assessments or simulated markets is advocated. There are three methods that are of special importance here – hedonic pricing, the travel cost method and contingent valuation. All of them assume that environmental values can be measured in monetary (commensurable) terms. They also assume that aggregate willingness to pay is a valid measure for the value of environmental goods.[4]

11.2.1 Hedonic Pricing

Hedonic pricing (HP) is based on the idea that environmental values can be elicited from prices of marketed goods whose value varies with some environmental characteristic. The idea is that a good can be characterized by a vector of attributes, **a**. Then, by keeping some of these attributes fixed, one may estimate the value of the other characteristics. Typically, the market for housing is utilized. The value of environmental attributes such as pollution, noise or scenery, can be assessed by looking at the price of similar houses in areas with, for example, clean air as opposed to areas with polluted air. The difference in price is a measure of the capitalized value of clean air as assessed by the buyer.

The method was developed by Lancaster (1966), Griliches (1971) and Rosen (1974). Its strength is that existing market data can be utilized. This certainly increases the realism of the analysis. It is actual willingness to pay that is measured.

There are some problems involved not least related to the fact that normally a rather large bundle of attributes may vary between the actual areas studied. Thus, it may be difficult to assess the effect of one single quality. There is certainly also a problem concerning whether the homeowners have knowledge about (all) quality parameters. The most important restriction is still the fact that the method can be used only for a subset of environmental goods – those whose values can be captured by the market price of a complementary marketable good such as a house. If we accept that WTP estimates are relevant, HP can be used for local pollution, local recreation and scenery, but not for environmental values that go beyond that of locality or for so-called 'non-use' values.[5]

11.2.2 Travel Cost Method

The travel cost method (TCM) is similar to HP in that actual behaviour is observed. In this case it is not the variation in actual prices, but the 'willingness to travel' to consume environmental goods or services that forms the basis for the method. Therefore it is typically used for assessing the value of recreational goods like visiting national parks and forests, going fishing, hunting and so on. The resources spent for being able to consume a good – for example, travel costs, entry fees, equipment costs, costs for accommodation – are used as a proxy for the value of the good. By observing these costs and the number of trips taking place, a demand curve for each site can be estimated.

The method was first proposed by Harold Hotelling in 1947 (Hanley and Spash 1993) and introduced into the literature by Trice and Wood (1958). Again the strength is that the method is based on observing actual behaviour. As with the HP method, it can only be used for a rather restricted set of environmental goods, though. Furthermore, there are problems related to estimating the value of time. Is the travel in itself a cost or maybe a pleasure? If a cost, what is then the price? Trips may have multiple purposes and so on.

11.2.3 Contingent Valuation

Both the problems related to the above methods and their rather restricted area of application, motivated the development of a method that could elicit prices of environmental goods more directly. CV implies asking respondents either for their willingness to pay or their willingness to accept compensation for a certain good. The distinction between WTP and WTA refers to existing or assumed property rights.[6]

The first step in a CV study is to define and describe the good. Next, one must decide upon a payment vehicle, so that the respondents know how contributions are to be collected: by general taxation, by payments to private or public funds and so on. One must also define a bidding procedure – that is, whether open-ended or closed-ended (such as dichotomous choice, double-bound dichotomous choice)[7] – and next choose respondents. The latter is a difficult issue because it is not readily given who 'consumes' or who has interests in the existence of the actual good. On the basis of interviews with respondents, average bids, bid curves and aggregate bids can be calculated.

The method was first proposed by Davis (1963), but the main advance took place in the 1970s and 1980s. An important summing up of this development is found in Mitchell and Carson (1989). The method has gained a strong position within environmental economics. In the early 1990s, CV seems to have been the most dominant activity within environmental economics (Vatn and Bromley 1994).

The strengths and weaknesses of the CV method are in many ways opposite to those of HP and the TCM. CV is very flexible. It can in principle be used to elicit any value – certainly under the assumption that it can be converted into a price. It can thus be used to estimate the value of the existence of a species, the existence of a protected area one will never visit (or 'consume') and so on. There are also problems involved. Do people take the bidding seriously since it is hypothetical? It is just about responding to a questionnaire, not actually making any payment. Do they understand what the issue is about? They may not have any own experience with the good. Moreover, a series of inconsistencies are observed in the literature, raising doubts about both the reliability and the validity of the method. This has also provoked a comprehensive debate among neoclassical economists themselves about the undertaking (see, for example, Hausman 1993).

11.3 THE PROCESS OF ENVIRONMENTAL VALUATION[8]

All methods presented above assume that a single metric such as a price – that is, price variations (HP), resource use measured in monetary terms (TCM), or price bids (CV) – is a good way to represent environmental values. In addition, they seem to assume that there are no fundamental information problems involved – that is, the individual knows and understands the issues as soon as they are framed. Finally, they imply that it is aggregates of individual preferences[9] or willingness to pay that should be used to decide over environmental choices.

To evaluate this practice, we need first to understand what is required to produce a single metric as a measure of value. This is the objective of the present section. Section 11.4 will focus on the wider issue – the relevance of such a single metric in collective environmental decision making.

Individual valuation of environmental goods is a complex undertaking. Referring not least to the discussions made in Chapter 9, environmental goods and services embody characteristics that represent important complications both for individual and collective choice. First, it may be difficult to define the goods – that is, give a precise description and demarcation of them given that they are complex and often are best characterized as processes. Second, there are important issues related to the process the individuals must go through when estimating their implicit or explicit willingness to pay, including the issue of compressing a complexity of value dimensions into one metric.

Of course, even the most ordinary commodity embodies a multitude of attributes. A loaf of bread is characterized by a constellation of calories, taste, smell, structure and texture. There is bread for everyday use, bread for feasts and bread for ceremonies. By purchasing bread on a routine basis one learns about the relations between price and those attributes of bread considered both desirable and undesirable.

Even with repeated transactions, there remains the problem of how consumers measure and value all of the relevant attributes, subsequently to transform them into a single metric. In other words, the problem is how do individuals map a multiplicity of attributes – mediated by preferences – into one measure? The verb 'valuing' describes precisely this information-processing activity, in which the final product is some 'reduced form metric'. In valuing, individuals must weight each attribute by some standard, and thereby compute one metric reflecting the multitude of characteristics of the object under consideration. Producing a single metric – that is, the value measure v^m – one must calculate the scalar product of two vectors. First we have the vector describing the attributes $\mathbf{a} = (a_1, \ldots, a_n)$ of the commodity that the person recognizes as pertinent and hence valuable, and next we have the one describing the weighting $\mathbf{w} = (w_1, \ldots, w_n)$ of each attribute reflecting the individual's preferences.

Despite its formal simplicity, this computation process is difficult for most goods. This was discussed in Chapter 5 not least in relation to satisficing strategies. Hence, long experience may be required for it to work quickly and well. Children sent to the bakery for their first purchase would not find the process simple. Adults are reminded of this when they undertake purchases that are rather rare – for example, an automobile.

The calculation process may break down and give rise to information losses due to three different reasons. First, we have those related to cognitive restrictions: the difficulty or cost of observing and weighting attributes of the object of choice. We may denote this the *cognition* or *information problem*. Second, losses will occur if different rationalities or classes of values are involved; if the value dimensions are incommensurable. This property means that the chooser cannot easily map disparate attributes – via $\mathbf{w}$ – into one dimension. If some value dimensions are non-comparable, one metric is unable to capture all relevant information. We call this the *problem of incommensurability*. Finally, there is the *composition problem*: the problem of demarcating parts of an environment, given that it is best understood as a system of interrelated parts or processes. If the various attributes – that is, the elements of $\mathbf{a}$ – are dynamically interrelated, either internally or with the attributes of other goods, the computation of v^m is distorted. In practical terms, information problems are created as soon as the value of one attribute depends upon the level of another. As we shall see, all the above points are of pre-eminent importance where environmental goods or services are concerned.

11.3.1 The Information Problem

The process of individual 'valuing' of goods and services entails the selective perception of certain data about the good or service, and a corresponding disregard of other data. There are two issues involved here. First, there is the issue of observing and understanding those attributes that define a particular good or service – recognizing the elements of vector **a**. Following from the perspective presented in Chapter 9, environmental goods are, to a large extent, characterized by their *functional invisibility*. This situation creates obvious problems for the valuation process. Second, there is the problem of weighting various distinct attributes – that is, transforming them into a common denominator for the vector **w**. There is evidence that people have restricted capabilities in making *comparisons across dimensions or scales*.

Perception and functional invisibility

The computation of a single value metric demands a precise understanding of which issues are at stake. As emphasized in Chapter 9, the environment is developed over an enormous time span where the various elements are created as complements. A continuous trial and error process has shaped a myriad of relations and feedback loops that can best be characterized by functional invisibility.

This means that the precise contribution of a particular functional element in the ecosystem is hard to know simply because when functioning, it is difficult to see all the factors that are necessary for the ecosystem to work. There may be thousands of processes keeping a system functioning, but what we observe is that it works, not why. It is often through failure that we learn about the critical ecosystem functions that, while working, are invisible. However, even when we observe that something has gone wrong, it is none the less difficult to say what went wrong. In the case of biodiversity, it may be very difficult to say anything precise about what will be the effect of a single species becoming extinct, reducing the amount of wetlands and so on. The importance of certain species or areas may exist at many different levels. They may be crucial in some situations while not in others.

Relating this to the valuation process, it becomes a challenge to choose what to describe, to define the problem, really. As will be discussed more systematically in Chapter 12, the main challenge in such situations is to agree on what should be valued. Doing the valuation in sequence so that (a) the analyst first defines the good and (b) the respondents next value what is thus defined, is not at all unproblematic. If a road or a dam for hydroelectric power generation is to be built, which consequences are really worth mentioning?

Given that we accept that the analyst can define the stakes or effects, these must next be described. Even this is challenging to do in such a way that the participants in the valuation study really have the same feature in mind when revealing their bids. In a situation where the good or service is nested in a set of relations, it is not stretching the point to say that the 'resource' in question can be practically anything the respondent wants it to be. The perception of the elements of the vector **a** may therefore vary across individuals because they demarcate and understand the goods differently. If people could meet and discuss which issues are pertinent, as in focus groups, the above problems would most probably be reduced as compared to a purely individualized process. This is an option that is being increasingly utilized.

Valuing across various dimensions

The calculation of a single value metric demands that the respondents have cognitive and normative beliefs about these values, making it possible for them to convert them into one value term. We shall here focus on the cognitive and leave the normative for the next section.

The logic of the vector approach discussed above is the presumption that individuals can make extensive comparisons across multiple dimensions or scales. As an example, they must be able to convert the visual beauty of a landscape and the various dimensions of its functionality – for example, biodiversity composition, hydrological capacity, carbon binding capacity – into one metric. Some may normally be measured in numbers, others in kilograms and so on.

Gregory et al. (1993) argue that individuals are not accustomed to interpreting environmental goods along one dimension – for example, in monetary terms. Tversky (1969) argues that it is much simpler for people to compare the alternatives dimension by dimension – scale by scale – than it is to evaluate each good across all dimensions and then compare these total assessments.[10]

We see this at work in regular commodity markets where individuals restrict their calculative comparisons to commodities embodying rather similar attributes. Choices between commodity groups seem, by cognitive and computational necessity, driven by other considerations. Such comparisons may be driven by learned behaviour over a previous constellation of attributes and prices. Price-based choices – that is, decisive comparisons – are largely confined to price changes within the same general group of goods or services.

11.3.2 The Issue of Incommensurable Preferences

The calculation of a single value metric demands, furthermore, that the respondents have normative beliefs about these values that are commensurable. Above, we suggested that people may have problems with transforming different dimensions or scales into one metric. This problem is vastly compounded if (some of) the values involved are incommensurable. Then one metric (price) will be unable to capture all relevant information since there are elements of the vector **w**, which resist being transformed into each other. In the language of mathematics, they are orthogonal.

The *ethical or moral dimensions* of environmental choices introduce one important basis for such incommensurability. Douglas (1966), Douglas and Wildavsky (1982), Kneese and Schulze (1985), Etzioni (1988) and Sagoff (1988, 1994) are scholarly explorations of the problem of restricted trade-offs, as discussed also in Chapters 5 and 6.

Commitment and moral judgements are concepts often attached to those domains where issues about life, quality of life, and integrity are at stake. These are areas where social norms very often restrict or reject the commodity perspective. In the case of environmental goods, moral issues are related to the presumed 'right' of wildlife to survive, and therefore the obvious impertinence of respondents being asked to pay for it individually. Environmental issues raise the question of the 'right' to life or to a certain quality of life for humans as well as for wildlife – be it now or in the future. The intergenerational question is certainly a core issue here. Future generations have no possibility of entering into a discussion about what their options or possibilities should be. It becomes the responsibility of those living now to choose for them – again emphasizing the moral dimension of environmental choices.

As suggested, involving people in discussions with, for example, experts, about the good or problem before the valuation takes place can reduce information problems. The problem of incommensurable preferences cannot be thus mediated. The character of the value issues can be clarified better through group processes. None the less, if values are of different classes, understanding more deeply the value issues involved does not make them any more commensurable.

11.3.3 The Problem of Composition

Finally, we have the composition problem. The calculation of a single value metric demands that the value of the good at stake can be calculated independently of the larger context of which it is a part. Consequently, in economic valuation studies, environmental goods and services are described in a manner that renders them commodity like. While this may seem merely an 'economistic reflex' – every good must be a commodity – the perspective may also follow from the practical aim. A precise valuation demands a precisely demarcated object. The essence of commodities is that conceptual and definitional boundaries can be drawn around them and individual property rights thereby attached. Here we are back to Polanyi ([1944] 1957) and the point he makes concerning the arbitrariness of commodity demarcations. Indeed, he talks of the 'commodity fiction'. As markets evolved in human history, Polanyi suggests that it became helpful, but also necessary to regard certain aspects of reality as 'mere' commodities. After all, markets can only operate where things are commodities.

For the most part, environmental goods can be looked upon as those that are not commoditized. This simple fact may in itself indicate something about their character. Demarcation may be technically impossible, it may be too costly to accomplish, or it may actually destroy what we are after in our pursuit of monetary values. The danger with commodity fiction is that the environment becomes simply a location for recreation – water to swim in, land on which hunting may occur, wildlife that may be slaughtered and brought home. The 'market' exchange becomes one of trading euros (or some other currency) for the opportunity to use the commodity for a certain period of time. The connection with modern labour markets is immediate and suggestive.

The perspective outlined in this book forces one to try to comprehend environmental goods and services in a more holistic way. Much of the hostility arising in the ecological community towards economics and economic valuation rests on this aspect of holism and I shall make three brief comments on valuing in *situations where functional issues* are core issues.

First, in a fully functionalized system, each part must actually be as 'valuable' as the whole. If one function ceases to work, the whole system stops functioning, and hence the value of any single component cannot be understood – or priced – separately from its contribution to the whole. In this situation, the idea of continuous trade-offs among various components has nothing to offer. The body gives a good illustration. It will stop functioning if single parts are damaged, such as the heart, the brain, the stomach and so on. These are fully functionalized elements and each must be as valuable as the whole body.

Second, ordinary commodities are characterized by their capacity to be exchanged, and their 'value' – as measured in prices – is an exchange value. In these circumstances, the commodity represents a distinct set of attributes over which the use and enjoyment can be defined and controlled by the buyer. The very process of production in an economy is one of transforming disparate factors of production (raw materials) into a constellation of attributes which, when taken together, offer usefulness and so command a certain price. For environmental functions this condition is not met. The value of most environmental goods and services is derived from the very act of *keeping them working within their existing functional relations*. Moreover, environmental goods and services most often do not exist as discrete units.

Third, from a systems perspective, individual components do not acquire their value from their uniqueness to us as humans, but rather from their uniqueness in relation to the whole system of which they are a part. In the standard approach, uniqueness seems to be captured by the concept of 'existence value' for distinct, often spectacular species. This is a reflex of the commodity perspective and its implicit itemizing. Cummings and Harrison (1992) document that it is this kind of uniqueness that actually dominates so-called 'non-use values' as that concept is normally used in the valuation literature. Kahneman and Knetsch (1992) conclude similarly.

Thus, the spectacular or the visual tends to dominate the systemic or functional. Bald eagles and grand vistas get much more attention – and hence become more 'valuable' – than an insignificant mussel or a muddy wetland. I am not saying that what is spectacular is not valuable. What I am suggesting is that the more the issue is about the common good, about systems' functions or features, the more urgent are the problems concerning information, preference incommensurability and functional indivisibility.

11.4 INDIVIDUAL VALUES, ETHICS AND SOCIAL CHOICE

11.4.1 Aggregation, Rights and the Weighting of Individual Values

The VAI not only influences the process of individual value articulation, but also determines whose interests are to count in the decision process. In the case of CV, the willingness to pay is certainly influenced by the capacity to pay. It is total WTP across the defined population of individuals that is decisive for the conclusion. Consequently the preferences of the rich will count more than those of the poor. This point is not only very important, but also straightforwardly evident and needs no more elaboration. Rather more demanding is the way the distribution of rights in itself will influence price bids both via the income effect and by the effect of so-called 'loss aversion'.

Under the CV institution, the rights issue is implicitly defined by the choice between willingness to pay and willingness to accept compensation. If the environmental good involved were of some size or importance, one would expect a certain difference between WTP and WTA due to the income effect implied by the right itself – see Chapter 8, Section 8.3 for a discussion of this issue. However, the same section of Chapter 8 also documented that WTA measures are normally much higher relative to WTP than the theory about income effects should imply. Hence, the implicit rights structure might have profound effects on the outcome of a CV study dependent on which rights structure is implied. Following Hanemann (1991) one should expect extra large differences for environmental goods, since here we may lack (perfect) substitutes.

Yet one sees a distinct preference for WTP measures in economic valuation studies. It is considered preferable because it is more 'realistic'. This stems from the argument that it is bound by the income of the respondent (Arrow et al. 1993). Respondents cannot just report an unreasonably high compensation bid. This is a weak argument. It mixes a technical problem with a normative one – the issue of fundamental rights. If what is really going on is a loss of some defined or presumed right, WTA is the only relevant measure. Why should people who once enjoyed good drinking water have to pay to prevent it from becoming even more contaminated? They will wonder why they should have to pay to obtain a state of nature that existed prior to the advent of chemical runoff caused by someone else. Indeed, they might legitimately wonder why they should not be asked about their necessary level of compensation while the contamination of groundwater continues. The above argument concerning what produces the most 'realistic' bids does not carry any weight in evaluating this issue.

An important point here is the fact that in the case of environmental goods and services, the rights structures are generally unclear. 'In the beginning' there was undamaged nature. Damages take place over time due to different activities, and what was is gradually lost. This implies that a resource or benefit stream that was previously abundant – non-rival – now becomes scarce. Therefore, in the beginning there was no need to define rights. When the scarcity becomes visible, rights do not exist or it is not clear who has the right. Owning the land may have given the owner the right to cut down the trees that stand there, draining its wetlands and so on. Does that right also give the right to the accompanied biodiversity when over the years that becomes a scarce good? What about emissions that have been going on for years before any problems become visible?

If we refer to the status quo at the time of action, giving the right to the victim implies that someone has to lose (for example, the polluter since pollution has been accepted up to that date), while giving the right to the offender or polluter just implies sanctioning the status quo 'rights' that were never really granted. Thus, the time lags involved in environmental degradation may be said to give some 'protection' to those degrading the environment. The general favouring of WTP both accentuates, and may be a reflex of, this tendency.

If the goods or services are considered a right for those experiencing environmental degradation, WTA would be the proper approach if monetary measurement is to be undertaken. But care must be taken even here. If an individual is asked to 'value' wildlife, this may be viewed as an issue concerning the right of a certain species to exist. In this situation, WTA suddenly smacks of bribery. This is because it is the 'right' of the species to exist that is at stake. The same reasoning applies to situations where the respondent finds it relevant to take the rights of other humans into consideration. But in situations like this, WTP is not a good measure either. Both WTA and WTP might imply forcing a conversion of moral commitments into money equivalents. Certainly, if moral issues are involved, there are strong arguments for shifting to a forum-type institution where arguments and judgements constitute the basis.

11.4.2 Prices or Norms: The Ethical Aspects

Since individual preferences are context relative, a fundamental problem becomes which of many institutional settings is relevant to a particular choice problem. Again we must emphasize the distinction between two kinds of choice processes in society. One concerns decisions from within a given institutional structure. The other is about choosing these sets of institutions, implying which norms and values then become accentuated.

As already indicated, transforming environmental values into commodities may create many ethical concerns or problems. It deprives it of much of its meaning and worth. In many cultures, the sense of sacredness is attached to the natural environment or parts of it. Even in our more secular type of societies, the natural environment is of great importance in creating identity and defining belonging. Many further view it as heritage – that is, primarily as something we inherit with the responsibility to hand it on to later generations in good shape (Burgess et al. 1995).

Going further into this, there are several issues that do not conform well to the commodity concept. I shall distinguish between the issues of 'nature's own right' and the moral concerns raised by the interconnections between humans through their common environment (Vatn 2000).

Holland (1997: 130) suggests that the first issue 'arises from the fact that the natural world contains many items which undeniably in the case of sentient animals, or arguably in the case of other animals and plants, have moral claims on us'. Being arguable implies, as I understand Holland, that conclusions about the nature and extent of the moral claim will vary between cultures and over time. This does not, however, eliminate the challenge.

Certainly, a definition of 'nature's right' has to be culturally or socially defined. It can in no way be given from nature itself. If we grant such a right, it is on the basis of reasoning about *socially constructed* ethical beliefs. Some have questioned why the utilitarian calculus is defined only over human interests and needs. Environmental economics has in a way tried to alleviate this problem by introducing the previously mentioned concept of 'existence value'. In this way the rights of nature – as each *individual* perceives it though – becomes part of the calculus. This reveals a misunderstanding about the character of moral claims, which have to go beyond *individual* evaluations, as ethics and morality are social phenomena. They belong to another category than those to which ordinary trade-off calculations are appropriate.

While the issue of 'nature's own right' is important, many of the ethical issues involved in environmental decision making concern the relation between humans. The issues stem from the *interdependencies* that the natural environment creates between humans. The preferences and acts of one must then by necessity influence the possibilities for others. This is the case with emissions as it is the case with more direct changes or destructions of a habitat. Services will be lost. In such a situation it seems more relevant for people to discuss what is reasonable to do, rather than to offer a price. It becomes not only acceptable, but also creditable to engage in a discussion about which preferences one should hold. In a situation of physical interdependencies, preferences lose much of their privacy. That is typically the reason why norms about what is right develop in society (Vatn 2000).

Concerning the interests of future generations, no communication or deliberation with those concerned is possible. However, offering a price may seem even less relevant in relation to that issue. Instead, people may want to meet and reason – as contemporary fellows – over how one best can secure the interests of future generations. The moral responsibility faced here is that of forming the future, not only its environmental basis, but also the values to be protected. Thus, again it is the reasoning over which values should be favoured that is at stake.

Once more this way of thinking illustrates the consequences of the model and concepts we use for the conclusions we reach. The commodity concept is developed within a model based on *independence* between goods and between agents. Interdependencies with ethical force are expelled from the core of the model and into its external sphere. And even here the ethical concerns are twisted or made largely invisible by proposing the same rule of calculation for the external environment as for the internal market.

Certainly the whole endeavour of the neoclassical revolution of economics was in the end to remove any questions about what is the right or good way to live (see Chapter 8). Either these issues are understood as purely private and not open to debate, or the debate is said to belong to other spheres of society than the economy. Due to the interdependencies explained here, they cannot be just private. They cannot be kept outside the sphere of economics, either. They certainly concern the formulation of economic institutions with their defined rights structures. Such a 'division of labour' between 'individual' and 'social' institutional arenas – between economics and politics – is not tenable.

11.5 ENVIRONMENTAL VALUATION AND RATIONAL CHOICE[11]

So far we have asked questions concerning what – from a theoretical point – characterizes environmental evaluations. We shall now look at what the existing valuation studies themselves say about environmental preferences and the rationalities involved. Certainly, since a value-articulating institution like CV in itself motivates respondents to think in monetary terms, one would expect respondents to exhibit a certain degree of conformity with what is demanded. Nevertheless, if the issues raised concerning information and plural value dimensions have any significance, one would also expect these issues to show up in the CV studies undertaken. The aim of this section is therefore to explore what environmental valuation studies themselves may reveal concerning the issues raised in Sections 11.3 and 11.4.

Certainly, over the years a long list of problems has surfaced in monetary valuation studies. This is not least visible in a whole host of concepts that define different deviations from the behavioural assumptions of the neoclassical model: for example, 'starting point bias', 'information bias', 'question order bias', 'yea-saying' and 'part–whole bias'. These concepts are developed to explain seemingly inconsistent behaviour or, more precisely, to construct explanations so that the observations may fit better to the core behavioural assumptions of the economic model. It is worrying, though, that most of them seem to be rather *ad hoc*. They do not stand up very well to the theoretical stringency of the underlying model itself. Indeed, the concepts often seem inconsistent with the model they are developed to defend. In discussing this, we shall again distinguish between the cognitive and the normative. We shall first look at the informational aspects and next at the aspects of preference formation.

In doing this, we shall review a series of studies done from the 1970s onwards. Certainly, over the years the CV method has been refined, and some of the problems encountered previously may now be avoided. Nevertheless, the material from CV studies using formats now abandoned – for example, bidding games – are also of interest to us. Our aim here is not to determine what might be a better CV format, but what the responses to various CV formats may tell us about (a) the information problem and (b) the issues raised concerning preferences and preference formation.

11.5.1 The Information Problem

A CV study is a particular type of communication. Concerning the information aspect, the problem for the analyst is – as perceived – to make up his/her mind concerning what is 'relevant', 'neutral' and 'enough' information. The respondent on the other hand, faces the problem of understanding what the information presented implies when trying to capture the logic of the choice situation (Fischhoff et al. 1999). Focusing on the information problem as previously discussed, I shall divide the discussion in two: the issues related to information and the character of the good; and questions about the communicative aspects of information and information transfers.

Information and the character of the good

The problem from a CV point of view is to ensure that the respondents' perception of the situation is the same as that intended by the analyst. This concerns both the type of transaction implied and the description of the good. Consequently, the concepts of 'amenity misspecification bias' or 'information bias' are introduced in the literature to describe situations where information transfer is felt to be in some way distorted (Cummings et al. 1986; Mitchell and Carson 1989). Certainly if information costs are zero, such problems will not exist. If they are positive, many different problems may appear.

As emphasized by Ajzen et al. (1996: 43): 'the nature of the information provided in CV surveys can profoundly affect WTP estimates'. Since the respondents are often faced with issues that are unfamiliar to them, the literature indicates that they seem to look for 'clues' that can help them to categorize the problem at hand under more familiar classes of issues. Hence, subtle contextual cues can seriously 'bias' the CV estimates under conditions especially of low personal relevance (ibid.). One would expect this problem to increase the more complex and novel the good is. WTP for a fishing experience may be informationally a much easier task than WTP for preserving the fish species. It is less complex or perhaps more precise: it is an activity that has been defined socially rather well. Fewer interpretations are left to the individual. To illustrate the information problems stemming from the character of the good, I shall focus on two issues, the so-called 'part–whole bias' and the 'question order bias'.

Part–whole bias Earlier, we discussed the problem of composition – that is, the problem of how to demarcate or define a good that is a systems property more than an item. In CV studies the issue appears through certain insensitivity in WTP estimates to the size of the good. Protecting one or several wilderness areas, different numbers of a species and so on give rather similar WTP figures. The observation has been termed the 'embedding effect' or 'part–whole bias' (Kahneman 1986; Kahneman and Knetsch 1992; Diamond et al. 1993; Boyle et al. 1994; Hanemann 1994; Smith and Osborne 1996; Carson 1997).

The literature on these issues is far from conclusive. Hanemann (1994) and Carson (1997) suggest that studies showing embedding effects are badly conducted or that the observations are not counter to what should be expected from economic theory. Substitution rates between different environmental goods or the marginal value of different levels of the same good may be such that observations still fit the standard core assumptions. Authors such as Kahneman and Knetsch (1992), on the other hand, dismiss this kind of argument on the basis that insensitivity is far too strong.

Two types of explanations related to the perspective of this book may be put forward.[12] First, given the existing complexity, respondents may search foremost for ways to classify the issue. Given a chosen *typification* of the issue involved – that is, whether it is about 'species protection', 'scenery protection' or more standard consumption aspects, the relevant response is defined by the classification. It is the typification, not primarily the size or amount of a good that matters.

The other type of explanation concerns the idea that respondents may have more sophisticated views of the good than the analyst assumes. Concerning species protection, Schulze et al. (1994) argue that the respondents may not only think of the specific species, but they may also include the ecosystem to which it belongs: 'Butterfly species in the Amazon are becoming extinct because of the loss of habitat. The only way to save one species is to save all of them by saving the forest as well' (p. 16). This explanation responds directly to the composition problem as previously defined.

The two explanations may appear to be contradictory. This is not the case. They are both based on the idea that the interpretation of the data presented lies in the *typification/classification* of the issues involved. If the issue is protection of a species, it also involves the protection of the ecosystem which sustains its living. Following the institutional view on rationality, classification or typification is the process by which people come to understand which issues and values are involved. The observed 'part–whole bias' supports such a view.

Question order bias Another regular observation is that a good is differently valued depending on what place it has in a sequence of goods presented to the respondent – that is, in a sequence of different environmental goods, the same good is more highly evaluated the earlier in the sequence it is mentioned. Thus, we observe 'question order bias' or 'sequencing effects'. Hoehn and Randall (1989), Hoehn (1991) and Carson et al. (1998) are among those arguing that some question order bias should be expected. If goods are substitutes, the WTP for an increase in one of the public goods should decrease the farther out in a sequence it is valued, since its contribution to welfare has been (partly) covered by goods listed earlier. The opposite is true if they are complements. Only if there is independence, should the sequence not matter. Carson et al. (1998) acknowledge that some observed differences in values under different contexts are still too large to be plausible.

Halvorsen's (1996) study may serve as an example. People were asked to value the human health and environmental effects of reduced air pollution. In a sequential valuation with the health effect focused on first, the mean WTP values were 1133 Norwegian Kroner for the health effect and 6 Norwegian Kroner for the environmental one. If both goods were valued simultaneously and the respondents were asked to split their bid on the two elements afterwards, the results were 278 and 862 Norwegian Kroner for the two effects, respectively. As we see, both the ordering and the mean WTP estimates for each good changed. The latter change was substantial not least for the environmental impact. Even though, for example, environmental and health effects may be substitutes to some degree, one may certainly ask whether results like the ones above are reasonable.

However, if we assume maximization and full information, the sequencing should not matter. People would know and remember all goods there are and make the necessary trade-offs instantaneously. Pollution also has effects beyond health and the environment. The analyst establishes that focus. Taking the model seriously implies that the respondent also considers these when the ones focused on by the study are priced. So whether substitutes or complements, whether mentioned early or sequenced late in a study, none of these issues should matter. If the assumption is imperfect information, sequencing may matter. Then, however, the problem is fundamental. In concrete valuation studies, important information will always be lacking. In such a setting the only reasonable thing to do for the individual may again be to use some sort of classification or 'mental accounts'[13] for each type of goods and solve the problems of choice as they appear, that is, sequentially. This is a type of satisficing strategy and may result in exactly the type of observations made by Halvorsen. Whether this is a rational basis for collective environmental choices is another issue.

Communication and the cost of information

CV studies show that responses are not independent of the technical characteristics of the elicitation procedure. According to one type of problem, the level of proposed payment in closed-ended elicitations formats influences the WTP estimates (Kealy and Turner 1993; McFadden and Leonard 1993; Holmes and Cramer 1995; Herriges and Shogren 1996; Boyle et al. 1997; Kahneman et al. 1999; O'Conor et al. 1999). The so-called 'starting-point bias' describes responses that therefore deviate from the model of rational choice.[14]

Again, if we knew our preferences and they were complete and continuous – that is, following the assumptions underpinning the neoclassical model, the levels of payment proposed in closed-ended arrangements should not influence final WTP estimates. Concerning the specific formats of double-bound dichotomous choice or double-bound referendum format, McFadden and Leonard (1993) conclude:

> Summarizing, we find that the distribution of stated WTP depends strongly on the elicitation format. . . . Thus, the 'starting-point bias' that led CV researchers to abandon repetitive bidding games is already a damaging effect in second response, and the double referendum elicitation format is internally inconsistent. (p. 191)

Holmes and Kramer (1995) even find that starting-point bias occurs in single referendum (dichotomous choice) formats, a conclusion supported also by Meade (1993).

In the literature we find two explanations for starting-point bias, both of which seem to be of an *ad hoc* nature. One is so-called 'yea-saying' (Mitchell and Carson 1989; O'Conor et al. 1999). The other relates to the possibility that respondents consider bids to carry information about the value of the good. While yea-saying directs the concern mainly towards the social dimension (see Section 11.5.2), the latter explanation makes us focus on the completeness of preferences and respondents' ability to handle information and compare across categories of goods.

Mitchell and Carson (1989: 240) suggest that 'confronted with a dollar figure in a situation where he is uncertain about an amenity's value, the respondent may regard the proposed amount as conveying an approximate value of the amenity's true value and anchor his WTP amount on the proposed amount'. Boyle et al. (1997) propose that 'bids may carry *unintended* cues to survey respondents regarding the *quality* of the item being valued' (p. 1496, added emphasis). They also refer to various marketing studies indicating that consumers use price as shorthand for quality. Actually, information costs may make it rational for the respondent to use the offered price bid as 'cheap information'. Certainly complexity and unfamiliarity may make this effect especially important within the fields we are looking at here.

Prices may thus be understood as the result of tests made by other consumers. It tells you what a good should be worth. If so, the cause–effect chain is turned around due to information costs and social dynamics: *prices inform preferences*. On the basis of this, starting-point bias should be expected particularly when valuing goods where respondents have no other easily available price information – that is, no class or type specific price, to which they can compare or relate.

In a study by Schkade and Payne (1993), verbal protocols were used to obtain insights into how people reached conclusions over their value bids. Their study shows that people tend to look at what they have spent on similar goods – that is, the classification aspect is again emphasized. They further show that the cost of producing a good also seems to influence what we are willing to pay, possibly also since it guards against the feeling of being cheated. Clark et al. (2000) document that some respondents explicitly want to anchor their bids to the costs. Schkade and Payne (1993) observe in their study that 41 per cent of the respondents searched for the appropriate or necessary bid, assuming that everybody paid their share. Similar arguments are made by Shabman and Stephenson (1996). The above observations fit nicely the concept of reciprocal behaviour. Since the environmental good is a common one, it makes sense to expect both reciprocity and fairness concerning the distribution of costs to be important.

The above observations turn the issue on its head. Instead of supporting the idea that information functions as neutral data to support people when applying a preference-maximizing calculus to determine choices, the hypothesis develops that the information delivered plays a crucial role in determining both what type of preferences become activated (the typification), and what in the end is considered to be the value of the good. Moreover, not only individual values are underlined. Instead what is a fair distribution of costs becomes important. This takes us to the realm of preference formation.

11.5.2 Preferences and Preference Formation

Consistent with the neoclassical position, most CV studies presume that people have individual-specific, well-defined preferences for different states of the world. The empirical evidence seems to be at odds with these assumptions: preferences may have to be learned; they may not be easily ordered; trade-offs may be blocked.

Preferences: individual or social?

The idea that people have predefined values for different goods has come under scrutiny not least as a function of problems observed in environmental valuation studies. Slovic et al. (1990) conclude that people have well-defined preferences only for familiar goods. Obtaining values for less familiar ones demands some kind of inferential process. This has facilitated a move towards a more *constructivist* perception of preferences. Hence, during the 1990s, several papers focused on the issue of preference construction – for example, Gregory et al. (1993), Slovic (1995) and Payne and Bettman (1999).

The perspective presented in these studies is still individualist. The role of the analyst is to support the individual in the process of clarifying his/her preferences when learning about the various elements and characteristics of the (unfamiliar) good involved. The perspective is therefore not necessarily at odds with the neoclassical position – see the reference to Hanemann in Chapter 6 and his emphasis that the real issue is not whether preferences are constructs, but whether they are *stable* constructs. Individuals merely learn about themselves. It becomes a challenge to the model first when preferences are also socially contingent.

From the perspective of *social* constructivism, the observations made would be understood very differently. As already emphasized, we perceive and understand on the basis of socially produced concepts. This not only influences the production of knowledge and the dynamics of cognition. According to this perspective, preferences reflect characteristics of both the individual and the society within which the individual is raised. This kind of perspective generates important and very different hypotheses concerning several observations surviving in the contingent valuation literature. The discussion around the concept of yea-saying is illustrative.

Yea-saying is certainly an *ad hoc* explanation for observed deviations from rational choice theory. The concept seems to have been used in both marketing and psychological research before it entered the CV literature. It is defined as 'the tendency of some respondents to agree with an interviewer's request regardless of *their true views*' (Mitchell and Carson 1989: 240–41, added emphasis).

Again some true, individual value is presupposed. It is, however, seen as twisted by the elicitation process so that something other than the 'true value' appears. If one instead shifts focus, and looks upon preferences formation as – at least partly – searching for a social norm, as expressing a social belonging and so on, the interpretation of what is going on is rather different. Yea-saying would not be understood as twisting a given, true value. It would rather be conceived as part of the preference development itself and conform to the idea that there is a communicative process going on between the interviewer and the interviewee. Certainly, the interview situation is not what would be thought of as a normal communicative process. However, there are strong enough similarities to expect respondents to be influenced if social mechanisms have any effect at all.

Some recent environmental valuation studies have combined CV with opportunities for the respondents to communicate. Brouwer et al. (1999) and Clark et al. (2000) have thus combined a CV study with a focus group and an in-depth group study, respectively. Brouwer et al. conclude that a combination of the two methods is most appropriate. Their respondents were happy to give CV estimates, while a majority also favoured the more participatory process involved.[15] In the Clark et al. study, the focus was more directly on how people went about constructing their WTP figures. They conclude: 'When deconstructed by the respondents themselves, their WTP figures proved to have little substance and they unequivocally rejected CV as an acceptable means of representing their values . . . valuing nature in monetary terms was incommensurable with deeply held cultural values' (p. 60). Clark et al. refer to a study by Vadjnal and O'Connor (1994), which obtained similar results. Also the above-mentioned study of Schkade and Payne (1993) concludes likewise.

While both the Brouwer et al. and Clark et al. studies obtain positive responses concerning the role of deliberation in value expression exercises, there are strong differences between the two studies concerning respondents' appreciation of the CV method. It is impossible to draw any definite conclusions about the reasons for this. The different formats concerning the studies may have had an effect. The type of good involved was simpler to respond to in monetary terms in Brouwer et al.'s study (flood alleviation scheme) compared to that of Clark et al. (nature conservation – the Pevensey Levels area in England). Therefore CV may have been found more appropriate in the first case, which involved issues of a more private kind where costs such as damage to buildings and so on were already perceived in monetary terms. This would not have applied in the nature conservation case.

Ethical concerns and plural preferences

While the issue of blocked trade-offs is observed as a feature of people's preferences more in general (Chapter 5), it is a very visible characteristic in the realm of the environment. Offered only the option to monetize, some respondents deliver 'protest bids' in the form of non-responses or zero bids in cases where it is still obvious that they think positively about the good. The standard reaction to protest bids in CV studies is to exclude them as irrelevant, implying that there is restricted evidence concerning what is really going on here.

In an evaluation of the environmental valuation debate, Milgrom (1993: 422) emphasizes that protest reactions are typical for environmental goods, but not for commodity surveys:

> Protest responses are extreme responses that express an objection to some aspect of the contingent valuation scenario, rather than reflecting the respondent's economic self-interest. These include reports of willingness to pay a large fraction of the respondent's income and probably also include some WTP responses of zero. The regularity of protest responses and the absence of similar responses in survey valuations of consumer products suggests that protest responses do reflect some kind of genuinely felt value for environmental goods.

Studies such as Stevens et al. (1991), Spash and Hanley (1994) and Spash (2000), document that approximately 25 per cent of the sample refused to state a WTP amount on the grounds that the environment had a right to be protected. These respondents protested about the survey on the basis of ethical arguments. They held preferences that were lexicographic. Given that the logic of the involved VAI draws the individual's attention towards monetary bids, 25 per cent is a rather high figure.[16]

Spash (2000) is an interesting study in relation to the latter. He documents that among people, who in his survey have responded in a rights-based/lexicographic way, there are some that still offer a monetary bid. Such behaviour seems inconsistent. However, the observation is not so remarkable if we bear in mind the effect of the institutional setting. Since monetary valuation is involved, we may make the interpretation that some still feel compelled to conform to the rules given.

Taken together, the various observations made in this section show that the environmental valuation studies themselves give results that actually support the institutionalist perspective on cognition, information and preferences. We have made several important observations. The type of information delivered influences price bids. Preferences concerning environmental goods are often not defined. Preferences are often plural – implying restricted trade-off possibilities. Finally, we have also observed that CV respondents question the very use of monetary valuations as a basis for at least some types of environmental decision making.

11.6 INVISIBILITY, UNCERTAINTY AND PRECAUTION

Returning to the insights from Chapter 9 concerning the uncertainties involved in changing environmental systems, one may ask whether valuing the environment in economic terms is really the right response to the question about what should be preserved. Since it is hard to assess what may happen if we destroy a habitat or change a certain environmental process, there is really no strict basis for evaluating the consequences. If it is not so much the elements in themselves, but the resilience of the system that is important to preserve, thinking in terms of concepts such as the 'precautionary principle' (European Environment Agency 2001) or the 'safe minimum standard' (Ciriacy-Wantrup 1968), may be highly relevant.

The precautionary principle is a general rule of public action to be used in situations of potentially serious or irreversible threats to health or the environment. More specifically, the focus is on situations where there is a 'need to act to reduce potential hazards *before* there is strong proof of harm, taking into account the likely costs and benefits of action and inaction' (European Environment Agency 2001: 13; original emphasis). The idea was first developed in Germany in the 1970s and was more universally adopted at the UN Rio summit in 1992 (see Chapter 10). In the Declaration on Environment and Development it was concluded that the precautionary principle should be used on all matters of environmental concern:

> In order to protect the environment the Precautionary Approach shall be widely applied by states according to their capabilities. Where there are threats of serious irreversible damage, lack of full scientific certainty shall not be used as a reason for postponing cost-effective measures to prevent environmental degradation. (UN 1992b: principle 15)

Despite the above reasoning, we still need to evaluate when being precautionary is too costly. The principle directs attention towards uncertainty and irreversibility. It demands that if a choice is characterized by uncertainty, this uncertainty should be used as an argument in favour of preserving the environment, thereby linking to the concept of critical natural capital and the preservation of resilience. There is thus a lot of progress implicit in this change of perspective, but the principle does not avoid the problem of deciding when it should apply – that is, when the uncertainty is large enough and the possible stakes are high enough.

Toman (1994) offers a perspective on the idea of a safe minimum standard (SMS), which may take us a step further in clarifying the involved issues. The concept relates to the same issue as the precautionary principle. It is about developing standards for the preservation of environmental goods to avoid future unwanted losses. The SMS framework is, however, not intended to offer a specific decision rule. Its purpose is rather to 'provide

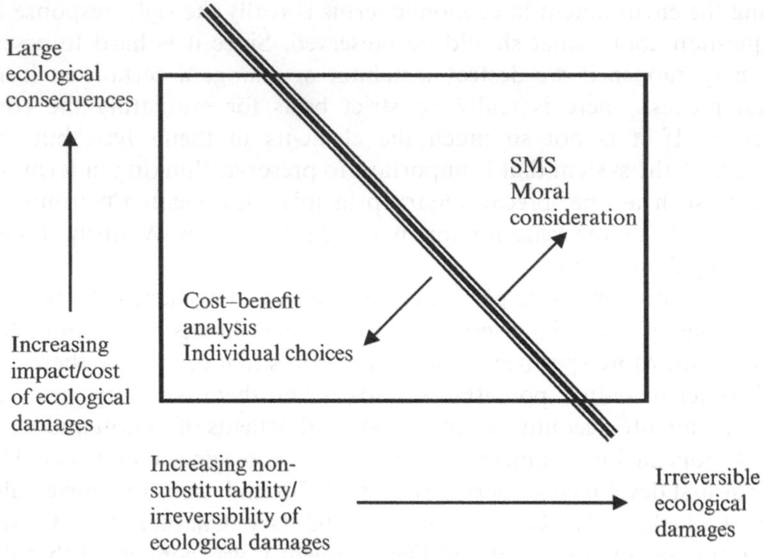

Source: After Toman (1994).

Figure 11.1 The safe minimum standard

some common ground for consideration of differences in conception among economists, ecologists and ethics' (ibid.: 405).

Toman suggests that in the case of reversible changes and small environmental stakes, individual preferences and CBA should reign. In the opposite situation – that is, that of irreversibility and large environmental stakes, SMS should be used. Figure 11.1 illustrates the idea. The basic choice to make is to define the borderline between the two decision spheres – the one governed by CBA/markets and the one governed by SMS. As Toman draws the line, he puts more emphasis on irreversibility than on the size of involved (yet uncertain) stakes.

The idea is an interesting one, again because of its focus on precaution, especially if irreversibility is involved. However, in this case a process is also needed to decide when SMS should apply and what it will imply in each case. Second, other ethically relevant dimensions than irreversibility and the size of stakes may be involved. Consequently, there may be other reasons to restrict the domain for CBA/markets beyond the ones defined in Figure 11.1. The perspective does, however, support a basic position of this book: if no more than individual issues are at stake, letting markets decide is certainly OK.

11.7 SUMMARY

There are different ways to make decisions over how to use the natural environment and what to protect. Methods for doing this are here understood as value articulating institutions. By applying this perspective we emphasize that the various methods for establishing the environmental priorities influence how values can be expressed and thus what gets priority. This perspective links back to the issue of resource regimes and their influence on the development and protection of values as developed in Chapter 10.

Based on the neoclassical perception of choice, a set of VAIs have been developed – of which contingent valuation is the most prominent. Here the individual WTP forms the basic value measure. These methods are based on the assumptions that choices are individually calculative and that the individual has been or can be offered the information necessary to make the calculation. They are furthermore based on the assumptions that environmental goods can be viewed as commodities and that while real markets 'fail' to capture their values, simulated markets will do so.

There are some fundamental problems related to using monetary valuation in the field of environmental goods, three of which have been highlighted. First, we have the information problem – the problem of perceiving which values are involved and then transforming them into one single scale. Environmental goods primarily take the form of functions, which are fundamentally characterized by invisibilities. In addition, people are confronted with the problem of transforming unfamiliar dimensions into one measure. Second, we have the issue of incommensurable value dimensions – that is, the fact that while some issues can be treated in monetary terms, ethical issues are also often involved – issues that go beyond the logic of individual calculation. Finally, since environmental goods are often parts of interrelated wholes – they are systems features – demarcating them into separately valuable objects is both difficult and highly problematic.

While the information problem is a fundamental problem independent of the VAI in place, the various types such as CV/CBA, MCA and deliberative institutions have different capacities concerning their treatment. This issue has been briefly mentioned in this chapter and will be reviewed more completely in Chapter 12.

Concerning the ethical questions involved, the distinction between the consumer and the citizen – between monetary bids and normative reasoning – is a core issue. First, we have the issue of the 'rights' of other species – a topic that is fundamentally ethical. Second, we have the fact that the actions of different individuals are linked via the functioning of the environment.

We therefore influence each other's opportunities quite directly. Then it becomes fundamental to ask whether decisions about the common environment should be based on people's individual willingness/ability to pay or on a communicative process where the quality of various arguments for and against can be evaluated.

Theoretical clarifications concerning what characterizes environmental values and choices are important. However, these kinds of analysis are of minor value if not backed by empirical evidence. While we have presented much empirical evidence for our position in previous chapters, we have here added to our knowledge base by looking into the findings of monetary valuation studies themselves. This analysis documents that there are a number of deviations from the standard model of rational individual choice. 'Information bias', 'question order bias' and 'starting-point bias' are all concepts developed to 'explain' deviations from the model concerning respondents' treatment of the information given. These 'biases' are all examples of how the information procedures themselves profoundly influence WTP estimates. They also demonstrate the difficulties many respondents have with handling the values involved under the CV institutional setting.

The problems observed are probably even greater if we turn to the issue of preferences and preference formation. The data available confirm that many respondents have difficulty in transforming their values into one dimension. Thus, CV studies themselves seem to support the view that preferences are plural, that they are influenced by the institutional setting under which they are elicited, and finally that they are anchored to the status quo rights – that is, the WTP–WTA discrepancy. All these observations are counter to the assumptions of the neoclassical model.

Certainly, some of the errors observed are due to badly conducted CV analyses. Nevertheless, this cannot fully explain the many problems observed. Development of alternatives to monetary VAIs may therefore seem warranted. Such developments are under way, and these will be discussed and contrasted with the CV/CBA in Chapter 12.

NOTES

1. This is a difficult issue since there is no way one can objectively define who 'are affected' or 'have interests'. This is an issue of importance for any kind of VAI, not just CV.
2. It should be emphasized that many CV practitioners use focus groups to pre-test the quality of the material to be used in the study. This reduces the information problem by ensuring that the questionnaire material is readily comprehensible. However, the content of the discussions in the focus group is not thought to have any other bearing on the evaluation. In the final CV study, new people are interviewed in an individual setting.
3. Following the discussions in Chapters 5 and 6, there is nothing in the (simulated) market

situation of a CV study that forces the individual to act only as a consumer. However, the basic perspective is on the consumer perspective or 'role'.

4. Here I shall give only a very brief introduction to each method. For a more comprehensive summary, see Hanley and Spash (1993).
5. The method also assumes weak separability, implying that the MRS between two goods appearing in the individual's utility function is independent of the quantity of other goods.
6. According to the literature, 'WTP measures give an estimate of compensating variation for welfare-improving moves and of equivalent variation of welfare decreasing moves . . . Likewise, WTA replies give information about compensating variation for welfare-decreasing moves and equivalent variation for welfare-increasing moves' (Hanley and Spash 1993: 53). This implies an inconsistent treatment of the rights issue. While consistent with the logic of the standard welfare measures, a loss can only be measured against a situation where one previously had a right to what is now lost – that is, WTA. Likewise, WTP implies gaining something that one did not possess before. Therefore it seems consistent to estimate only compensating variation in both cases.
7. 'Open-ended' implies that individuals are asked to give their maximum WTP (WTA). 'Closed-ended' implies saying yes or no to a payment level proposed by the researcher – a dichotomous choice. If a higher (if yes) or lower (if no) price is then proposed, the procedure is called 'double bounded'. A so-called 'bidding game' – implying that increased amounts are suggested until a maximum accepted bid is reached – seems now to be less used.
8. This section uses material from Vatn and Bromley (1994).
9. The emphasis on the influence the social sphere has on preferences does not make it inconsistent to call them individual. Even in cases where a preference is attributed 100 per cent to social process, they are still also held by individuals. The more important the social element is, the more I tend to term something a value, though.
10. Multicriteria analysis is based on similar ideas – see Chapter 12.
11. This section is based on Vatn (2004).
12. Kahneman and Knetsch argue that the embedding effect is a result of the purchase of moral satisfaction – a 'warm glow' effect. The issue of 'warm glow' concerns the character of preferences, and will be discussed in Section 11.5.2.
13. The idea of 'mental accounts' is developed by Deaton and Muellbauer (1980), and deviates clearly from standard rationality assumptions.
14. In relation to this, note that the eliciting procedures used over the years in CV studies have been changed, in an attempt to avoid new problems. The move from open- to closed-ended formats was motivated partly by the fact that it would reduce the possibility for the bidder to act strategically and partly to better mimic the type of decisions we make in ordinary markets – that is, we buy or do not buy given the price offered (Arrow et al. 1993). Similarly, concerning closed-ended formats, there has been a shift from dichotomous 'bidding games' to single- or double-bound dichotomous choice. If the choice is single bound, the respondent is asked to say 'yes' or 'no' to only one price proposal for a certain good. If it is double bound, a 'yes' ('no') to the first price proposed is followed by a higher (lower) price proposal. In this way, some inconsistencies in the form of 'starting-point biases' in a bidding game with several steps (repetitive bidding game) is avoided. However, the material from this latter type of bidding game is of interest for us when evaluating people's behaviour. It shows how people reacted to that kind of structure. Furthermore, if standard rationality assumptions hold, the formats chosen should not have any effect. Finally, shortening the chain of successive bids, as in the case of single- or double-bound dichotomous choice, has not eliminated the starting-point bias – see the text where such bias is also observed in double-, and even in single-bound dichotomous choice.
15. Note that in this case, the same people were participating in the focus group and the CV study. It therefore deviated from the standard use of focus groups in CV studies where these are used only to carry out pre-tests of questionnaires and so on.
16. One should be aware that lexicographic preference orderings are not necessarily

absolute. Preferences that are thus bounded may only be bound within certain limits. This implies that preferences that are hierarchically structured may be traded off against other preferences if minimum levels are available. Michael Lockwood (see Spash 2000) distinguishes between four categories: a strong lexicographic ordering; a modified lexicographic ordering operating within thresholds; weak comparability where choices are made between goods without attributing a common value to them; and commensurability as in standard rational choice theory. This structure is parallel to that of O'Neill (1993) with incommensurability equating with lexicographic preferences (see Chapter 6). The difference is that O'Neill did not include the possibility that incommensurability was bound by a threshold.

12. Comparing value articulating institutions

This chapter will clarify the differences between a set of value articulating institutions (VAIs): cost–benefit analysis (CBA) (Section 12.1), multicriteria analysis (MCA) (Section 12.2) and deliberative institutions (DI) such as citizens' juries and consensus conferences (Section 12.3).[1] The aim is to focus on the principal differences. Those intending to use some of the methods should refer to more specialized and comprehensive sources. Some references are given in the text.

With regard to the clarification of differences, we shall look at two sets of dimensions: the first set concerns the *rationality* assumed and the kind of understanding of the *process of value assessment* and *articulation* that the method is based on. The other concerns the *aggregation* of individual preferences. Regarding the first set, we shall consider what assumptions are made concerning the values involved – that is, to what degree these are assumed to be commensurable and compensable.[2] We shall also look at how the VAI treats the complexities involved. This concerns specifically which assumptions are made concerning the capacity of individuals to clarify the issues involved and to do the necessary evaluations and computations. The second set of issues concerns how aggregation of individual preferences or priorities is handled. The VAIs are based on different understandings concerning interpersonal comparison and interaction. While CBA focuses on maximizing some aggregate of individual preferences, the other methods focus more on finding a reasonable or tolerable solution to a conflict, or developing a consensus concerning what the common good is or should be.

Certainly, CBA, MCA and the set of deliberative institutions to be presented here, do not cover all types of VAIs. However, those selected should give a good coverage of the main issues and positions involved. While some VAIs are more relevant for environmental decision making than others, no (existing) VAI can be said to be ideal. We shall therefore also look at ways of combining VAIs to counteract problems that are observed (Section 12.4).

Some of the problems encountered follow from the character of societal or public choices, and are thus common to all appraisals. Dominantly the VAIs discussed here are used to assess 'projects' such as transport issues, building of dams, regulation of air pollution, protection of some species, release of a genetically modified product and so on. In each case one has to define what are the alternatives, whose interests should count, which consequences are relevant, and how they should be accounted for. These challenges are common to all the VAIs, while they are solved differently. Our aim is to clarify how the basic assumptions underlying the various VAIs influence the chosen solutions and how well they fit the characteristics of environmental problems.

12.1 COST–BENEFIT ANALYSIS[3]

CBA is based on neoclassical welfare theory. The fundamental idea is very clear and rather simple. A project is to be prioritized according to how much benefit it gives less its costs. Thus it supports the maximization of the total benefits for society. In principle, CBA can be used for both public and private issues, but it is developed mainly for public decision making – that is, when markets (as perceived) fail – implying that environmental decision making is one of its core areas.

According to Boardman et al. (2001: 7) a CBA has the following steps:

1. specify the set of alternative projects;
2. decide whose benefits count (standing);
3. catalogue the impacts and select measurement indicators;
4. predict the impacts quantitatively over the life of the project;
5. monetize (attach dollar values to) all impacts;
6. discount benefits and costs to obtain present values;
7. compute the net present value (NPV) of each alternative;
8. perform sensitivity analysis; and
9. make a recommendation based on the NPV and sensitivity analysis.

This is a structure generally accepted across textbooks in the field.

While the basic idea is straightforward and strong, there are several challenges confronting the CBA analyst. First, who should define the alternatives? In the case of a transport problem many alternative solutions may be of interest. Defining these and selecting which should be studied in more detail is a crucial issue. However, this is a problem common to all VAIs and concerns the authority relationship among the decision maker, the analyst and the consumer/citizen. In a CBA, the alternatives are normally defined by the relevant decision maker – for example, a public body like a ministry, a county or local council – or sometimes by the analyst.

Concerning step 2 – whose benefits counts – there is no simple answer, either. A transport project may involve the destruction of habitats. Should only locals have a say or should the interests of all inhabitants in the whole country be considered? What about foreign tourists who might like to visit the spot? Existence values might be involved. Then maybe every citizen in the world should have a say? Again the basic problem is, in principle, the same for all VAIs, although it is tackled differently. In CBA those with a standing are perceived as consumers. They are thought to represent only themselves and their personal preferences.

The next steps concern the definition (3) and measurement (4) of relevant impacts. These issues are related to the vector of attributes, and the problem is not least to define which systems boundaries are relevant. This is especially crucial in the case of environmental issues. As earlier emphasized, these goods are often linked. They take the form of a set of processes rather than being separate items. In the CBA framework, both steps 3 and 4 are taken care of by the analyst (Boardman et al. 2001). Hence, the definition of impacts is thought to be a technical issue. This is consistent with the general perspective of neoclassical economics – that the world around us is directly comprehendible independent of various social constructs. Defining and measuring impacts is therefore not considered to be a value-laden issue, and can safely be taken care of by the analyst/experts.

Step 5 concerns the valuation – the weighting of the different impacts or attributes. In CBA this is done by assigning monetary values based on individuals' willingness to pay as revealed in real markets or by various valuation techniques such as CV. This is a core feature of CBA and has several implications. First of all, social choices are to be based on individual preferences, more specifically the intensity of individual preferences as measured by the willingness to pay. Second, it implies that all values involved are considered commensurable. They can be transformed into one measurement scale. Values are also *compensable*, implying that a loss observed in one attribute or good can be compensated by a gain in another. Finally, it is assumed that the individual – *qua* individual – is the ultimate judge and has well-informed preferences concerning the goods involved.

As discussed throughout this book, objections can be raised against all these assumptions. To some extent the assumptions are also problematic within the current neoclassical model itself. As mentioned in Chapter 6, neoclassical economics has abandoned interpersonal comparison and holds that preferences should be looked at merely as a ranking of goods. By summing WTP estimates, CBA actually undertakes interpersonal comparison.[4]

In addition, willingness to pay depends not only on preferences, but also on ability to pay. Consequently, allocations will depend on the distribution of income. There are ways for CBA to deal with this latter problem. One can assign different weights to the various individuals. Dasgupta and Pearce (1972) discuss some of the philosophical questions involved and practical ways of dealing with this. Weisbrod (1968) proposed using the weights implicit in past governmental decisions to obtain distributional weights. Krutilla and Eckstein (1958) pointed specifically to the information implicit in the (progressivity of the) income tax schedule.

The assumption of commensurability and compensability implies that choosing is always about making trade-offs.[5] These are to be made by the individual. This demands that the individual is informed and has preferences about the goods involved. It also demands that there is no problem for the individual to undertake the necessary calculations and that the issues involved are well understood by the various individuals. Since CBA is often used for issues concerning the common good, as in the case of environmental problems, one may question the emphasis on individual preferences more fundamentally. It may be argued that, for example, communicative processes based on arguments about what is right and wrong might seem more relevant.

A CBA may to a substantial degree be undertaken on the basis of market prices. This will dominantly be the case for costs related to investments and running costs of a project, since market prices for these usually exist. In some cases these costs have to be corrected due to imperfect competition and government 'interventions' since it is the 'perfect market' that forms the reference point. Environmental costs and benefits must also be estimated. Normally these do not exist in the form of market prices, and various methods such as hedonic pricing, travel costs or contingent valuation must be applied. In some cases these costs and benefits are estimated directly for the actual project. To an increasing degree, so-called 'benefit transfer' (Navrud 2004; Navrud and Ready 2004) is used. This implies that benefit estimates made for one situation – for example, estimating the value of a certain wetland – are used for a project where another wetland area is at stake. The reason for this practice is to reduce the costs of data collection. Normally the 'transfer' implies some correction due to changes in socio-economic conditions.[6]

Steps 6 and 7 concern aggregation and are also crucial characteristics for CBA. This implies the calculation of the difference in costs and benefits in NPV terms:

$$NPV = \sum_{t=0}^{T} \frac{B_t}{(1+r)^t} - \sum_{t=0}^{T} \frac{C_t}{(1+r)^t} \tag{12.1}$$

where B_t are benefits and C_t are costs in time period t, r is the discount rate and T is the time horizon of the project.

There are two important value issues involved here. While some may experience the benefits, the costs may fall on others. More precisely, net benefits may be positive for some and negative for others. In any project it is unlikely that all persons involved gain. Thus, the NPV calculation of CBA does not secure Pareto optimality. Instead CBA is based on the Kaldor–Hicks rule, the rule of potential Pareto improvement.

The second value issue concerns discounting future effects. A social discount rate is used, and the choice of this rate raises an ethical question (Hanley and Spash 1993). There are some unresolved issues in economic theory about the choice – specifically the relationship between private (market) and social discount rates (see Portney and Weyant 1999). The private discount rate is assumed to equal the marginal rate of time preference, and in perfect markets it will be equal to the marginal rate of return on capital. Marglin (1963) offers three arguments for the social rate being less than the private one. First, society might want to save more collectively than what comes out of the sum of individual decisions. Second, individuals, as members of a society (citizens), may have other time preferences than as consumers. This may especially apply to environmental benefits and costs. Finally, individuals have a time horizon related to their life expectancy, while society must also consider future generations. It is interesting to see that Marglin accounts for the differences between the role of the consumer and that of the citizen. Since CBA is based on the perspective of the individual consumer, this might be viewed as inconsistent – as mixing two different principles.

The effect of a positive discount rate is that future benefits and costs count less than present ones.[7] The choice of discounting and the choice of the rate are highly contested issues. Especially in the case of environmental problems like climate change and biodiversity loss, the dominant part of the costs will not arise until long into the future, while the benefits of economic activity endangering the climate and the species are immediate. This may make it hard for long-term environmental projects to pass the NPV test even with low discount rates. Spash (2002) offers some instructive examples and evaluations in the case of greenhouse gas emissions.

Returning to the issue of long-run sustainability, we see from the above that the only sustainability rule consistent with CBA is that of weak sustainability (Chapter 9). This is based on the rule of compensatory decisions, implying that the various goods are considered fully substitutable. In cases where critical natural capital is involved – that is, natural capital which cannot be replaced by produced capital – discounting becomes inconsistent. No discount rate – not even a negative one – can consistently handle this kind of problem. Bromley (1989) and Page (1997) argue that future interests have to be taken account of by formulating rights for future generations in, for example, natural resources. In relation to this, one should recognize that discounting future *production* may not endanger the utility of coming generations if all goods are substitutable and technological progress is at least as high as the discount rate. Discounting the *utility* of future generations is something different. It discounts their welfare and goes directly against the ethical rule of sustainable development – both weak and strong. It simply implies that their utility counts less than ours.

Step 8 is the sensitivity analysis. The impacts of a project may be uncertain. The same goes for the monetary values and the distributional weights assigned. Finally, the discount rate can have a crucial impact on the ranking of the projects. To evaluate these uncertainties, sensitivity analysis is demanded and the various types of software available have made this fairly straightforward – for example, Monte Carlo simulations. However, as Boardman et al. (2001: 16) argue, there are certainly practical limits to such analysis:

> Potentially, every assumption in a CBA can be varied infinitely. In practice, one has to use judgement and focus on the potentially most important assumptions. Although this can mean that CBA is vulnerable to the judgment biases of the analyst, carefully thought-out scenarios are usually more informative than mindless varying of assumptions.

CBA can coherently handle uncertainties in the form of risk as defined in Chapter 9. That is easily integrated into the NPV calculation itself and further evaluated via the sensitivity analysis. Certainly, if risks are involved, sensitivity analysis becomes very important. If there is uncertainty, especially if radical uncertainty is involved, the method faces problems. There is one option available – to formulate constraints for some impacts, and remove alternatives not conforming to these constraints in the final evaluation. As an example, if important habitats are involved, an option is to formulate a restriction on how much habitat loss can be accepted, if any. Alternatives not passing this extra test should not be accepted despite high NPVs. This is a practice in line with the ideas of Toman (1994) as presented in Chapter 11. It is, however, an option that is rarely used, perhaps because it is outside of the 'world' of substitution and trade-off so basic to CBA. One should also acknowledge that there is no specific procedure defined in CBA for how to formulate such restrictions.

The final step – step 9 – is to recommend which solution should be adopted. In principle, this is simple. The analyst should propose the project with the highest NPV (Boardman et al. 2001). In cases with restricted budgets, however, this may not result in the right priorities. Choosing on the basis of the benefit–cost ratio is then favoured. In this way, maximum benefit is secured within the given budget. If constraints are set for any resources involved, these should certainly be considered when making the final choice.

Summing up, CBA focuses on finding the optimal solution to a decision problem based on the potential Pareto improvement rule (Kaldor–Hicks). It assumes value commensurability and compensability, and priorities are made on the basis of individual preference intensity as measured through willingness to pay measures. Individuals are assumed to be utility maximizers. Trade-offs over time are made on the basis of net present value calculations.

12.2 MULTICRITERIA ANALYSIS[8]

An environmental issue may foremost be viewed as a conflict between various interests or values (Schmid 1987; Paavola 2002). Decision making is then about *conflict resolution* – about identifying the best *compromise* between competing ends or interests. Whether the gainers gain more than the losers lose – that is, the Kaldor–Hicks criterion – may not persuade the losers. Instead the CBA may create a hostile environment for the final decision process.

MCA has a structure that may offer useful support when analysing conflicts. There are two fundamental aspects involved: conflicts between different interests and values held by different individuals or groups; and conflicts between different interests or values held by the same person.[9] The concept of a conflict may imply that interests and values are multidimensional and not easily traded against each other. MCA is formulated to handle values or criteria that are not easily transformed into one dimension such as a monetary measure. This is actually the core of MCA as the name also indicates: criteria are *multidimensional*, and the method allows for handling criteria that are *incommensurable* (Martinez-Alier et al. 1998). It can also handle the fact that weights may only be considered *coefficients of importance*, not signalling trade-offs/compensability (Munda 1996).

Another principal difference between CBA and MCA relates to the fact that MCA puts much emphasis on *the process*. Decision making, especially when environmental issues are involved, is a complex process. Problems are often ill-defined. MCA offers a distinct response to that challenge; not least by the way it fosters learning.[10] From this perspective, MCA can be described as a structured search process where the analyst supports the decision maker or the actual interest group(s) in defining the problem, looking for alternatives, assessing their consequences, ranking the alternatives, perhaps going back and formulating new alternatives and so on.

MCA is thus a response to *bounded rationality*. The problem of restricted cognitive capacity becomes especially important in situations characterized by complexity and unfamiliarity. In Chapter 11 we referred to Tversky (1969) and the point that it is simpler to compare alternatives dimension by dimension than it is to evaluate each good across all dimensions and then compare these total assessments.

Hence, there are three reasons why we should not expect any form of optimal solution to a decision problem: interests and perspectives are conflicting, problems are ill-defined, and decision makers have restricted cognitive capacity. The socially constructed perception of the problem influences in a crucial sense what is defined as a problem, whom it concerns, and the ways in which it can be solved. Done properly, MCA can make these issues more transparent and secure more self-reflection in the decision process. It cannot remove the problems.

The above may give the reader the impression that MCA is just one method. Here another difference to CBA appears. While the latter is based on one theoretical foundation, MCA is actually a group of methods whose theoretical foundation can be split into two main classes. First, we have a class of MCA methods which are utility based, with multiattribute utility theory (MAUT) having the core position. Here the focus is on utility functions/cardinal weightings. MAUT-based methods are therefore quite similar to CBA concerning the assumptions about commensurable value dimensions and the potential for making trade-offs (compensability). These methods use aggregating procedures where a single value of the different alternatives involved is computed and thus a ranking of them can be made according to a one-dimensional criterion (Nijkamp et al. 1990). The main difference compared to CBA therefore concerns how utility information is produced.

The second class of MCA emphasizes the aspect of incommensurable impact and value dimensions, lacking or restricted trade-off possibilities. These are focused on either avoiding aggregation or structuring it very differently from that of CBA/MAUT. Hence, the distinction made above between CBA as focusing on commensurability and trade-offs, and MCA as focusing on weak commensurability/incommensurability and non-compensability is actually relevant only for this second group of MCA procedures. Since the assumptions underpinning the latter methods are the most interesting, given our purpose and understanding of environmental problems, most weight will be put on these. This choice is also motivated by the fact that MAUT-based methods are characterized by many of the same strengths and weaknesses as CBA and these have already been discussed.

It is quite logical that MCA is not one, but a set of methods. This follows from the fact that there are many ways of ranking or excluding projects when we move away from commensurability and one-dimensional aggregation procedures. Different logics may then apply. Therefore, in Section 12.2.3 we shall look at some of the main principles concerning the ranking of projects. The literature is overwhelming concerning aggregation procedures. We shall just look at a few to illustrate some principal aspects concerning decision making when goods with multiple attributes are involved.

12.2.1 The Basic Structure of MCA

A core structure in a multicriteria analysis is the MCA matrix, illustrated in Table 12.1.

The problem may be to solve a transport issue or convert a forest into a residential area. A set of alternative solutions must first be defined. A transport problem may be solved by building a railway, setting up a bus system or by building a motorway. Next, a set of criteria is defined, where monetary costs (normally investments and running costs measured in market terms), landscape changes, time saved, accidents, pollution and so on may be relevant. The impacts of each alternative for each criterion are called scores. They are measured in the most relevant dimension, such as money for the costs, hours for time saved and so on. While these are all cardinal measures, qualitative scores may have to be used for some criteria. From the list, such ordinal ranking may typically be the case for many landscape effects since they may be difficult to quantify. It is possible to say that a landscape is more beautiful than another, but not how much more beautiful.[11]

Finally, the procedure may involve ranking various criteria – that is, to define the weights as emphasized in Table 12.1. Whether weights are used and what form they take depends on the assumptions made concerning compensability and commensurability. They may signal trade-offs (compensability) or be coefficients of importance.

From the above we observe that the MCA is actually mirroring the process of value calculation as discussed in Section 11.3 quite directly. It focuses on criteria (= attributes **a**) and on weighting these (**w**). It can

Table 12.1 The basic structure of an MCA matrix

Criteria	Alternatives				Weights
	a	b	. . .	*x*	
1	(scores)	(scores)	(scores)	(scores)	(weights)
2	”	”	”	”	”
3	”	”	”	”	”
. . .	”	”	”	”	”
y	”	”	”	”	”

therefore be seen as a response to many of the challenges mentioned in Chapter 11. It tries to treat the information problem explicitly and it allows incommensurabilities.

12.2.2 The Steps of MCA

While Table 12.1 gives the main components, an MCA will typically include the following set of steps:

1. define and structure the problem;
2. define the alternatives (the possible solutions);
3. define the set of evaluation criteria:
   - how many
   - type – whether comparable or not (cardinality, ordinality and so on);
4. characterize the alternatives – that is, assess the scores;
5. identify the preferences (weights) of the decision maker or different interest groups involved;
6. compare the alternatives – if relevant, choose aggregation procedures and aggregate; and
7. evaluate the result – including sensitivity analysis – and choose or propose the best compromise (often involving going back to (1), (2) or (3) and run the process for a second round).

The list of steps is not very different from that of the CBA. The differences may more typically be found in the way the steps are performed and what role they play in the entire process. Comparing with CBA, the focus on problem definition is more explicit in the case of MCA. This is due to the greater focus on the complexity or 'fuzziness' of decision problems – that a problem can be understood in many different ways. Defining it is therefore a fundamental step. The way the problem is formulated determines the conclusions that can be drawn. However, while there is a difference in practice, which may seem to flow from divergences in the theoretical underpinning, there is nothing that prevents CBA practitioners from putting as much weight on this issue as do most MCA analysts.

A more important difference concerns the role of aggregation. While CBA focuses on trade-offs as measured by prices/WTP estimates, MCA methods exist that can treat incommensurable criteria or weakly commensurable (that is, ordinal) weights. Cardinal weights may also be used, but do then not necessarily signal trade-offs. They may instead be considered only as coefficients of importance. They measure how much more important a criterion is compared to another without implying that an increased amount of the less-valued criterion can compensate for the loss related to the higher-valued one. As an example, some may argue that it is three times as important to protect wolves as bears. However, tripling the number of bears may not compensate for not protecting any wolves.

Finally, compared to CBA, the MCA procedure focuses more on an iterative process where alternatives are also changed as an effect of the learning implied by the process. In some situations this may involve reformulating the whole problem. In other situations what is learned has less fundamental importance, resulting only in a reformulation of the alternatives and/or the criteria. In the above list of sequences, such iterations are motivated at the stage of evaluation of results – stage 7. Certainly, returning to higher levels of the process can take place at any stage if found preferable.

Bana e Costa and Vincke (1990: 5) formulate the learning aspect and the necessary openness of the process in the following way:

> Thus the activity of multicriteria decision aiding can not be only restricted to the resolution of a problem where one has to aggregate *given* preferences to a *given* set of potential actions. The identification of the set of actions (and of the fuzziness of its frontier), the construction of the criteria and the preference modelling are fundamental and often difficult aspects of decision-aid. (original emphasis)

While the foundation of CBA is the rational agent with given preferences, MCA tries to capture the alternative view that preferences are not clarified. They may rather develop or become clarified as part of the decision making process itself.

12.2.3 The Participants of MCA

The *classic* MCA situation contains a decision maker (DM) and an analyst. The DM may be a political body like a city council or a ministry, or it may be the board of a firm. This is similar to CBA. The roles and interaction of the involved parties are, however, somewhat differently interpreted. Consistent use of MCA implies that the DM is strongly involved in the problem formulation, and in defining both the alternatives and the criteria. Traditionally, the scores have been assessed by the analyst who may seek assistance from specialist expertise. If weights are used, the DM again sets these. This is different compared with the CBA where consumers provide these in the form of WTP estimates. Finally, the analyst undertakes the aggregation (if applicable) and the sensitivity analyses.

There are alternatives to involve the DM directly in the assessments of the various alternatives. We shall briefly look at two: involving stakeholders and involving citizens. This involvement does not imply that there is no

DM in these cases. Somebody will still have to make the final decision. The point is that the various alternative solutions are first evaluated by either stakeholders or citizens.

Concerning a stakeholder-based MCA, the idea is that the issues at stake involve a set of specific interests. Banville et al. (1998) define stakeholders as those having a vested interest in an issue. They may be causing the problem, they may be affected by it or they may be both affecting it and affected by it. Defining who this is, is not a simple issue. As clarified in the section on CBA, who has interests in an issue depends not least on how it is understood. It is thus at the level of problem formulation that the structure is set which next defines who should be involved. Thus, while stakeholders invited to participate may reformulate the problem as part of the later process, the initial problem definition becomes important since it defines who becomes stakeholders in the first place.

The focus of a 'stakeholder MCA' is on generating the 'best' compromise between the involved and often conflicting interests. This part of the process may take different forms. One solution is to let each group of stakeholder representatives assess the various alternatives and reach a conclusion concerning their priorities. Normally, stakeholders will prefer different solutions. To develop a solution that can act as 'best compromises' between the involved interests, the analyst may, on the basis of the information from this first round, develop a new set of alternatives. Then the different stakeholder groups assess these alternatives in a second round (Stewart and Scott 1995). In principle, several rounds can be included. The basic point is that the development of the final solution – the compromise – lies in the formulation and reformulation of the problem and then the alternatives. At some stage the stakeholder assessment is closed and the material is presented to the DM who makes the final choice. Stakeholders may have agreed on a proposal, giving it much weight in the final evaluation. Alternatively, if no compromise is reached, much more is ultimately left with the DM.

Instead of involving stakeholders, a representative set of citizens can be invited to participate in the MCA assessments. In this solution, the focus is more on the 'general interest' as perceived by these representatives than the special interests as defined by stakeholders. Renn et al. (1993) suggest that a combination of stakeholders and citizens' participation may be preferable in that stakeholders may more clearly define what the stakes are (the formulation of the problem, alternatives and criteria) while the citizens may do the evaluation of which stakes or interest should be protected (the weighting).

In recent years we see a tendency to problematize the setting of scores. An increasing number of disputes over the assessment and understanding of the effects of various solutions to a problem has reduced the confidence in expert assessments. This is not least typical for environmental issues.

A good example is the controversies surrounding the use or release of GMOs, where the main conflict is more about assessing and evaluating the consequences of their use than over weighting these consequences (Stirling and Mayer 2001). One consequence of this is an increased tendency to involve stakeholders or citizens directly in the process of assessing the scores, and not treat that process as a purely technical issue.[12]

12.2.4 Different Procedures for Comparing Alternatives

With regard to the procedures for comparing alternatives, the assumptions made about the commensurability and compensability of the values involved are crucial. These form the basis for a set of different MCA methods – especially various types of weighting and aggregation procedures. For the purpose of this analysis I shall divide them into four:

1. no explicit weighting of criteria and therefore no aggregation (incommensurability);
2. commensurability and compensability is assumed;
3. commensurability is assumed, but not compensability; and
4. neither full commensurability nor compensability is assumed.

In the following, I shall give a very brief overview of the core characteristics of each type and a representative method. The purpose is to illustrate various ways of handling the more technical issues as defined. From the above we have seen that MCA can involve people in different capacities or roles. In the following discussion I shall refer to only one capacity, the DM supported by an analyst. This simplification is made because who expresses values or priorities – be it the DM, the stakeholder on the citizen – does not influence the technical issues concerning their aggregation, which is the focus here.

No explicit weighting and no aggregation

No explicit weighting implies that the decision maker makes the choice of the alternative s/he wants to prioritize directly on the basis of the information included in scores such as those in Table 12.1. From that perspective the MCA becomes very similar to a so-called environmental impact assessment (see, for example, Edwards-Jones et al. 2000). A simple example of a table of this kind is given in Table 12.2.

In an MCA context, the analyst can provide more help to the DM even though s/he does not want to offer any weights for the criteria. A *concordance set* can be established, based on a pairwise comparison of the alternatives and lists for which criteria alternative a is better than

Table 12.2 A scores table of a transport problem

Criteria	Units/scales	Alternatives		
		Highway (a)	Train (b)	Bus (c)
1. Costs	Million euros	20	40	15
2. Time reductions (per person)	Minutes/day	25	15	10
3. Emissions	Tons/year	1000	120	350
4. Landscape effects	+ + +/− − −	− − −	−	− −

alternative b; a is better than c; b is better than c and so on (Nijkamp et al. 1990; Munda 1995). This list may help to give an overview to the DM and simplify the more implicit evaluation of how the importance of each criterion influences the final conclusion. If an alternative is better than all other alternatives on all criteria, we have a so-called 'ideal point'. The alternative dominates all other alternatives and normally the choice is then fairly simple. Uncertainty as to which criteria should be involved and the problem with uncertainty about the scores may complicate the conclusion, though.

One alternative *dominates* another if it scores at least as well as another on all criteria and is strictly better at least on one. This opens up for further sorting without including weights. In principle, all alternatives that are dominated by at least one other alternative can be excluded for further investigation. Again, uncertainties concerning the scores warrant some caution. When all dominated alternatives are excluded, we are left with the so-called 'efficient set'. We observe that if this set contains only one alternative, we are back to the ideal point.

Figure 12.1 illustrates this in a simple situation with only two criteria. Alternatives c, d and e constitute the efficient set. An ideal point alternative would have had to score better than alternative e on criterion 1 and alternative c on criterion 2.

If such a sorting results in an efficient set containing more than one alternative – that is, no ideal point – then some further discrimination between these is needed. This is the situation with the example in Table 12.2. Actually no alternative can be excluded on the basis of domination in that case. All are in the efficient set as defined. Given no explicit weighting, the DM must chose among all the non-dominated alternatives on the basis of the structure of the scores. If the DM is not able to choose on the basis of this information, turning to some sort of explicit weighting is an alternative. In the following we shall differentiate between three options, depending whether commensurability and compensability is assumed or not.

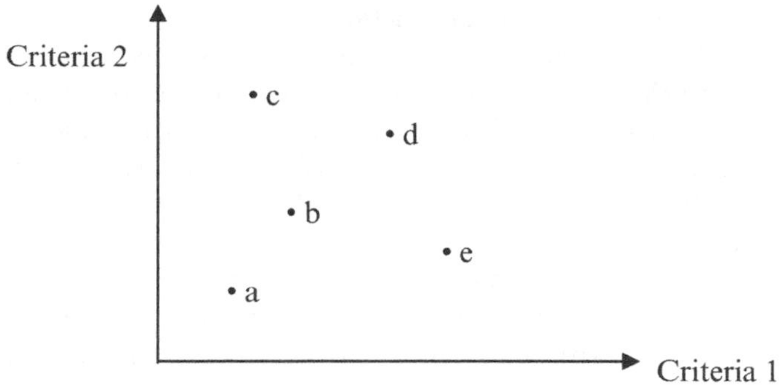

Figure 12.1 Efficient solutions in a situation with two criteria

Commensurability and compensability is assumed: MAUT

Multiattribute utility theory (Keeney and Raiffa 1976; Nijkamp et al. 1990) covers a set of methods which make the aggregation on the basis of the assumption that values or weights are commensurable and compensable. These methods thus demand that scores are cardinal.

MAUT-based methods can in principle take two forms:

$$\text{General additive form: } U_j = \sum_{i=1}^{n} u_i[g_{ij}] \tag{12.2}$$

where U_j is the utility of alternative j, u_i is a utility or value function which is a non-decreasing function of the scores g_{ij} of criterion i on alternative j.

$$\text{Weighted summation: } U_j = \sum_{i=1}^{n} w_i \cdot p_{ij} \tag{12.3}$$

where w_i is the non-negative weight given to each criteria i and p_{ij} is the standardized scores of criteria i on alternative j.

Weighted summation (12.3) is a special case of the general form (12.2) and is technically rather simple. It demands that scores are standardized. Since scores are normally measured along different scales (see Table 12.2), the computation is meaningless without this procedure. It implies making the scores on each criterion relative to one another, typically by dividing all scores for a criterion by the largest score for that same criterion. Standardized scores hence get values between 0 and 1. Various methods for standardizing are found in Nijkamp et al. (1990) and Munda (1995). Since the way the standardization is done may influence the conclusions, extra uncertainty is involved.

The above methods are defined for the deterministic case. MAUT-based aggregation routines are also developed for the probabilistic case.

Non-compensatory values: outranking methods

The weighting of the various criteria may be understood as coefficients of importance. The DM considers them to be non-compensatory, while still in this case commensurable. Also for this situation, a set of methods is developed. The so-called 'outranking methods' like ELECTRE are typical representatives (Roy 1990; Munda 1995).[13] These are based on two elements taking explicit account of the fact that trade-offs are not allowed.

First a measure of the dominance of one alternative over the others is computed. The outranking relation aSb implies that alternative (a) is preferred over alternative (b). This relation is accepted if the concordant coalition

$$C(aSb) = [g_i \in G: g_i(a) \geq g_i(b)] \tag{12.4}$$

is, 'sufficiently' large. $g_i(a)$ is the score of criterion i for alternative (a). This coalition is based on the concordance set for each pair of alternatives. It is measured in the following way:

$$c(a, b) = \frac{\sum_{i \in (aSb)} w_i}{\sum_{i=1}^{n} w_i}. \tag{12.5}$$

The sum $\Sigma_{i \in (aSb)} w_i$ is called a *concordance index* for (a, b) – that is, the sum of weights for all criteria where in this case alternative (a) dominates alternative (b). The measure c(a, b) is the concordance index divided by the sum of all weights. Implicit in this, the weights must be cardinal (commensurable). The DM must define thresholds for when alternatives are distinctly better or worse on each criterion. This follows from the fact that uncertainties may exist and this implies that scores have to differ somewhat before one can conclude that they are really different. Furthermore, a level of c(a, b) must be defined for demanding that (a) is better than (b) – that is, the concordance threshold $c(a, b) \geq s$, where s normally $\geq 1/2$. So if more than 50 per cent of the weights belong to criteria where an alternative dominates another, it is said to be better.

This is still not sufficient given the fact that weights are non-compensatory. A second step is established to secure that the alternative prioritized on basis of (12.5) does not score too badly on any of the criteria for which it is dominated by other alternatives. Thus, for these a so-called 'discordance threshold' is established:

$$g_i(a) - g_i(b) \leq v_i. \tag{12.6}$$

This threshold takes the form of a 'veto'. Since trade-offs are not allowed, the logic becomes to ensure that no chosen alternative scores too low on criteria where this is considered problematic. The outranking methods tries to find an alternative that both scores well on prioritized criteria and does not perform too badly on criteria where it is still dominated by other alternatives.

It should be emphasized that the function of scores in the case of outranking methods is only to determine for which criteria one alternative is considered better than another. Scores do not count directly in the procedure of 'aggregation' as we saw was the case in, for example, weighted summation. Thus, these methods demand that *weights are set taking the level of scores into consideration*. A criterion of generally high importance, but with scores that are almost equal across all alternatives, may consequently get a lower weight than a generally less important criterion where alternatives score very differently. As an example: loss of habitat may be considered the environmental issue of highest priority for a DM. However, if the alternatives considered all 'consume' approximately the same amount of valuable habitats – for example, in the case of a transport problem – distinguishing among these on that criterion does not become important and the weight will be low.

A main challenge when using the ELECTRE methods seems to be the rather complex structure of thresholds that is often needed. Certainly, the DM may have problems with assigning cardinal weights too. If so, methods that can handle weakly commensurable weights are an alternative.

Weakly commensurable and non-compensatory values

If the DM cannot assign cardinal weights to the various criteria, several options are available. If s/he is unable to give a ranking either, we are in principle back to the situation depicted above with direct choice based on the fact that weights were incommensurable. If s/he is able to rank (implying ordinal weights or weak commensurability), it is possible to perform an analysis on the basis of the scores and these weights. REGIME is one such method (Hinloopen et al. 1983; Hinloopen and Nijkamp 1990).

To explain how the method works, let us use the example in Table 12.2 reduced to a situation where there are just two alternatives and three criteria involved as in Table 12.3. Given the scores for the alternatives on the various criteria, the DM offer weighs in the following order: $w_3 > w_1 > w_2$. We see that alternative (a) (highway) is better than (b) (train) for criteria 1 and 2. For criterion 3 the situation is the reverse. None of the two alternatives is dominated by the other.

All scores are cardinal in Table 12.3. It does not matter for REGIME whether they are ordinal, cardinal or mixed. The core of REGIME is to

Table 12.3 A simplified version of the transport problem

Criteria	Units/scales	Alternatives		Weights
		Highway(a)	Train(b)	
1. Costs	Million euros	20	40	++
2. Time reductions (per person)	Minutes/day	25	15	+
3. Emissions	Tons/year	1000	120	+++

calculate so-called 'success indices'. These give the *probability* that each pair of alternatives will be ranked in a specific way, dependending on all cardinal values the weights w can take which are consistent with the ordinal ranking – that is, those that in our example are consistent with $w_3 > w_1 > w_2$. As an example, $w_1 = 0.02$, $w_2 = 0.01$ and $w_3 = 0.97$, and $w_1 = 0.33$, $w_2 = 0.32$ and $w_3 = 0.35$ are both consistent with that.

Let $C_{a,b}$ $(= \Sigma_{i \in (aSb)} w_i)$ denote the concordance index for the pair of alternatives (a, b) and $C_{b,a}$ $(= \Sigma_{i \in (bSa)} w_i)$ the similar index for (b, a). The important issue in REGIME is *the sign* of the indicator:

$$\mu_{jj'} = C_{jj'} - C_{j'j} \tag{12.7}$$

or in our case $\mu_{a,b} = C_{a,b} - C_{b,a}$. The sign is interesting simply because it indicates when one alternative is better than (dominates) another.

From Table 12.3, we see that the concordance index $C_{a,b} = w_1 + w_2$ and the concordance index $C_{b,a} = w_3$. From this it follows that the indicator $\mu_{a,b} = w_1 + w_2 - w_3$ and $\mu_{b,a} = w_3 - w_1 - w_2$. The probability that alternative a (highway) is better than b (train) can then be computed by an algorithm as illustrated by Figure 12.2.

The corners of the triangles represent the extreme points – that is, points where μ has its maximum absolute value or where it is zero. In the bottom left-hand corner we have the situation where criterion 3 gets all weight (that is, $w_3 = 1$ and $w_1 = w_2 = 0$). In this case $\mu_{a,b} = -1$ ($\mu_{b,a} = 1$), which is the highest absolute value of this indicator. Alternative b dominates alternative a. In the bottom right-hand corner $w_3 = w_1 = 0.50$ while $w_2 = 0$ the two alternatives are equal – that is, $\mu_{a,b} = \mu_{b,a} = 0$. All points on the line dividing the triangle in two define a situation where the two alternatives are equally ranked. Finally, the upper corner shows a situation where it is alternative a that dominates – that is, $\mu_{a,b} = 0.33$, which is the highest value the indicator can take given the concordance index and the fact that $w_3 > w_1 > w_2$. The success indices can be understood as the relative size of each subtriangle

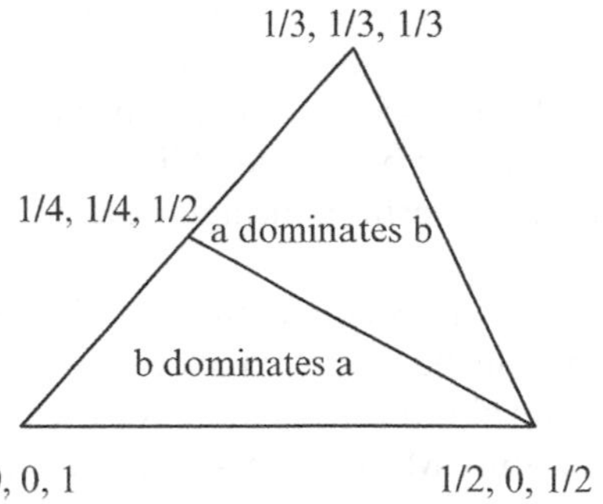

Figure 12.2 Illustration showing how success indices are calculated in REGIME

compared to the complete triangle in Figure 12.2. These fractions give the probability that alternative a dominates alternative b and vice versa. If the probability is 1.00, one alternative dominates another with certainty as the problem is formulated.

Normally there are more than two alternatives. Success indices are then computed for each pair of alternatives. Thereby a ranked order of the alternatives can be established. As in the case of outranking methods, when setting weights, the DM must take the level of scores into consideration. The arguments for that are similar, since scores play the same role in these two methods. They establish dominance relations only.

The algorithm used in REGIME implicitly assumes uniform distribution of all weights in the interval [0,1]. This may be the least prejudiced solution, but it makes the results vulnerable to the number of criteria. In other words, if there are few criteria for which an alternative scores positively, it may not be prioritized even if these criteria are given high priority by the DM and are actually much more important than the other criteria. The method does not capture this since it considers only ordinal information. If some alternative ranks positively on very few criteria, while these are still considered very important, splitting these criteria into a larger set is an option for providing more information. If each part of a criterion split in, for example, two, still both get a high rank compared to the other criteria, this emphasizes their importance, and the results from the REGIME analysis are more reliable.

Some concluding remarks

The above presentation represents a very simple introduction to some of the basic issues involved when systematizing the discrimination between alternatives as this is done in a selected set of MCA methods. The presentation covers the most typical assumptions concerning the use and understanding of weights. There are pitfalls with all the methods, and only some have been mentioned. While MCA may seem to fit the principal characteristics of environmental decision making better than CBA, there are still many problems involved. Especially concerning the aggregation procedures, there are issues that warrant great care.

The DM may have clear views concerning whether the criteria are commensurable or compensable, but more typically this is not the case. Instead it is a question of learning about the character of the problem involved and to the DM's own perceptions. MCA fosters such learning. The analyst must still support the DM in this process.

In trying to avoid the problems of CBA with its assumptions of commensurability and compensability, MCA practitioners have established algorithms that are themselves vulnerable to criticism. While there is obvious potential in MCA, it does not offer simple and unquestionable solutions. The problems should not be seen as a sign of bad methodology. Rather, they indicate various consequences of the kind of complexities involved not least in environmental decision making.

12.3 DELIBERATIVE EVALUATION PROCESSES

While both CBA and MCA are based on a calculative or mathematical logic, a third main position in the literature is to accentuate the role of communication – that is, the role of *the argument* and of potential preference changes following from communication about what should be done. This takes us to the arena of deliberative institutions or methods. The fundamental idea is that through deliberation people can reach agreement on the basis of the better argument, on the basis of mutual understanding and trust. The choice is made through *communicative interaction*. Out of such a process a potential consensus over what to do may arise.

Like MCA, deliberative evaluation can also be seen as a response to the *bounded rationality* of individuals. The process of deliberation not only makes the persons involved aware of the needs and perspectives of others, which is the core of the communicative aspect. The citizens also help each other to clarify what the issue is about and how they themselves (should) think about the problem.

In this section I shall give a brief introduction to the basic ideas behind deliberation, explain its theoretical foundation and give some insights into its practical applicability. Thereafter a set of deliberative VAIs will be briefly characterized. One should be aware that the thinking about deliberative methods goes far beyond that of project appraisals. It is more correct to say that its perspective is basically to turn democracy by representation into a more direct participatory system – that is, participation going beyond that of elections and voting.[14] The treatment of this more fundamental issue concerning democracy and its forms will be rather superfluous due to the format of this book. For those interested in overviews of the broader discussions on deliberative democracy, see Dryzek (1990, 2002), Bohman and Rehg (1997), Elster (1998) and Smith (2003).

12.3.1 Characteristics of Deliberation

Deliberation implies communication that induces 'reflection upon preferences in non-coercive fashion' (Dryzek 2002: 2). Thus, the idea goes to the heart of the issue of collective or social choice; the question whether preferences are fixed, and whether consensus can be found concerning which preferences or norms should govern a certain issue. While CBA and welfare theory is about *aggregating* individual preferences, deliberation is about individuals *agreeing* over preferences. In that process, arguments form the core inputs.

The role of coercion-free communication

The emphasis on deliberation has its roots in ancient Greece and its *polis*. In modern, twentieth century developments, names such as Dewey (1927), Arendt (1958), Habermas (1984, 1995), and also Rawls (1993) are central. The core idea is that of communicative rationality (Habermas 1984) with its focus on the creation of understanding through dialogue and the force of the better argument. It is a form of common reasoning where consensus is obtained by mutual learning, understanding and changed preferences.[15] In the ideal Habermasian form, communication is thought to be free of coercion, strategic action and manipulation.

Habermas does not disregard the role of instrumental or strategic rationality. Markets and instrumental reason have their legitimate place in society. The problem he focuses on is that in modern societies instrumental rationality, in the form of technocratization, has invaded spheres of social life to which it does not belong. It implies a scientization and commercialization of politics and social issues and thereby of the conversation about the common good.

One may certainly argue that the idea of communicative rationality is a 'naïve ideal'. It may be vulnerable to strategic manipulation, for example, by those in power who have the capacity to set the agenda, to suppress arguments and constrain access to the debate. Agreement comes about on the premises of some powerful actors. There are two main arguments against this, which underpin the idea of deliberative democracy.

The first relates to the observation that dialogue in itself discourages strategic or instrumental behaviour. Rather, it accentuates the focus on the common good since the dialogue logically is about 'We' not 'I'. It motivates people to think in public interest terms (Goodin 1992). While there might still be an interest in bringing forward arguments in favour of pure individual gains, one is still forced to couch these in the form of something that can support the common good. This claim is then open to tests and counterarguments and if not found valid, the argument can be discredited as the disguised strategic action it really is. In this way the advocate is also discredited. Elster (1998) deals with a special aspect of this in his focus on the 'civilizing force of hypocrisy'. When one is stating that something is good for the society, it binds the individual to what is said and obstructs the realization of the potential selfish motive behind the argument. Dialogue and public statements civilize behaviour (Dryzek 2002; Smith 2003).

The second point relates to the way that communication is instituted. Several remedies are available to strengthen the internal drives of communication towards neutralizing special interests, to secure open agenda setting, the involvement of those with a 'weaker voice' and so on. One may advocate that agendas should be public, that their formulation is to follow certain rules securing equal access. Meetings may be explicitly defined as public, and socially weak groups may be given special support so that their voice might be heard. This is all about creating conditions for a coercion-free communication in Habermasian terms.

However, there is a crucial issue involved related to who formulates these rules and how they should be defined. This issue is about the basic constitution of a system – 'the rules for making the rules'.[16] Certainly, there is no way to secure the development of good constitutional rules. However, the creation of a public sphere – like freedom of speech, organization freedom and a free press – has also resulted in arenas for voicing arguments concerning the legitimacy of the constitution. Thus, while these spheres may still be biased towards certain interests, their mere existence has fostered a civilizing process just because of the role it plays in the testing of arguments. Thus one can observe a continuous process over the last few hundred years in granting wider groups access to the public sphere and to public decision making. The 'rules of making the rules' have become more democratic.

In the beginning this process, as observed in western societies, focused on the right to vote, which as we know, was first given only to a subgroup of men – predominantly those owning property. Later this right was granted to all men and finally to all women. The public arena opened up for testing the arguments in favour of existing voting privileges, and they were unable to stand against the arguments of equal access.

This is therefore a good example of how the institutionalization of a public sphere almost by its own logic created a process leading to general access for all groups. However, systems are rarely unbiased. Resourceful groups can 'buy' favourable treatment either by buying good communicative expertise, or by 'buying' politicians and so on. However, the existence of a public sphere in itself may provide the opportunity to question and then restrict such practices. Nevertheless, many systems of unequal access survive and are even formally accepted. The opportunity to argue against a practice is just a part of the process of eventually changing it. In that process the power granted by existing rules is certainly a challenge to overcome.

In today's world a core problem is evolving rapidly which seems to be reversing the trend towards more open communication in society. That is the development in the media for communication itself. In mass media, concentration of ownership and the heavy reliance on advertisements for financing the media business has narrowed down the focus, reduced access and resulted in an increased amount of distorted information. Information that might reduce financial support (from advertisers) or is against the interests of the owners more directly, runs the danger of being suppressed. Information is turning into a commodity and is not guarded as a common good. There is great need for reversing that process so that the media for communication itself becomes much more neutral.

The idea that *uncoerced* communication by itself will foster consensus also has some problematic aspects. As emphasized, environmental issues often involve conflicts between different interests. Certainly, in a public debate these interests have to explain themselves, implying that their arguments will be tested. This is a very different process from just documenting costs and eliciting individual WTP estimates. However, communication – the use of arguments – will not in itself eliminate the fundamental conflict. The type of conflict may play a role concerning how serious this problem is. To illustrate the issues involved, I shall distinguish between interest and value conflicts.

Interest versus value conflicts[17]

In the case of an *interest conflict* – as the concept will be used here – the parties involved have the same understanding of the problem. A common set of values are involved. What is at stake is the distribution of losses and gains. A typical example may be that of building a small dam. All people involved observe the benefits created through easier access to irrigation water and so on, but they cannot decide on whose land it should be placed due to the different distributional effects of that choice. The force of the better argument – for example, that some location involves fewer environmental damaging effects, that the construction and running costs are lowest and so on – may result in a consensus about where to site the dam. A consensus about how to deal with those losing land may also evolve on the basis of a common understanding of how these issues – compensations and so on – should be treated.

It is possible that in such a situation the dialogue may turn into *negotiations* between those gaining and those losing. Negotiating – in contrast to deliberating – brings us to the instrumental rationality of bargaining. This indicates the dual potential of the process. It may end in a 'game' between gainers and losers where the focus is not on the overall rules of distributing gains and losses, but on what compensation the individual losers demand before they will accept the concrete solution. This is not a problem, but if conflicts are always transformed into pure negotiations, there is a danger that the sense of community will deteriorate and the chance of creating a suspicious climate with escalating conflicts increases.

A *value conflict* is much more fundamental. In this case the parties involved do not agree on the basic understanding of the problem, what values are at stake and which should be given priority. There is no community across interests. Typically we observe this in many preservation processes. The environmental movement may want to preserve some species; an issue which may imply changed land use. A forest that previously provided jobs and income, may now have to be set aside for protection purposes. It may even be that one environmentally positive value – such as protection – conflicts with another, namely the use of a renewable resource. The parties to such a conflict often have difficulty in acknowledging the values of the opposing party. The perspectives involved are incommensurable.

The role of dialogue is very important in such a situation. It may have the force to change the perspectives of the involved parties. Actually, what deliberative VAIs offer is the possibility of creating consensus, even in a situation with value conflicts, through changing the understanding and the preferences of those concerned. However, this may fail, and the counterpart to a deliberative resolution of a value conflict is that of a *war*. There is nothing even to bargain about. The interests cannot be reconciled via compensations.

The way the process is formulated seems to influence the likelihood that an interest conflict will take the form of pure negotiations or a value conflict will end in war. One important aspect is the *role* that is given to people. Another is the *time perspective*. Interests are often defined by the existing position a person holds in society. If a person is asked to step out of that position – especially if the issue is formulated in terms of choosing for future generations – the principal issues are put more in the centre and the contemporary position of each individual is given reduced influence. What is considered a common good and what is considered a private one may then shift.

While the above underlines both the potential of and challenges to deliberative processes, it must be emphasized how different the basic thinking is compared to that of a market solution and CBA. In Chapter 8 we drew a distinction between defining what is best in subjectivistic and objectivistic terms. While CBA is based on a subjectivist understanding of value, deliberative methods are based on the idea that values can be reasoned over and some common understanding of what is right or good to do may be created. They are objective in that sense. Given that environmental goods are common to us, the potential for reasoning about what is best to do, creates opportunities for the societal choice processes that are very important to explore, even though deliberative methods are not a simple panacea.

12.3.2 Some Deliberative Value Articulating Institutions

We shall now move from the more general and broad issues to the more specific – that of presenting concrete project-orientated deliberative VAIs. Again, the field we enter is a rather complex one, and a large number of deliberative VAIs exist. Furthermore, the field is compounded by the fact that the same VAI may be named differently in different countries/research traditions. The interest in such institutional solutions is still in its infancy, but it is growing rapidly, a fact that may explain the situation. I shall restrict the presentation to three deliberative VAIs – that is, focus groups, citizens' juries and consensus conferences.

Focus groups

A focus group is a small randomly selected citizens' discussion group – of about ten people – led by a moderator. The aim is to explore the views of the involved persons in an environment that is supportive of bringing forward views and arguments. There are three distinct characteristics of a focus group. First, it is normally based on the knowledge of those participating – that is, no experts or stakeholders are called in. Second, it does not conclude or propose a conclusion. The material is brought from the focus group to the decision maker in the form of a summary of arguments. Finally, the subject area to be discussed is defined by the problem definition given by the organizers/DMs.

Focus groups seem to have originated in market research and are also being utilized in environmental research: in CV studies to develop and test questionnaires, and in research that is aimed at clarifying the arguments that people voice concerning certain environmental issues (Burgess et al. 1988a,

1988b; Kerr et al. 1998; Brouwer et al. 1999). Finally, as mentioned above, focus groups may also make a direct input into decision making processes (for example, Lenaghan 2001).

With regard to the last point, it is a problem that the focus group procedure offers relatively little power to the participants. One issue is the lack of influence on the problem definition. Even more important is the fact that focus groups are removed from the decision making process. They merely deliver arguments and usually have no control over the conclusions that the organizers draw from the group process.

Variants of focus groups are 'in-depth groups' (Clark et al. 2000) and 'deliberative focus groups' (Wakeford 2001). The latter also involve external expertise. These forms establish more comprehensive processes – often several consecutive meetings.

Citizens' juries

A citizens' jury is also a small group of citizens – 10–20 jurors – normally selected at random. It is led by a moderator, but deviates from a focus group in several ways. First, a citizens' jury is expected to reach a conclusion on the actual matter in the form of a proposal to the commissioning body/DM. Second, its discussions are supported by 'witnesses', dominantly experts but also stakeholders, who present additional information to the panel of jurors to facilitate the deliberative process. The jury is given the power to define what information it requires, and the procedure normally takes 3–5 days. The method is presented more completely in Stewart et al. (1994) and Smith and Wales (1999).

Citizens' juries have been used since the 1970s both in Germany (where they have been called 'planning cells') and in the United States (Smith 2003). Peter Dienel, Ortwin Renn and Ned Crosby (Dienel and Renn 1995; Crosby 1995) have been important for its development. The method has been especially successful in Germany, but it has also been taken up in many other countries: in the UK there have been more than 100 citizens' juries since 1995 (Delap 2001). Several of these have been related to environmental issues, such as evaluation of the release of GMOs, wetland creation, waste management and so on (Kenyon et al. 2001).

The idea behind the citizens' jury is to develop a consensus proposal. The style of moderation and the way the agenda is structured reflects this objective. Nevertheless, consensus may not be possible, and then a voting procedure might be applied.

Consensus conferences

A consensus conference has many of the same features as the citizens' jury. Again they concern a small group of randomly selected citizens who deliberate over an issue under the leadership of a moderator. The method originated in Denmark in the 1980s. It was developed by the Danish Board of Technology as a means of incorporating the perspectives of the lay public within the assessment of new and/or controversial scientific and technological developments, raising serious social and ethical concerns (Joss and Durant 1995; Joss 1998; Smith 2003).

In Denmark, a consensus conference involves 10–16 lay citizens in a four-day inquiry to evaluate a sensitive scientifical or technological issue. The laypeople are given the opportunity to question a panel of experts and interest-group representatives, assess the information brought forward and discuss the issues among themselves. Joss (1998: 5) emphasizes: 'The active involvement, in dialogical way, of lay people, experts and interest-group representative allows for the subjects under consideration to be evaluated, beyond a purely scientific context, to include economic, legal, ethical and other social considerations'.

The main difference from a citizens' jury is that of degree. A consensus conference focuses even more strongly on consensus than does a citizens' jury. Thus, disagreement or the recommendation of a diversity of options is not encouraged (Wakeford 2001).

Challenges to deliberative VAIs

This very brief presentation raises several issues that need consideration and elaboration. All the above deliberative institutions comprise small groups. This seems to be a necessity to make deliberation work well. The problem then becomes that of selecting a representative group of people – that is, a group able to represent the variety of perspectives, interests and arguments of relevance to the case. All deliberative institutions presented are based on a random sample. However, from which domain they should be selected is no simple issue. Moreover, if a specific domain is accepted, random selection of groups of from 10–20 people from that domain also raises concerns of representativeness. To counteract that problem, some kind of stratification may be used – that is, the organizers ensure that persons from all relevant groups of a society are given access (sex, colour, age and so on). Nevertheless, someone has to define what dimensions are important for such stratification. O'Neill (2001) gives a good overview of the problems related to that issue. Such problems have resulted in a great variety of recruiting procedures to deliberative VAIs (Niemeyer and Spash 2001).

A way to counteract some of these problems is to enlarge the group. One such deliberative institution is the opinion poll (Fishkin 1997) which normally involves 200–500 persons. This, however, restricts the depth of deliberation and the opinion poll is not the result of a common conclusion, but of an individual poll or vote. The opinion poll is thus a pre-poll deliberation.

All deliberative institutions presented above are based on people being involved as citizens. Some authors argue for involving stakeholders instead. The people involved in the deliberation are then directly engaged and have therefore both the necessary interest and knowledge to participate. The problem with this is that the idea of open dialogue and deliberation is difficult to envisage. Instead, there is an increased likelihood that the process will develop into negotiations or even 'war'. However, the fact that stakeholders disagree is a fundamental issue, and conflict resolution will not be achieved by excluding stakeholder representatives from the deliberative process. The proposal of the participants of the DI then has to be decided on in the relevant political body. Stakeholders will certainly try to get access to that process through either lobbying or more directly, representation. However, the wider public can then ascertain the conclusion of the deliberative institution, how that conclusion stands in relation to the position of important stakeholders, and finally how these various elements influenced the final decision. In relation to this specific issue there is no difference from CBA- or MCA-based proposals.

One may argue that using voting – possibly as an option for citizens' juries – is counter to the whole idea of deliberative VAIs. Certainly, it implies that the process of reaching a consensus is relinquished. However, there is a difference between the standard electorate voting procedure of representative democratic systems – that is, isolated individuals voting in elections – and voting in a jury after a process of deliberation has occurred. Dryzek (2002) argues for the capability of deliberation to create a stronger sense of community which narrows down the conflicts and also restricts the possibility of various voting paradoxes appearing.

The latter refers to Arrow's so-called 'impossibility theorem' (Arrow 1963), which says that it is impossible for any mechanism to aggregate individual preferences into a collective choice that satisfies a set of standard criteria like unanimity (any unopposed individual choice should be incorporated into the collective choice), non-dictatorship, transitivity, unrestricted domain (no restriction on individual preferences), and independence of irrelevant alternatives. Arrow's conclusion has been used to support the idea of 'minimal democracy' since democratic procedures – that is, voting – will tend to create inconsistencies, arbitrariness and instability. Creating markets and focusing on individual choice as in the case of CV is better as it circumvents these problems.

This conclusion has, as we have seen, its own problems, not least concerning common goods. Furthermore, it denies the dynamics of the social. Dryzek (2002) argues that the capacity of deliberative institutions to develop common values and preferences, in technical terms resulting in so-called 'domain restriction', substantially reduces the aggregation problem. The potential of deliberative institutions is to cut the knot that Arrow tied for liberal democracies. Dryzek uses the experience from various deliberative institutions to conclude that there is enough empirical evidence to postulate that the mechanism of domain restriction implying some clustering of preferences across individuals follows from deliberation.[18] This is really what *the creation of a society* is all about.

12.3.3 Radical Uncertainty and Deliberation: The Cognitive and the Normative

We have already focused on the common good aspect in relation to the capacities of various VAIs, but there is one more feature of environmental issues that should be mentioned – the issue of radical uncertainty (see Chapter 9). Not least in modern societies – that is, societies with rapid and comprehensive changes in technology and resource-use patterns – the role of expertise becomes rather unclear. The content of what knowledge means and the traditional distinction between expertise and citizens' laypeople's evaluation has thus become challenged. We observe this in the lessening of trust in pure scientific advice and in the development concepts such as that of post-normal science directed at capturing this new relation between citizens and experts (for example, Funtowicz and Ravetz 1993, 1994).

The tradition of 'normal' science has been based on experts' ability to determine, with high certainty, the relations between the different variables of a system. It is based on the assumption that ignorance can be reduced at least to risk. In line with this, the problem has been to avoid accepting a false statement as true – that is, to avoid a type I error. Given radical uncertainty, the role of science changes. Two issues are of importance. First, radical uncertainty changes the focus of error treatment. Second, the distinction between facts and values becomes blurred in a way fundamentally different from when dealing with certainty and ordinary risk. The cognitive becomes in a way normative.

We have already mentioned (Chapter 10) that in the case of radical uncertainty it is the type II error that becomes crucial. Given that ignorance here is irreducible, one will have to make type II errors when working on the basis of standard practices that seek to avoid type I errors. The question is then: who should decide on these matters? The point is that science does not offer much help in situations with radical uncertainty. It can hardly be proven with the necessary certainty that damage will occur. However, the opposite cannot be proven either. Consequently, the issue becomes a normative one. What chances we are prepared to take is an issue for the citizen, not the expert. And it is not least for these reasons that deliberative institutions have become increasingly popular. As we have seen, the consensus conference developed in Denmark as an acknowledgement of this. Core issues taken up in focus groups, citizens' juries and consensus conferences relate to such difficult decisions – that is, nuclear power, GMOs, biodiversity protection and so on. They have been used to regain some legitimacy for the decision process. Beck (1992), with his focus on the need for new institutional structures in the 'risk society', is an important reference in relation to this development.

12.4 COMBINING THE METHODS?

No VAI is ideal. Combining VAIs may counteract some of the problems encountered for each of them used alone. Combining CBA or MCA with a deliberative VAI has been proposed. We shall look at the arguments in favour of both types of combination.

12.4.1 Combining CBA and Deliberation

Niemeyer and Spash (2001) observe a recent trend towards what they term 'deliberative monetary valuation'. This implies that economic valuation (CV) is combined with a more or less formal deliberative process before eliciting the WTP estimates – for example, Brown et al. (1995); Sagoff (1998); Brouwer et al. (1999); Ward (1999); Kenyon et al. (2001). This trend is a response to the critique of the assumption underlying economic valuation that people have predefined preferences for the goods involved. If they do not have such preferences, deliberation may help people sort out their views about an issue. It supports preference 'construction'.

Sagoff (1998: 227) comments:

> The introduction of a more discursive approach to value elicitation . . . makes intuitive sense. If individuals do not come to CV surveys with predefined preferences but must construct them, then the process of construction may legitimately involve social learning, since this is precisely what occurs in other contexts in which people work out their values.

Combining, for example, a focus group with a CV appraisal makes sense when focusing on the learning aspect. This is a justifiable argument and the solution represents progress in so far as the problem is (only) that of learning and understanding better what is at stake and what are one's own views on the issues involved.

Problems occur if one goes beyond learning about own preferences. If the reason for using deliberation is that of fostering communicative rationality – of evaluating and defending arguments – then combining deliberative institutions and CV/CBA seems to be based on a contradiction. It mixes collective reasoning and consensus building over principles and norms with individual trade-off calculations. It combines a VAI based on capturing incommensurability with one that is focused on commensurability. It mixes a VAI directed towards the 'We' with one based on an 'I' perspective. So while combining a deliberative institution with CV may produce figures that are easy to use in calculating an optimal solution, it may come out of a process where such calculation has really no logical support.

12.4.2 Combining MCA and Deliberation

There are also some evolving initiatives concerning combining MCA with deliberative processes. While this is a very recent development, one can observe two directions. First, there is work where the MCA is the core element with participation added to it – for example, Stewart and Scott (1995); Banville et al. (1998). Second, we have a solution where the deliberative process is the core method, but MCA is included to structure and document the deliberation – for example, Renn (1999); Clark (2002); Stagl (2003).

The first type of solution acknowledges that several interest groups or stakeholders are involved and that the MCA can be developed to include these in a communicative process around alternatives, criteria and potentially the weighting of the criteria. As mentioned earlier (Section 12.2), it may take the form of separated processes among each stakeholder group where the analysts inform each group about the priorities of the other stakeholders. On the basis of this information, the analysts can also propose new alternatives that are viewed as potentially best compromises between conflicting interests. As an example, Stewart and Scott (1995) did this for a dam project in South Africa, where the dominant mechanism to resolve the conflict consisted in reformulating the alternatives once the most controversial issues had been clarified.

The second solution starts from the other end – from the deliberative institution and its problems. Here a combination is a response to the claim that the deliberative process is very complex and verifying why a specific consensus is reached for the wider public is a problem. Stagl (2003) documents a study where a deliberative process to review the UK energy policy was supported by a rather simple form of multicriteria evaluation. She suggests that the MCA not only supports verification, but it also supported the participants in clarifying what they meant about the various issues. It fostered focus and structuring.

In general, there is a greater potential for combining deliberation with MCA than with CBA. At least some of the MCA methods are based on much the same ideas concerning incommensurability and the normative aspects of decision making as the deliberative institutions. What MCA may offer deliberation is to support it in handling complexity and bounded rationality. It also offers a way to document the results both for the different participants in the process and for the wider public.

Deliberation also has much to offer MCA. It is a way to involve the public – as stakeholders or citizens – participating in multicriteria assessment. The narrow focus of MCAs centred only on the decision maker and on all kinds of technical issues related to aggregating criteria and so on is opened up. More knowledge and broader perspectives are brought into the MCA and the credibility of the results is increased.

There is also a price to this. The costs of doing the assessment increase when methods are combined. More time may be needed. Those assisting the process must have a wider competence. Thus, with regard to combining methods, the issues involved should be clarified and an evaluation made as to whether such combinations are likely to result in significantly better outcomes.

12.5 SUMMARY

Different value articulating institutions have been developed. We have looked at three different groups of such institutions, namely cost–benefit analysis, multicriteria analysis and deliberative institutions. We have put emphasis on clarifying the assumptions underlying the different VAIs. This is of fundamental importance for evaluating their relevance for different types of assessments and problem solving.

CBA is based on welfare theory. It is founded on *rationality as maximizing* and the idea that a decision problem has an optimal solution. In this lies the assumption that values are commensurable and compensable. CBA is based on aggregating individual preferences expressed via WTP estimates. It is assumed that individuals have given preferences for the goods involved and the capacity necessary to calculate the consequences of different alternatives for their preference satisfaction.

MCA is a term covering a set of tools. The basis for all MCA methods is that goods are multidimensional and that decisions are viewed as very complex. The methods are based on *bounded rationality*, as they are concentrated on supporting the decision maker in a difficult decision process. MCA is furthermore focused on establishing the best compromise between conflicting criteria or interests. An important group of MCA methods is based on incommensurability and/or non-compensability, implying that the method may help in analyses when weights are incommensurable or when they are coefficients of importance rather than measures of trade-offs.

Finally, deliberative institutions are built on *communicative rationality*. The ideal solution is the consensus, which is based on coercion-free communication and evaluation of arguments. As in the case of MCA, cognitive capacity is considered restricted. The dialogue supports the individual in understanding what the issue is about. However, the main rationale for the communication is that people through learning about various arguments and changing own perceptions may be able to reach agreement on what solution to a problem to prioritize.

All methods are idealizations, while it should be noted that MCA is explicit on the existence of capacity problems of individuals. Because they are idealizations, combinations of the methods may be a reasonable way to counteract deviations from the ideal observed in real-world decision processes. We have argued that this is an important option. However, experience in this is still limited. The main argument made here is that combinations must be built on consistency concerning assumptions. While CBA is based on a set of assumptions that are very different from those of most MCA and deliberative methods, the possibility for a combination of MCA and deliberative institutions seems to have more potential.

Choosing who should participate is a challenge in CBA, MCA and deliberative institutions. Certainly the problem is less severe the larger the selected group. Therefore it can be viewed to be less of a problem in CBA/CV. However, there is clearly some balancing to be done between the width and depth of an analysis, between involving many people and ensuring that the work is thorough. Involving more people cannot compensate for a wrong or illegitimate type of assessment. Given that what is good can be evaluated in objective terms – that is, accepting that one argument can be evaluated as better than another *in the public domain* – reduces the conflict between width and depth. The capacity of deliberative processes to create consensus, or at least some domain restriction, is perhaps its greatest value.

NOTES

1. Strictly speaking, in CBA it is for example, the CV element that is the value articulating institution, while the CBA in total rather may be classified as a 'decision-recommending institution' (Jacobs 1997). In the case of MCA and even more for the different deliberative institutions, the value articulation is an integrated part of the whole assessment. I have therefore come to the conclusion that distinguishing between value articulating and decision-recommending institutions will raise more complications than benefits.

2. As will be explained later, commensurability and compensability are different issues.
3. For more complete expositions, see Dasgupta and Pearce (1972); Pearce (1983); Hanley and Spash (1993); and Boardman et al. (2001).
4. One might ask whether this is not also the case in market allocations. A cardinal WTP measure – the price – is decisive here too. In this case it is not, however, necessary to assume that a kilogram of strawberries for £1.50 gives the same utility to person A as to person B even if they both buy it at that price. It only says that a kilogram of strawberries is *at least* as much worth spending £1.50 on as a kilogram of raspberries (for instance), if that was the alternative. If this transaction enters a CBA, the price of £1.50 is used as a measure of the benefit it gives to both A and B. Their benefits or 'utilities' are compared.
5. Trading off one good for another implies full compensability. Munda (1996) argues that there is an inconsistency involved in assigning different weights to the different individuals. Distributional weights must be seen as coefficients *of importance*. They define the relative importance of each individual. Munda argues that it is a problem when including these in a framework otherwise based on trade-offs. He concludes: 'Unfortunately, since CBA is based on a complete compensatory mathematical model, the (distributional) weights can only have the meaning of a trade-off ratio, as a consequence theoretical inconsistency exists' (p. 163).
6. If the corrections follow from changes in income levels between the study site and the policy site – that is, the site of the project – changing values due to the effect of this seems consistent with the basic assumptions underpinning the neoclassical model. It may cause us to ask why the income situation at a certain point in time should be decisive especially for irreversible reallocation of resources, but that is not a particular issue for benefit transfer. My main question, however, concerns the tendency to use more elaborate models for value transfers that also involve cultural, social and educational factors. While I believe that this practice is highly relevant, it actually implies accepting that the social sphere influences the individual and is contray to the very basis of the undertaking itself – that is, that preferences are context independent.
7. Dasgupta et al. (1999) show that social discount rates will in general be lower than private ones. They may also be both zero and even negative.
8. For more complete descriptions of MCA, see Bana e Costa (1990); Nijkamp et al. (1990); Janssen (1994); Munda (1995). It should be emphasized that many authors talk about multicriteria decision *aid* rather than multicriteria *analysis.* The distinction is important in clarifying that the final decision will always be more than just to pick the solution prioritized by the analysis. However, this applies for any of the tools described in this chapter. Therefore I have decided not to use the concept. All methods presented in this chapter are considered as decision-supporting tools.
9. Bogetoft and Pruzan (1997) thus talk about intra- and inter-person conflicts. They furthermore identify a third level – systemic conflict. While inter-person conflict is defined as conflicts between members of a group that are to make a decision, systemic conflicts are those between decision makers and 'those who are at the receiving end of the decision' (p. 8). To exemplify, 'receivers' are, in the case of a firm decision, employees, customers, suppliers and so on.
10. Because of this, there is a distinction in the literature between 'decision making' and 'decision aid', with the latter referring to the learning process as the role of the methods or VAIs involved.
11. For a very instructive discussion of the characteristics of value dimensions, see O'Neill (1993).
12. This is not an insight that is shared by all MCA theorists and practitioners. In many studies the assessment of scores is considered a technical issue. One should be aware that the degree of conflict over scores is dependent on the involved issues.
13. ELECTRE is actually a set of methods (ELECTRE I, II, III, IV and A) with some variation in the assumptions concerning how strict the outranking relation is (see Roy 1990).
14. A distinction is normally made between participation and deliberation – that is, deliberation can be combined with different participatory principles.

15. The term 'deliberative democracy' has different meanings in different literatures. Some equate it with the standard representative democracy of the liberal state, where the basic idea is that of individuals with given and stable preferences that do not change as a function of the context of interaction. The single individual, through his/her personal reasoning, draws the conclusion about what s/he finds to be the right solution. Rawls is the most prominent representative of this position. Dryzek (2002: 15; original emphasis) suggests that 'Rawls downplays the *social* or interactive aspect of deliberation, meaning that public reason can be undertaken by the solitary thinker. This is deliberation of a sort – but only in terms of the weighting of arguments in the mind, not testing them in real political interaction'. There is thus great variation between different deliberative schools concerning the implication of deliberation, the role of the group and the role offered for changed preferences. This will not be dealt with here; see Dryzek (2002) and Bohman and Rehg (1997) for an exposition of various stances. The way the concept of deliberation will be developed here, implies the possibility for a discussion with others about which preferences to hold.
16. The point is not only related to the constitution of a society at large. While this is the most fundamental level, the issue of 'the rules for defining the rules' is also of great importance to the creation and choice of various deliberative VAIs as these are defined in this book.
17. The distinction is very much in accordance with Aubert (1979).
18. Boulding (1970) has argued similarly that 'a public requires some sort of organization, an organization implies community, a community implies some kind of clustering of the benevolence function . . . which denies the assumption of independent utilities' (cited from Schmid 1987: 30).

13. Policy and policy measures

Given that we have decided what we want to accomplish – Chapters 11 and 12 – we need to motivate changes in behaviour that secure the achievement of our goals. This is attained by institutional change. Policy measures are a common term for the various institutional arrangements involved when trying to reach societal goals. While this term normally relates to taxes and subsidies, legal regulations or information, the basic issue is the choice of which resource regime should be installed to treat specific resource allocation questions. The issue concerning which particular instruments are relevant to use depends on the regime in place.

Most environmental economics textbooks discuss policy measures in a principal–agent context. The regime referred to is therefore normally taken as given – that is, that of a market economy with private enterprises and a state. Effects that are external to the market are internalized by economic instruments or legal regulations as formulated by the principal – the state. These policy measures establish new constraints on the behaviour of the firms so that they have to take into consideration the full costs of their activities. The model is developed to a high level of sophistication. However, there are some important inconsistencies and a certain 'narrowness' involved in the treatment that needs further consideration.

Chapter 10 showed how the type of regime has a great impact on both which externalities appear and what principal–agent relations we face. We shall therefore start this chapter with a short discussion of the regime issue (Section 13.1). Following this with a more detailed analysis under all possible regimes is far too demanding. In the next two sections we shall therefore focus on the formulation of policy measures in the standard private property/market situation: partly in the setting of contracting firms (or firms and households), and partly in the principal–agent setting – that of a state regulating firms'/households' activities. The perspective of these sections will be two-sided. On the one hand we shall look at how the choice of institutional framework – that is, the contracting or a principal–agent setting – influences what becomes efficient. This is the task for Section 13.2. In Section 13.3 we shall on the other hand look at what an analysis based on the matter flow perspective from Chapter 9 implies for the choice of policy instruments. That discussion will be restricted to the principal–agent framework. Both these sections will focus on the role of transaction costs.

Thus, while the analyses in Sections 13.2–3 focus on the more technical questions, Section 13.4 returns to the broader issues concerning interest protection, values and rationality. Following the perspective of this book, the choice of policy measures may evoke different rationalities, and bring about or foster certain values. The choice of policy measures is not just about reducing transaction costs (the coordination issue). It is also about understanding the motivations involved and forming the society with its individuals.

The discussions in this chapter will to a large extent be carried out 'as if' environmental costs can be estimated in monetary terms and 'optimal damage' can be calculated. As should be clear by now, I am critical of this assumption in many situations. For expositional reasons, I shall here not question the way by which society reaches the environmental targets it supports – whether they are formulated on the basis of willingness to pay measures, in the terms of a safe minimum standard or on the basis of some other kind of precaution threshold. Finally, I shall mostly focus on negative externalities with pollution as the exemplar. This is necessary to limit the analysis. The conclusions can easily be translated to a situation with positive externalities.

13.1 POLICY MEASURES AND REGIMES

Again, the issue is about the collective and the individual. It is about how individual choices may interact and form collectively irrational results. It is about how the society affects the individual, his/her values, and how it frames individual choice sets to avoid such unwanted situations. At the heart of the problem lie the *interdependencies* between individual choices that by necessity must follow from the physical characteristics of the world we live in.

The basic units of an economy – that is, firms, common properties and the state – are all command and communicative structures. These units are given the power to regulate interconnections within their realms. In the case of a firm/private property, interrelations in the process of making products are at the centre. In the case of a common property it is the interconnections between various uses of a common resource that are at stake. As we saw in Chapter 10, an important reason for establishing common property was to handle the external effects of individual use of a common resource. This suggests that by changing the relationships between the economic agents – by shifting between property regimes or allowing firms to merge – changed opportunities concerning the treatment of physical interconnections is established. Given that we start off from a situation where individual agents – firms or households – exist and produce external effects, the following five strategies are of special interest:

1. The establishment of *common property*. This implies, as we have seen, a system of common governance for resources where individual use of the resource by separate units (households, firms and so on) produces external effects for other users.
2. The strategy of *merger*. This is most relevant for firms, but in principle may also cover common property entities and even states. The point is that since the action of one unit has external influence on the other – for example, one firm is damaging the production possibilities for another firm – combining them into one unit internalizes the problem. It can then be treated within the common command structure of that firm. The typical case is the paper mill upstream of a brewery in a river whose emissions pollute the water for the brewery.
3. The strategy of *negotiations*. This is the Coasean solution. Instead of a merger, the firms (states or common property entities) may regulate the externalities via agreements. In the above case, the paper mill may compensate the brewery for the damages, or the brewery may pay the paper mill for reducing its outlets. This situation demands the existence of a rights structure stipulating who is free to shift costs onto whom – that is, from the basis of which rights structure the negotiations should be undertaken.
4. The establishment of *state property* of both the emitting and the receiving activity. From the perspective taken here, this strategy resembles in many respects the establishment of common property or a merger. It is again about 'internalizing the externality' by bringing it under one common set of goals and one common governance structure.
5. The *execution of state power on the conditions* for private economic enterprises, be it firms or common property entities. In this case, the state establishes regulations such as taxes, emission quotas and so on, signalling the costs of external effects to those creating them. This is the Pigovian solution.

The above forms have many features in common. But they also have different implications concerning two main issues: that of value articulation and that of transaction costs. Concerning value articulation, the three first solutions are characterized by a direct expression among the parties involved. The value articulation is an integrated part of the regime or the choice process. In the case of a common property, the decisions that establish the internal rules will be based on the arguments produced by the members concerning the common problem they face. In that process they must decide which resources are most important to protect. In the merger setting, the issue will also be handled by a structure of value articulation defined by the board of the new firm. In the case of a negotiation, the parties decide about the costs or losses involved as they develop their claims for compensations.

Under state ownership or direct state regulations, the situation may look slightly different. The state needs to develop a procedure to ask its constituencies about what should be done in the case of, for example, a pollution problem or an issue of species protection. How important is it to protect a specific lake? What should the level of forest protection be? In this case a CBA, an MCA, or deliberative procedures may be used, depending on how the state representatives and the citizens view the problem. What foremost distinguishes the state from a firm in this respect is its relations to the citizens. Firms are responsible to their owners, states to their citizens.

Different regimes are characterized by different types and levels of transaction costs. As is implicit in the idea about firm mergers, the cost of transacting is influenced by the number of firms involved when considering an externality issue. As we saw in Chapter 8 on the subject of welfare theorems, efficiency is obtained in a situation where firms operate in a competitive environment. This requires individualized private property rights. But the imperative to divide control over resources among atomistic agents is, at the same time, the mechanism responsible for creating coordination problems between firms. Through atomization, the number of borders among economic agents increases, thereby amplifying transaction costs and hence contributing to the generation of externalities. Bromley concludes similarly:

> The individualization of the world – its atomization really – is argued to be the very best means of individuals to be made better off and, by simple aggregation, for the collection of all individuals (call it society) to be better off. Now, if externalities arise at the boundary of decision units, and if theory and policy celebrate and sanctify atomization, then theory and policy would seem to advocate the maximization of decision units and, *ipso facto*, the number of boundaries across which costs might travel. Bluntly put, atomization ensures potential externalities. (1991: 60)

Therefore what is efficient concerning the internal becomes in a way the cause of inefficiencies regarding the external. There is thus an intricate trade-off problem here, which becomes invisible for a model based on the assumption of zero transaction costs. The problem is enhanced if one accepts that preferences also depend on the kind of regime that is set up. Choosing regimes is not neutral to the issues of either the cost of internalizing externalities, or how highly an externality is valued. The rest of the chapter will discuss these problems at different levels.

13.2 RIGHTS, TRANSACTION COSTS AND EFFICIENCY

As mentioned in Chapter 9, Coase (1960) criticized the Pigovian solution of environmental taxes. He suggested that if TCs are zero, no state regulation is needed except that of defining the rights structure. The parties would arrive at the optimal allocation via individual, costless bargains. Coase's attack may be interpreted as just demanding consistency. The standard neoclassical position (Pigou) did not take account of the full consequences of its own assumptions – that is, that of zero TCs.

However, in this way the problem was turned upside-down. It was not primarily the irrelevancy of assuming zero transaction costs that was emphasized, but inconsistent use of the model.[1] It may seem even more curious that the response by the Pigovians was not to counter by shifting towards studying the effect of positive TCs, where the Pigovian model of state regulations could develop a strong defence. However, the Pigovian or neoclassical analyses were continuously refined, still assuming TCs to be zero.

The aim of Section 13.2 is to clarify the differences in 'what becomes efficient' given that we (a) design a situation where we let the agents negotiate (Coase) or (b) establish a principal–agent framework (as implicit in Pigou). In doing this we shall focus on the role of transaction costs and the role of rights. As will be made clear: the basic argument in favour of the Pigovian solution is the capacity to reduce TCs.

13.2.1 Transaction Costs and Efficiency

The bargaining – the Coasean – situation

From the neoclassical perspective, the existence of externalities is considered a 'market failure' (Bator 1958). Given the Coasean perspective, it is still not a failure of the market. It is a rational result facing high TCs. Dahlman (1979) claims that positive TCs are the reason for the existence of externalities. If transaction costs are zero and rights are defined, optimal allocation will be reached via direct bargains between the involved parties – for example, the polluter(s) and the victim(s). This is the Coase theorem as explained in Chapter 8. Assuming that rights are defined, Figure 13.1 depicts the situation where the emissions of a polluting substance are involved and TCs are zero.[2]

The optimum is reached where the marginal abatement cost (MAC) equals the marginal environmental cost (MEC). This situation will be obtained via costless bargains independent of the right being with the victim(s) (R_v) or the polluter(s) (R_p) – that is, that the negotiations start off from R_v or R_p in the figure. The emissions from q^* to R_p are so-called

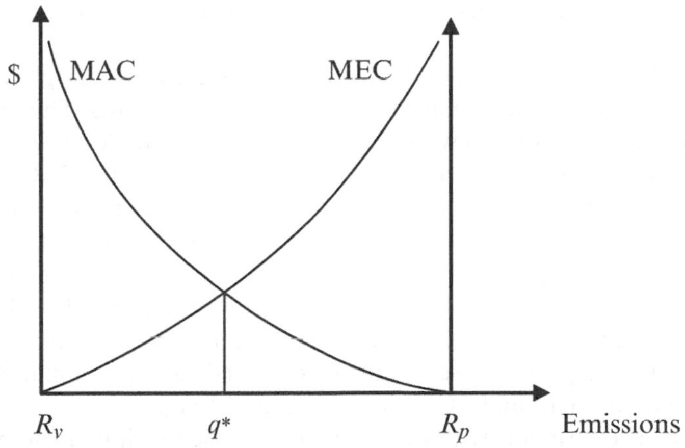

Figure 13.1 The efficient level of emissions with zero TCs

'Pareto-relevant externalities'. This implies that if these emissions are reduced, both parties will gain. The emissions from R_v to q^* are – following the same language – 'Pareto irrelevant'.

As soon as we shift to the more realistic assumption – that is, that TCs are positive – we get the situation depicted in Figure 13.2, assuming that the defined right is protected by a property rule.[3] Panel I can be used to analyse a situation with fixed TCs (only). If these are less than area A – in the case of victims having the right to an unpolluted environment (R_v) – or less than area B – in the case of the polluter have the right to pollute (R_p) – the optimum will still be q^*. If the fixed TCs are greater than A and B, respectively, the cost of bargaining is greater than the gains and R_v and R_p become

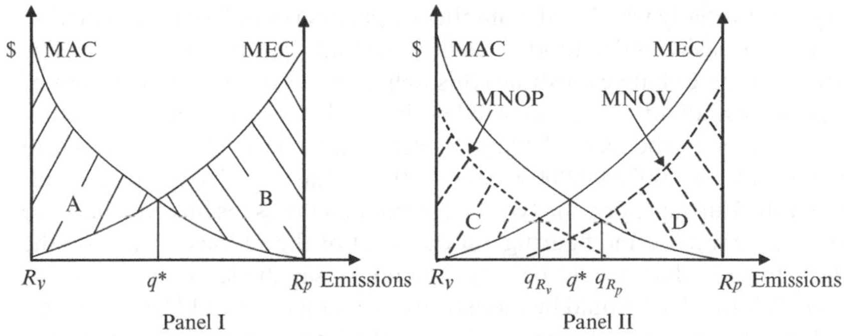

Figure 13.2 The efficient level of emissions under bargaining given positive fixed (panel I) and variable (panel II) TCs

the optimal allocations dependent on whether the rights are with the victim or the polluter.

In panel II, a situation with variable TCs is depicted. If the right is with the victim (R_v), the polluter must carry the transaction costs. This implies that the polluters' maximum willingness to pay to avoid undertaking abatement – which in the case with zero marginal TCs will equal the MAC curve – is reduced to MNOP (marginal net offer for polluters – the dotted line). This is the new offer curve for the polluters. The difference between the two curves is the marginal TCs. The victims' willingness to accept compensation – as measured by the MEC curve – will not be influenced by the TCs under these rights conditions. Thus, the optimum (= the negotiated equilibrium) will be q_{R_v}. As we see, emissions are reduced as compared to a situation with no marginal TCs – that is, from q^* to q_{R_v}.

Similarly, if the right is with the polluter (R_p), the maximum willingness to pay by the victim will be reduced equal to the marginal TCs, and the optimal allocation (the negotiated equilibrium) will be q_{R_p} since the offer curve now is shifted to MNOV (marginal net offer curve for victims – the dotted line). The polluters' willingness to accept compensation is, under this rights structure, equal to the original MAC. We see how the Pareto-relevant externality shifts systematically between rights structures dependent on the level of TCs. If there are also some fixed TCs involved, these must not be greater than areas C and D, respectively, to obtain a bargained result that differs from the initial situation – R_v or R_p. If they are greater, no emissions will be Pareto relevant in the case of R_v and no abatement will be the Pareto optimal solution in the case of R_p.

As emphasized, all previous figures cover a bargaining/market environment. The negotiations may be between firms, between different common properties, between firms and households, between states and so on. The reasoning has equal importance in all cases. However, the level of TCs may vary substantially whether the number of parties is small or large, whether the issue is easily understood or environmentally complex and so on. The issue of climate change illustrates this well: millions of firms and billions of households/individuals are involved both as polluters and victims. If action is to be taken in the form of bargains between the individual emitters and victims, at least 6,000,000,000 × 6,000,000,000 deals[4] need to be made simultaneously. Furthermore, the type of environmental issues involved here are extremely complex and figuring out the effect of the actions of each on the effects for all others would be very difficult, to say the least, and the combined TCs involved would be monstrous. No bargains would be made – no deals would be struck. Aggregate TCs would far outweigh potential gains.[5]

Moving to a scene where the bargains are undertaken by states – as in the UN process concerning the Koyto Protocol – the number of parties and hence TCs will be reduced.[6] Results in the form of reduced emissions may be possible – that is, TCs are now brought down to a much lower level. Negotiations become feasible and will most likely be justified by the gains. Nevertheless, the problem with getting signatures from enough parties (see Box 10.4) shows that there are substantial difficulties. These, however, are mainly related to disagreements about the distribution of costs. We are in a bargaining environment where no common authority structure exists, and therefore no rights structure is established from which basis the bargains may start. Instead the bargains also involve sorting out what rights should exist. This is certainly problematic.

The principal–agent – the Pigovian – framework

Moving back to the level of one state, the necessary authority structure normally exists for determining the rights. Thus, individual bargains have a basis from where to start (for example, Coase). However, too many parties will often be involved and direct state regulations are often the only realistic option for any environmental protection to take place. Certainly, it may be possible to handle some situations via individual bargains, some might be solved on the basis of consumer boycotts, some might be handled via firm mergers and others more through common property solutions. Nevertheless, in the standard situations with markets between firms and households, state 'interventions' will continue to be important in making any externality become Pareto relevant.

A core issue when comparing this principal–agent framework with that of a bargained solution is the differences in the levels of TCs. The question of which instruments to use is also of importance. Principally, the situation can be described as in Figure 13.3.

Let us start with a simplification and assume that state regulations imply only fixed TCs. If these are less than areas A (given R_v) or B (given R_p) in the figure, it is optimal to regulate. If the right is with the victim, the principal establishes a tax at level T. If the right is with the polluter the principal has to buy emission reductions from the polluter in the form of a subsidy $S\,(=T)$. This establishes emissions at the level q^* as in the bargained case without TCs.[7]

The likelihood that the TCs in the case of a principal–agent framework are much lower than the TCs in the bargained solution is rather substantial. This follows from the simple fact that the principal 'represents' the victims, implying that in most cases the number of interactions goes down substantially. Furthermore, the solution is not a bargained one. Instead, the state sets the tax or subsidy on the basis of its knowledge concerning the MAC and MEC curves (or on the basis of a level of emissions set in a form of an SMS). Hence, the interactions are not only fewer. They also entail

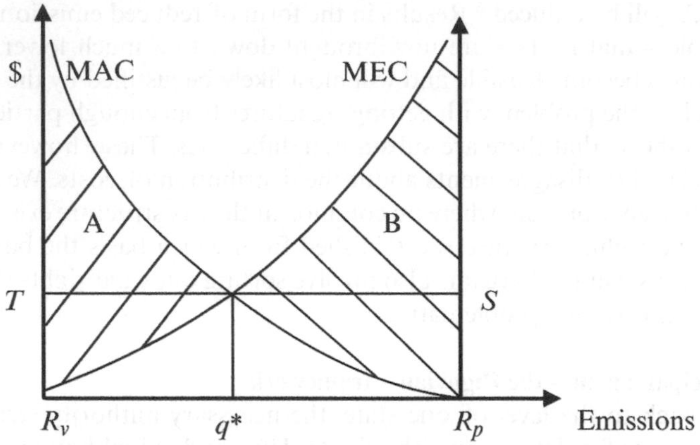

Figure 13.3 The efficient level of emissions within a principal– agent framework with fixed transaction costs only

only controlling that the polluters pay taxes on all emissions (R_v), or that they do not receive subsidies on anything more than actual emission reductions. The implication of this is that as TCs decrease, many more externalities become Pareto relevant as compared with the bargained case.

The state may find that it will not tax all emissions, because some are too small or they affect a recipient where the environmental damage is low. In such situations the state finds that the TCs are larger than the gains. Since TCs will vary with the number of firms involved in the regulation, a variable TC component appears. The basic reasoning about its effect is parallel to that of the bargained situation – see Figure 13.2, panel II. With positive TCs and, for example, victims' rights (R_v), fewer emissions become Pareto relevant. The main difference to the bargaining framework is that variable TCs will also generally be (much) lower in the case of a state regulation as compared with a bargained solution. Lower marginal TCs imply that the effect of who has the right will be smaller – the distance between q_{R_v} and q_{R_p} becomes much less. It will still not vanish as long as there are variable TCs.[8]

While in the bargained solution victims are compensated if they have the right, the situation is different in the case of state regulations. Normally the victims do not receive compensation. The tax revenues go to the state. According to public finance theory, such revenues should be used where they offer the highest return. One may argue that such tax revenues reduce the need for other taxes and that the compensation thus 'trickles back down' to the victims. However, the distributional effects would normally be different since it is unlikely that this tax relief would fall exactly on those experiencing the externality.[9]

While I believe that the above presentation of the Pigovian solution is the consistent one, it is somewhat different from what appears in most neoclassical expositions. First, it is argued that negative externalities like pollution should be regulated only by taxes (or tradable emission quotas). Subsidies should not be used – for example, Baumol and Oates (1975); Kolstad (2000). Second, while victims' rights are protected (see also the polluter pays principle (PPP) as advocated in this literature), the way such rights are instituted differs from the analysis behind Figure 13.3. Both kinds of deviation actually originate from what appears to be an unclear understanding of the rights issue.

Concerning the first question – only accepting taxes in the case of a negative externality (for example, pollution) – the argument goes as follows: if one subsidizes firms to motivate them to reduce emissions, this will also create an incentive for firms to become polluters. It will give incentives to an incorrect dynamic, and itself be a source of increased future environmental damages. While this is correct, it is nevertheless still consistent to use subsidies if the polluters have the right.

Following Coase (1960) there is actually a symmetry concerning the rights and incentives involved. He suggested that victims' rights would similarly motivate people to become victims. If they moved into a polluted area, they would experience damages for which they were then entitled to compensation. In the bargaining environment this would follow directly and there is a dynamic allocation problem appearing independent of whether the right is with the polluter or the victim. Also in the case of state regulations, victims' movements would result in increased environmental costs and environmental taxes should be increased. Emissions would not increase, but the effect of them would be greater due to the fact that more victims are exposed. Following the Pigovian rule, taxes should then rise and victims' movements influence what becomes optimal. Vatn and Bromley (1997) give a more systematic treatment of these issues. While the above issues give rise to many practical problems for policy makers, the more fundamental point concerns the fact that making clear-cut distinctions between 'distribution'/rights issues and 'efficiency' becomes impossible. I have nothing against environmental economics taking a stand for the victim, but I fail to see that this can be supported by anything other than ethical judgements. The argument concerning dynamic efficiency, so often used, is at least not consistently utilized.[10]

The second issue concerns the way the tax is instituted – that is, the PPP as it is formulated in practice and presented in standard expositions. Figure 13.4 shows the solution where a tax $T = \text{MEC} = \text{MAC}$ is instituted.

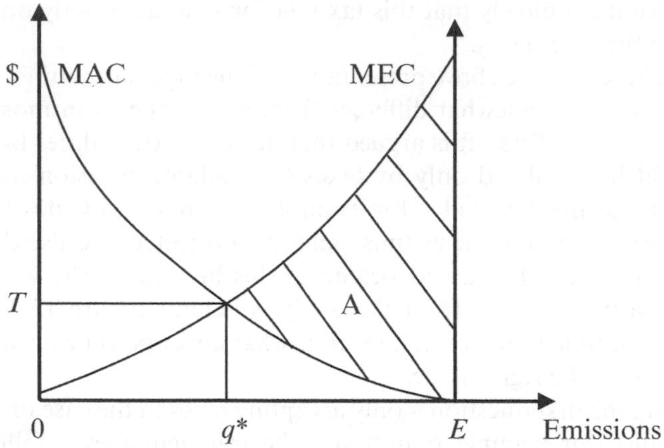

Figure 13.4 The polluter pays principle, environmental taxes and the efficient level of emissions

This is equal to the situation presented in Figure 13.3 with R_v as the basis. However, there is a difference concerning how one goes about formulating this solution. With consistent treatment of victims' rights, one would start off from R_v in Figure 13.3 or 0 in Figure 13.4, but this is not how policies are normally formulated. The victims are not protected by some right to a clean environment that cannot be interfered with. Instead, regulation occurs first when some negative consequences are observed – when emissions are at, for example, *E*. This implies that we start off from a situation described by some emissions. We evaluate whether a regulation is worthwhile – that is, if the MEC is larger than the MAC at that state. If so, a tax is instituted. Certainly, in practice we also have to evaluate the level of TCs. If we start from *E* and these costs are larger than area A, no regulation suddenly becomes optimal. The effect of this is that high TCs actually 'protect' the polluters and not the victims as would be the case with a consistent use of R_v.

The distinction involved here is basically about who has the responsibility of *proving that harm is inflicted.* A consistent use of R_v would imply that before people are given the opportunity to engage in, for example, some productive undertaking, they must document the (likely) environmental effects of their planned activity. If damaging emissions follow, they will be taxed from day one and their emission levels will be regulated so that marginal abatement costs equal the tax. This is the situation in the case of a bargained solution where the right is protected by a property rule. As the PPP is normally practised, the burden of proof is instead with the regulator. In this sense we actually observe a kind of a mix between

R_v and R_p. In other words, the responsibility to pay (PPP) and the burden of proof issues are separated in the way the Pigovian solution is normally instituted.[11]

13.2.2 Rights and Environmental Costs

So far the MEC and MAC curves themselves have been treated as uninfluenced by the existing rights structure. This is not a correct treatment of the issues involved. We shall start by looking at the marginal environmental costs. This issue applies equally to both the bargained and to the principal–agent settings, and no distinctions will be made concerning the two here.

We have previously demonstrated the great difference between WTP and WTA measures not least for environmental goods. The MEC curve of Figures 13.1–4 would have to be based on either WTP or WTA assessments. This choice depends on the rights structure. In our case the following situation can be depicted (see Figure 13.5).

Given victims' rights (R_v) the MEC should be established on the basis of willingness to accept compensation – $MEC_{R_v}^{WTA}$. However, WTP estimates are normally used when applying R_v/the PPP. I find this inconsistent. It is only when polluters' rights are assumed (R_p), that the MEC should be established on willingness to pay – that is, $MEC_{R_p}^{WTP}$. As previously observed (Chapters 8 and 11) WTA estimates are normally much higher than those based on WTP. Therefore 'optimal emission

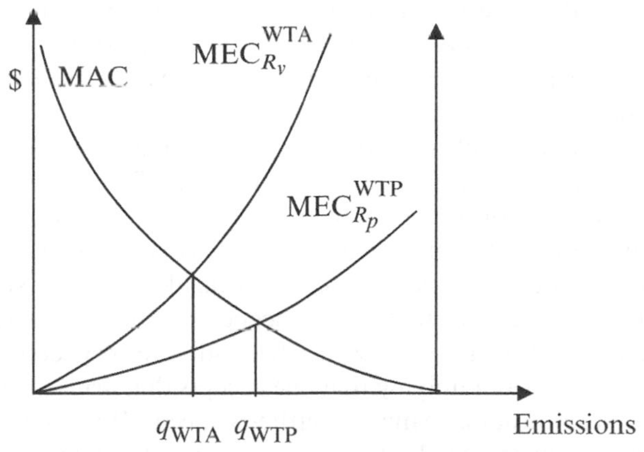

Figure 13.5 The efficient level of emissions given WTP and WTA estimates of environmental damages

levels' would be clearly higher in a situation given polluters' rights (R_p) than in a case with victims' rights (R_v). The difference would certainly depend both on the steepness of the MAC curve and the difference between WTA and WTP.

13.2.3 Rights and Abatement Costs

As previously emphasized, it normally takes (a long) time from the start of an environmentally damaging activity to the point when the harm becomes visible. Therefore the rule concerning who must prove harm is also important for the level of abatement costs. In a bargaining framework where the right is with the victim and the victim is protected by a property rule, the polluter has the burden of proof. S/he does not have the right to emit anything before the victim is approached and a deal is made concerning the level of compensation for (potential) future harms. In this situation the costs of abatement will be taken into account from day one, and investments will be made taking the (potential) environmental costs into consideration.

Turning to the situation with state regulation and the PPP, the burden of proof is normally with the state. This implies that the (later to be observed) polluter may make whatever investments s/he wants, and considerable investments in various productions will be undertaken under the assumption that no harm will occur. At the time when harm is proven or accepted and action is to be taken, the costs of abatement are to be calculated. Abatement might imply reducing production or changing production processes and so on. These costs will certainly depend on the investments already undertaken under the assumption that no harm was involved. Given the structure of the problem, the abatement costs will be higher than under a consistent treatment of victims' rights. This is illustrated in Figure 13.6.

The PPP – in its conventional formulation – will actually be similar to a polluter's right[12] concerning this specific issue. Thus, MAC^{PPP} will equal MAC_{R_p} (see Section 13.2.1 on the principal–agent framework). If we instead formulate the PPP on the basis of full victims' rights, two solutions – that is, *ex ante* and *ex post* – become available. First, the state could demand that the polluter has to prove innocence *ex ante*. This would equal a victim's right based on a property rule. Second, a retrospective (*ex post*) rule could be established, implying that one accepts that firms act under the assumption that some damage may nevertheless occur. If it occurs – that is, when harm is proven/accepted – the reference point for calculating abatement costs is still not the costs for the firms at that point in time, but the costs that would have occurred if one had known from day one what effects

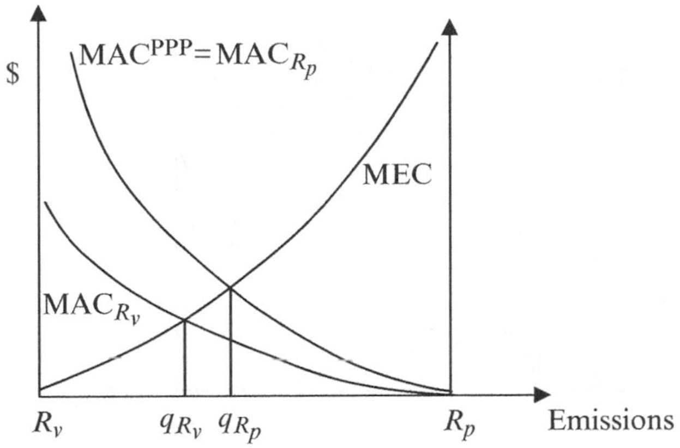

Figure 13.6 The efficient level of emissions with MAC valued according to different rights structures: full victims' rights (R$_v$) or the PPP as practised

would later appear. This secures that the *ex post* solution will, in principle, be determined on the basis of the same MAC = MAC_{R_v} as the *ex ante* one and it produces the same optimal abatement. The difference lies in the procedure involved.[13] In the *ex ante* case a property rule is instituted. In the *ex post* case a liability rule is used.

13.2.4 A Consistent Treatment of Rights and Costs

The final step in this analysis is to put together the various issues discussed in Sections 13.2.1–3 in a consistent way. First we shall do this under the bargaining framework of Coase. Next, we shall do the same with reference to the principal–agent model.

The bargaining environment of Coase

If we combine the various arguments above for the bargaining situation in a consistent way, we get a picture as presented in Figure 13.7. The solid lines depict a situation under victims' rights (R_v) while the dotted lines depict the case with R_p rights.

In the case with victims' rights (R_v), the environmental costs ($\text{MEC}_{R_v}^{\text{WTA}}$) are measured as WTA. The marginal net offer curve for the polluters – MNOP_{R_v} – can be consistently defined as MAC_{R_v} (as defined in Figure 13.6) minus the marginal TCs, which in this case fall on the polluters. The optimum in a bargained situation with victims' rights would then be q_{R_v} – assuming that total TCs are less than the gain from bargaining.

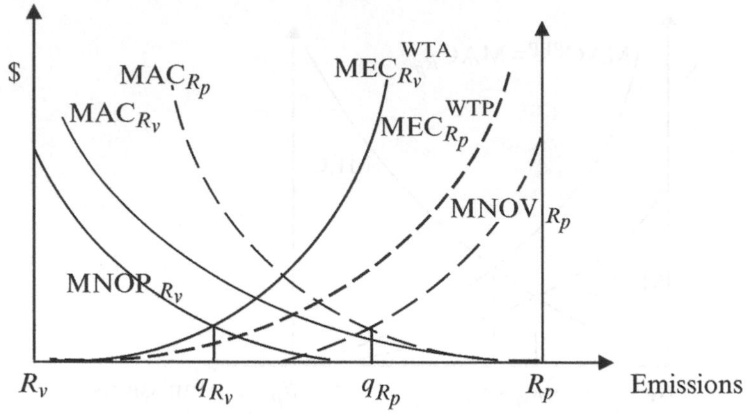

Figure 13.7 The efficient level of emissions given a bargaining situation

In the case of polluters' rights, environmental costs are measured as WTP and we get a $MEC^{WTP}_{R_p}$ which is lower than the environmental cost curve in the case of R_v. In this case transaction costs fall on the victims and their net offer curve will be $MNOV_{R_p}$. The marginal abatement cost will on the other hand be higher than in the previous case – that is, it is now equal to MAC_{R_p} of Figure 13.6. The optimum in a bargained situation with polluters' rights would then be q_{R_p} – again given that total TCs are less than the gain from bargaining. Taking all effects into account, we see that the difference between the 'optimal' abatement levels may be substantial depending on whether the rights structure is R_v or R_p.

The principal–agent framework of Pigou

As suggested, the main technical difference between the bargained and the principal–agent solution is the level of TCs. Using the regulatory power of the state reduces overall TCs and makes many more external effects Pareto relevant. While this is the main message, the understanding of the rights structure also in this case influences 'what becomes optimal'. Generally, the rights structure adhered to in the case of state regulations – both in the environmental economics literature and in practical policy – is the polluter pays principle. As we have seen above, this is not a full recognition of victims' rights. Figure 13.8 captures the main points. The way the PPP is normally instituted is described by the dotted lines, while the solid ones capture a consistent use of R_v.

The PPP rule is normally formulated as a mix between R_v and R_p. Polluters must pay (R_v), but costs and gains are measured as if polluters rights are

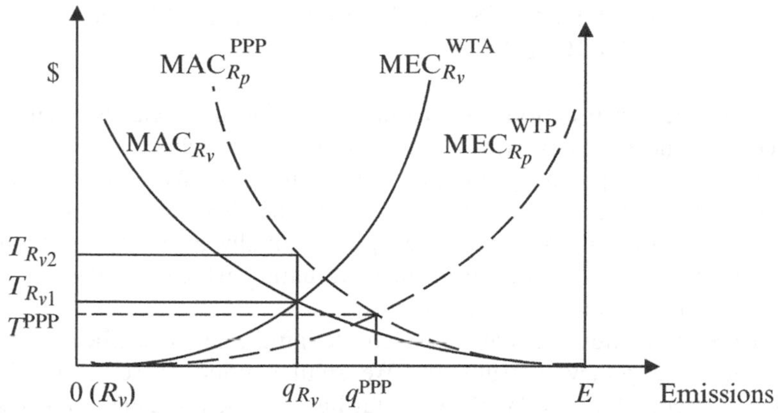

Figure 13.8 The efficient level of emissions given a principal–agent framework

governing – that is, R_p. Therefore environmental costs are estimated on the basis of WTP – $MEC_{R_p}^{WTP}$ – and the abatement costs are calculated on the basis of costs as faced at the time when the regulation is established. It is measured as $MAC_{R_p}^{PPP}$ – that is, as if R_p is the rights structure. Assuming insignificant marginal TCs, optimum will then be q^{PPP} with tax T^{PPP}.

If victims' rights are treated consistently, environmental costs should be measured as WTA ($MEC_{R_v}^{WTA}$), and the abatement costs to be taken into account ought only to be those that would have occurred if the regulation had taken place from day one of emissions (MAC_{R_v}). We observe that the optimum defined this way – q_{R_v} – will give lower emissions than the standard interpretation of the PPP. Thus, even in the principal–agent situation the definition of rights has implications for what becomes efficient. We also observe that when applying victims' rights consistently, two different tax levels are relevant depending on how that right is instituted. If the tax is based on *ex ante* regulation, the optimal tax is $T_{R_{v1}}$. If the regulation is implemented *ex post* – that is, is retrospective, as previously explained – the abatement costs of the firm is at that time $MAC_{R_p}^{PPP}$, and the tax must be set to $T_{R_{v2}}$ to motivate for attaining what is then defined as optimal abatement – that is, q_{R_v}.[14]

A similar reasoning as the above can be made for a situation where state regulations are based on consistent use of polluters' rights (R_p) compared with either a consistent use of victims' rights or the present way of treating the PPP. These analyses will be omitted, partly because they are less relevant, and partly because they would not provide any principally new insights beyond those already captured by Figure 13.8.

13.3 MATTER FLOWS, TRANSACTION COSTS AND EFFICIENCY

As we have seen, transaction costs are important for the efficiency of environmental policies. They are of importance when choosing between regimes – for example, bargaining or state regulations. We have also seen that they are of importance when we look at the effect of various rights structures applied within each of the regimes. Finally, as we shall show in this section, they are important when evaluating which specific policy instruments to use and how to apply them.

Here we shall study the role of TCs for defining what is an efficient way of formulating policy instruments. We shall confine ourselves to the principal–agent situation. While we so far have focused on emission taxes, the focus will now be on the following issue: given that TCs are positive, is it always best to regulate emissions?

13.3.1 The Systems Perspective

In Chapter 9 we described environmental disruptions as disturbances in natural or existing matter and energy flows. Focusing on matter, we observed that the economy was dependent on inputs from the biosphere and that all matter entering the economy would eventually return to the environment as waste. To regulate the negative effect of waste emissions, there seem to be three options or points at which instruments can be applied:

- We can regulate directly on the emissions or their consequences, as is the basis for the Pigovian position. All generally specified pollution models like those of Baumol and Oates (1975) and Fisher (1990), focus on regulating *emissions*.
- The second option is instead to regulate *inputs*. If all inputs still become waste in the end – the laws of thermodynamics – this may be an interesting option in many cases.
- Finally, regulating *production/consumption processes* is also an option.

Taking full account of the fact that we have two main spheres in the economy – the production sphere and the consumption sphere/the households – we can sketch the situation as in Figure 13.9.

Inputs (1) go first of all from the biosphere to the production sector – more precisely the extraction and processing sectors. In production (2) some matter is lost (3) and returns to the environment. Some matter is

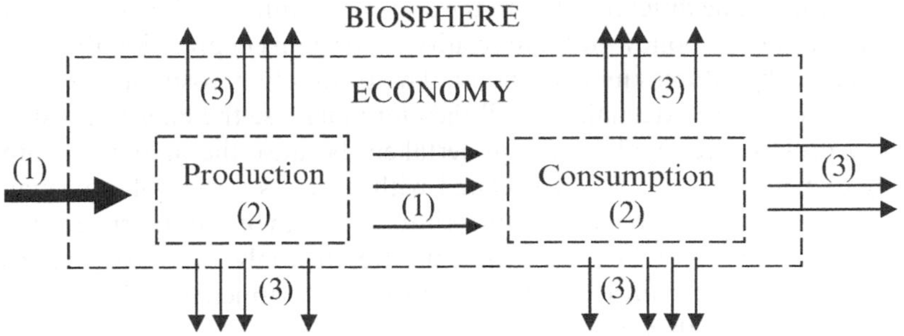

Figure 13.9 Materials flow, pollution and the three main points of instrument application

processed to become part of the final product, which then becomes an input (1) into the consumption sector. Consumption processes (2) might be of different kinds. Finally, the matter taken into the consumption sphere is also lost to the environment (3). Therefore moving from (1) to (2) increases order in the form of final products for consumption. But following from the second law of thermodynamics, this comes at the cost of increased entropy in the system as a whole (waste production).

The figure also illustrates another relation, that the number of bordering points between the economy and the biosphere – the dispersion of waste – is increasing from the first inputs of matter (1) until final emissions (3). The point is that the number of extractors of, for example, oil is fairly low. The number of producers of various oil products is much larger, and the number of points from where matter might be lost to the environment increases. Next, gasoline and so on is sold to retailers, which are increasingly more numerous than the producers. Finally, the products are consumed by a vast number of consumers and all that is left from what was taken into the economy is then transferred back to the biosphere. All matter is now so dispersed that it has no positive economic value – that is, it has become waste.

Materials flow characteristics are in the literature normally apprehended by a distinction between point- and nonpoint-source pollution (for example, Baumol and Oates 1975; Hanley et al. 1997; Kolstad 2000). Theoretically the distinction is not very precise, since any discharge has to be located at some point in the landscape. What is important seems to be implicit in the treatment – that is, whether there are few easily demarcated points with rather high discharges each, or many points on a more or less continuous borderline towards the environment where each point is negligible. The economic relevance of this is not whether the pollution source is a point or not. The difference lies in the costs of acquiring information and administrating the policy measure relative to the gains in precision that can be obtained by addressing each point. The transaction costs of internalizing the occurring externalities will therefore increase the later in the dispersion process the regulation is undertaken, because the matter is more highly dispersed. High TCs associated with an emission regulation may therefore even sweep many externalities into the realm of Pareto irrelevancy as long as emission-related instruments are the only ones considered. Again, we observe the importance of a consistent treatment of TCs and the trade-off problems they cause.

On the basis of this, one may be tempted to conclude that it might be less costly to reduce pollution by taxing inputs rather than emissions. This statement certainly needs some qualification. When matter changes form, or the location of losses influences the level of damages, there is a trade-off between this way of reducing TCs and finding a tax or quota scheme that reflects the damage at the margin as matter is emitted. This trade-off problem will be explored in the following sections.

13.3.2 Emission Characteristics and Precision

In a world of zero TCs, the following two conclusions are drawn concerning optimality:

$$\text{MEC}_\text{A} = \text{MAC}_\text{A} \tag{13.1}$$

$$\text{MAC}^1_\text{A} = \text{MAC}^2_\text{A} = \text{MAC}^3_\text{A}. \tag{13.2}$$

For each recipient A, marginal environmental costs should equal marginal abatement costs (13.1). Furthermore, if there is more than one emitter to this recipient, marginal abatement costs should be equal across these. In (13.2) it is assumed that there are three emitters. If the equi-marginal abatement cost condition of (13.2) does not hold, it would be possible to reallocate abatement between the emitters (firms/households) and the same abatement could be reached at less cost.

An emission or ambient tax T_A where

$$T_\text{A} = \text{MEC}_\text{A} = \text{MAC}_\text{A} \tag{13.3}$$

has the capacity to secure (13.2) as illustrated in Figure 13.10. For firm 1 it becomes optimal to abate q_1, since it is cheaper to abate than to pay the tax for this amount. Firms 2 and 3 have lower abatement costs and will find it profitable to abate more. If the tax is set as in (13.3), the sum of q_1, q_2 and q_3 will equal the optimal total abatement level, q^*.

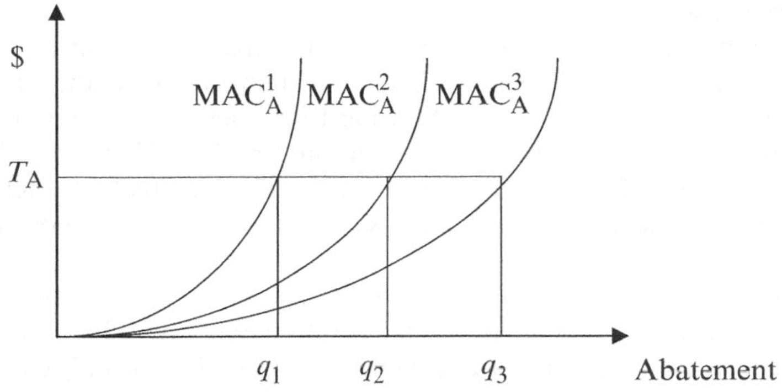

Figure 13.10 Optimal abatement levels obtained by a uniform tax

The higher the abatement cost for a firm, the lower the optimal abatement volume for that firm becomes. Conditions (13.1) and (13.2) give what is termed a socially optimal solution. Remember, however, that this conclusion demands zero TCs. Section 13.2 discussed the effects of positive TCs for defining optimal abatement levels – that is, the condition in (13.1). The problem to be discussed here concerns the effect of positive TCs for the status of the equi-marginal principle.

Let us denote a situation where this principle is satisfied for a *precise* solution. If TCs are positive, it may be so that the gain in precision obtained by establishing (13.2) – by using standard Pigovian emission taxes – could be more than offset by the extra costs in monitoring and so on.[15]

13.3.3 Three Illustrative Cases

To analyse the trade-off between precision and TCs in our setting, we need to define one more concept – that of a homogeneous emission. An emission is *homogeneous* if (a) the same input gives rise to only one type of damaging discharge (only one element or several elements in a fixed proportion) and (b) the discharges are uniformly mixed. The latter implies that the effect of a discharge is the same independently of where it takes place (Tietenberg 1985).

Given this, we can in principle specify four situations: (i) zero TCs and homogeneous emissions; (ii) zero TCs and non-homogeneous emissions; (iii) positive TCs and homogeneous emissions; and (iv) positive TCs and non-homogeneous emissions. In the following we shall look at (i) and (ii) together – since in a world of zero TCs it does not matter if emissions are homogeneous or not. The extra information necessary in case (ii) can still be acquired without costs. The following presentation is rather intuitive. For a more complete treatment, see Vatn (1998).

Zero transaction costs

As we know, in situations with zero TCs, no taxes (either on inputs or emissions) are necessary, since direct bargaining between the affected parties is by definition cost free. Starting from such an assumption may therefore seem rather contradictory given our problem. However, it helps to gain some initial insights, as it also covers the only situation where the Pigovian solution of emission taxes is guaranteed to be as good as an input tax.

Let us start with a situation with homogeneous emissions. A resource like carbon (in coal, oil or nature gas) is taken into the economy. Let us further imagine that it is used only for energy purposes and that the only emission is in the form of CO_2. Since the problem with CO_2 emissions is global, it does not matter where losses take place. This is therefore a situation where emissions are homogeneous.

If a tax on the input of carbon into the economy is instituted, the increased input price in the extraction sector following from this tax will be passed over to the processing sector and finally to the consumers. This implies that consumers meet a higher price on, for example, gasoline. If the tax is set right on the input of carbon, it will have the same effect on the consumer's decision problem as a tax on the emitted CO_2 gas following from his/her use of gasoline. With the given assumption of homogeneous emissions, the only way to reduce emissions and their effects is by reducing the input of the substance, as a certain amount of gasoline results in a given amount of CO_2. A tax on fuels or a tax on CO_2 emissions both result in the same changes in consumer behaviour: reduced energy use or substitution with another energy input that is less harmful to the environment and thus less taxed. The incentives to undertake these actions are identical for both an input and an emission tax given zero TCs. Since the emissions are homogeneous, the effects for the environment are also the same.

If emissions are non-homogeneous, the situation is in principle not different as long as TCs are zero. If the carbon in the previous example in some instances was not used for energy consumption and as an example was just emitted as harmless carbon, the environmental effect would be negligible. However, if TCs are zero, this does not matter. Remember that in this (curious) situation information is complete (costless) as is contracting and control. If some carbon is eventually used for purposes other than energy production, this will be known already at the time of extraction, given zero information costs. Taxing each carbon atom on the basis of its future use is then cost free. In this world, non-homogeneity does not matter.[16] Therefore input and emission taxes have equal efficiency characteristics.

Positive transaction costs and homogeneous emissions

In the case of positive TCs, the point of instrument application will matter. Let us continue with our carbon example and assume that carbon will be used only for energy purposes – that is, emissions will be homogeneous. If we assume that TCs are proportional to the number of points that we have to tax and monitor, it is easy to see that regulating on the final emissions is the option with the highest TCs. The lowest level of TCs is to regulate on the input to the economy/the extraction points. Due to the increasing number of agents throughout the chain from extraction to emission, we can produce the following typical relationships concerning regulations at different points as the matter moves through the economy:

$$\begin{aligned} &\text{TCextraction} < \text{TCprocessing} < \text{TCretailing} \\ &\quad < \text{TCemissions} < \text{TCeffect}. \end{aligned} \tag{13.4}$$

An ambient tax – a tax on the *effect of* an emission – will be the most costly in TC terms. Taxing emissions will be less costly and so on. However, precision will be the same in all cases since the emission is homogeneous, and the conclusion is straightforward: taxing inputs will always be the least costly solution in the case of homogeneous emissions since it gives the lowest TCs.

Positive transaction costs and non-homogeneous emissions

If emissions are non-homogeneous, the situation is more complex. We shall have to make a trade-off between TCs and precision. A typical example of a non-homogeneous emission problem is the release of various nitrogen compounds in to the environment following the use of nitrogen fertilizers in agriculture. First, not all nitrogen in fertilizers ends as polluting substances. Depending on the conditions in the soil and the waterways, some may be transformed back to nitrogen gas (N_2) and released to the atmosphere. As N_2, nitrogen is not a pollutant. Some may be lost as nitrates (NO_3^-), which pollutes some waterways and groundwater. Some may be lost as ammonia (NH_4^+), which may cause acidification. Finally, some may be lost as laughing gas (N_2O), which is a climate gas with global effects.

Losses may therefore have a different composition of polluting and non-polluting substances depending on the involved technology and locality. Moreover, nitrate losses have different effects in different recipients. Losses to the Barents Sea may have negligible impacts, while we know that losses to the coastal waters of Western Europe or the Baltic Sea are very damaging.

Taken together, there will certainly be a loss of precision (PR) moving from an ambient tax to a tax on the extraction of N_2 from the atmosphere to make fertilizers when emissions are non-homogeneous. We have:

$$\begin{aligned} &\text{PRextraction} < \text{PRprocessing} < \text{PRretailing} \\ &< \text{PRemissions} < \text{PReffect}. \end{aligned} \tag{13.5}$$

To strike a balance between losses in precision and increased TCs (see (13.4)) is an empirical issue. The point here is to illustrate which cost elements and which dynamics are involved. In the case of emission of nitrogen compounds from agriculture, TCs related to ambient/emission control are likely to be very high. To my knowledge no country has established a regime based on this type of regulation. Instead we have some examples of input taxes, but only at the wholesale level – that is, there are nationally defined taxes where all wholesalers in a country have to pay a uniform tax (see also Box 13.1). This increases precision since the situation may vary between countries, but it does not take account of variations within a country.[17] Having different input taxes in different regions of a country (that is, regulating on the retail level) would have increased precision with most probably a low increase in direct TCs. The problem of appearing black markets and the necessary control thus involved, seems to block this as an option. Taking all TCs into account, it is unlikely to be a reasonable solution.

An alternative to regulate on inputs as opposed to emissions is to regulate on the production process. One could establish management taxes for environmentally damaging practices (Griffin and Bromley 1982), or subsidies for environmentally friendly practices. In the case of nitrate losses from agriculture, one may observe subsidies for 'green fields' in the autumn (see Box 13.1). TCs will be much reduced compared to an ambient tax, while precision is lowered too. Which solution is in the end the best, can only be assessed empirically.

13.3.4 Optimal Point of Instrument Application

From the above discussion we can formulate a synthesis as depicted in Figure 13.11 (see also Box 13.1). While input regulations will offer the best option when emissions are (rather) homogeneous and TCs related to emission regulations are high, the situation is opposite for emission regulations. In many situations input regulations will be too imprecise and emission regulations will be too costly in TC terms, that is, they are administratively too costly. In such a situation, regulation of the production processes is an important alternative. Prescribing a specific technology in vulnerable areas enables

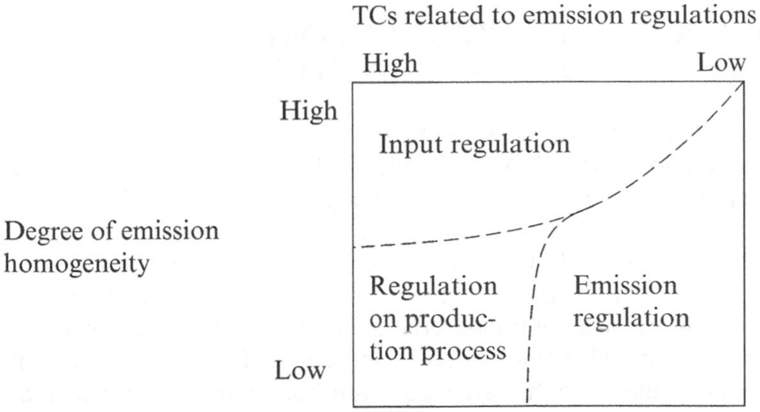

Figure 13.11 Optimal point of instrument application given different degrees of emission homogeneity and TC levels

the goals set concerning emission levels to be obtained. TCs may not be that much higher compared to input regulations, while precision is increased since the standard can be varied between regions and so on. Certainly, it may be argued that regulating production processes/technology is more costly than to regulate emissions. It is simply less flexible. However, this conclusion depends on zero TCs to generally hold. Certainly, this regulation procedure is less flexible, but the losses implicit in that may often be more than offset by reduced TCs compared to ambient taxes/emission taxes.

Recycling will often be an important way to reduce environmental damages. We observe that input and emission taxes are in principle equal regarding the incentives given for such a practice. In the case of an input tax, the substance taken into the economy increases in 'value', and this increases the motive for avoiding losses and in the end recycles the matter if technically possible. Similarly, an emission tax will also increase the cost of losing the compound and motivate recycling.

The way the problem is formulated here may also have influence beyond that of trading off TCs and precision. Focusing on the matter cycle as such forces both the principal and the agents to ask the following questions up front. What will happen if we bring this substance into the economy? Will it result in potential environmental damage? The systems view helps to sort out potential future damages and can be seen as an element in instituting the precautionary principle.

The question of optimal point of instrument application is not much focused on in the environmental economics literature. Input taxes are

BOX 13.1 THE CHOICE OF POINT OF INSTRUMENT APPLICATION

The evaluation of what is the best point of instrument application is a complex issue. Only in situations where precision is the same independent of point of instrument application does one not need to do further evaluations. Carbon is the typical case, and we see that to the degree that its use is regulated, some sort of input regulation is favoured. Even in the case of so-called CO_2 emission quotas, these are not based on measuring emissions but on measuring inputs. However, even in the case of carbon it may not always be the case that emissions are homogeneous. Injection of CO_2 into the ground is an opportunity in the case of some energy-producing plants. Dominantly this will imply returning CO_2 to the geological formations from where oil or gas was extracted. As a result, use of carbon will lead to fewer or no emissions. Taxing all extraction of hydrocarbons would become imprecise and in practice defeat such a solution. From a regulatory point of view, this could be handled by moving the point of instrument application one step down the chain to the point where the energy is sold or energy companies could get a tax reimbursement if they could document incineration. The increase in TCs would still be much lower than the loss in precision.

One issue in relation to this is that the costs of incineration are rather high. Thus to make oil extraction profitable, using the CO_2 incineration to extract more hydrocarbons from the ground is necessary. The net effect of incineration on total CO_2 emissions is therefore uncertain.

Coal and hydrocarbons used in energy production also emit SO_2, which causes acidification. Also in this case, input taxes rather than emission-based regulations may be appropriate. This tax could be set equal to the sulphur content of the energy carrier and the incentive would be in place to encourage energy producers to shift to less damaging coal or oil sources. However, two issues are involved. First, the effect of SO_2 is not global in the same sense as CO_2, and taxing inputs will result in some loss of precision even though the tax can be differentiated at the level of countries. Second, technology exists to clean SO_2 emissions. In some cases – larger energy plants – it may be less costly to clean than to shift to coal with less sulphur. To the degree that this sulphur can be stored in an acceptable way, this option should be made available. A general input tax could be combined with a system for reimbursement based on documentation of how much SO_2 was withheld through cleaning in the instances where this option was utilized.

In the case of nitrogen fertilizers, an input regulation is imprecise. However, it may be the best solution. This depends on the type of agriculture that dominates – for example, arable versus animal husbandry – and the kind of recipients involved. An alternative to a tax on inputs could be a regulation based on calculated emissions as in the form of an *N* surplus (*N* imported to the farm minus *N* in products sold). This is used in different forms in EU countries. However, this is also an imprecise solution, since it does not distinguish between the form the losses may take and the fact that not all surplus ends as a loss (for example, the building up of organic matter in soils). Finally, regulation on production methods such as using catch crops and abandoning autumn tillage, is an interesting alternative. A more detailed discussion of this is found in Vatn et al. (1997, 1999).

discussed, but dominantly in a negative way. It is interpreted as the only possible solution in some cases, but still seen as inefficient. Incorporating TCs in a meaningful way, changes the picture. TCs are as real as other costs and in a complete treatment of the involved costs they must be duly considered. The failure to treat TCs in a systematic way is the source of much confusion.[18]

Analysing the above issues, one should acknowledge that situations may exist where the point of instrument application excludes some instruments as applicable. In addition to taking account of the point of instrument application, a complete evaluation of TCs entails analysing the TCs of using one instrument instead of another. Therefore a system of tradable emission permits will normally be more costly in terms of TCs than an emission tax. However, tradable permits are favourable in some settings. One example of this is the Weitzman proposition on taxes versus quantities: if abatement costs are uncertain and the marginal environmental cost curve is steeper than the marginal abatement cost curve – that is, in situations where it is important to hit the environmental target fairly precisely – tradable permits are preferred over taxes (Weitzman 1974). In this case the cost of not hitting the target is high and the quota system actually guarantees higher accuracy. It may also be favoured due to some differences in the distributional characteristics of a tradable quota system as compared with a tax. This is an issue we shall comment on in the next section.

13.4 POLICY MEASURES, MOTIVATION AND INTERESTS

So far we have discussed the issues of regimes and how regimes and their implied rights and TCs will influence what will become efficient resource use. We have also discussed how the choice of point of instrument application influences regulation costs and thus efficiency. Finally we are at a stage where we can focus more specifically on the choice of instruments or policy measures themselves.

Three issues of special importance when choosing between instruments will be covered: (a) the motivational aspect – the issue of how the principal influences the motivation of the agents; (b) the distributional effect of various policy measures; and (c) transaction costs issues. This section focuses mainly on motivation and distribution. While there are some important TC-related questions that go beyond those emphasized in Sections 13.2 and 13.3, the basic thinking is the same, and we shall restrict ourselves to a few comments in relation to an analysis of the other two issues. However, before we can start looking at the motivational and distributional issues, we need to categorize the main groups of policy measures.

13.4.1 Types of Policy Measures

We shall distinguish between three main categories of policy measures – that is, economic, legal and informational. Each category is characterized by different assumptions about the rationality of the agents involved and assumptions about TCs.

Economic instruments are based on the assumption that agents are *rationally calculative.* Preferences are given and choices are assumed to be based on individual interests. The effect of the instrument – be it taxes, subsidies or tradable quotas – is to shift the payoffs of different actions and make changed behaviour more desirable for the agent than the existing behaviour. If this is successful, the agents produce the outcomes that the principal wants. Economic instruments may certainly affect behaviour in the desired direction even though respondents do not fit the model very well, for example if agents are boundedly rational. They are, however, dependent on individual profit/utility maximization as a dominant motivation in order to work properly. It is important to note that if people think more in *normative terms*, economic instruments may not be viewed by agents as just external and neutral incentives. They may instead be perceived as moral signals influencing self-esteem and accentuating certain values. Being taxed does not – according to this latter view – merely change trade-off structures. It may define the act as anti-social. It may also turn something that was previously thought of as a normative issue – that of right behaviour – into a pecuniary or calculative one.

Legal instruments may also evoke or embody two different types of rationalities.[19] The law may be seen as an external restriction on individual behaviour – as an external cost – again working on the motivation of the maximizing agent. The agent is then supposed to calculate the risks involved in breaking the law and comply if expected costs of doing so are higher than the expected gains. However, legal instruments may also be seen as following a more social or normative type of reasoning. Here, the idea is to define what is *right* or *legitimate* behaviour. The rationality is that of norm following, not that of least cost. If consistently applied, all actors must comply with the standard set. This implies that no actor should be allowed to buy him-/herself free from the restriction by paying someone else to correspondingly oversatisfy the claim. To be a legitimate agent everybody must comply it. Understood this way, the equi-marginal principle becomes irrelevant for the policy formulation.

Informational instruments – including communication – may also be understood on the basis of the two different perspectives on rationality. They may operate through *cognitive* and/or *normative* processes. As a cognitive measure, information may especially support economic instruments through making agents aware of what is best to do, given their preferences. Then it is assumed that information is costly – bounded rationality – and that a successful policy based on, for example, economic instruments demands some support from information campaigns and so on to support agents in their process of determining what is now profitable to do. However, information and communication can also work normatively by appealing to humans' care for others or by changing their perception of what is the right thing to do. Information may go beyond operating as pure knowledge and influence the norms evoked.

From the above, we see that each kind of instrument can be understood within an individualistic calculative framework or within the wider framework of social interaction and construction. Understanding the implications of this is crucial. However, we are only at the beginning of a process to perceive what goes on when an individual is confronted with different motivational structures such as the above categories of policy measures.

13.4.2 Policy Measures and Motivation

Neoclassical rationality

Neoclassical regulation theory – of which environmental economics is a sub-branch – emphasizes the calculative agent. The agent maximizes profits (firms) or utility (households and individuals). The goals of the agents may diverge from those of society (the principal) due to, for example, external effects. By taxing those external effects, correspondence can be established between the societal goals (the goals of the principal) and the goals of the individual firm, household or person. A tax $T_A = MEC_A = MAC_A$ secures equality between the defined social optimum ($MEC_A = MAC_A$) and the incentive the agent faces via the tax T_A. As shown in Figure 13.10, this tax also has the capacity to distribute abatement efficiently between firms with different abatement cost structures.

A system of tradable quotas has in principle the same incentive effect. An optimal emission quota (q^*) is defined – again where $MEC_A = MAC_A$. This total quota is next distributed between the involved firms. Figure 13.12 illustrates this in a situation with only two firms. The total quota is split, with q_1 given to firm 1 and q_2 to firm 2. If this initial quota distribution diverges from what is optimal given abatement cost structures, quota trade can restore equality at the margin.

The initial distribution of quotas in Figure 13.12 implies that the marginal abatement cost for firm 1 is much higher than that of firm 2 – that is, $MC_1 > MC_2$. Hence, costs can be reduced if emission rights are sold from firm 2 to firm 1 with final distribution q_1^* and q_2^*. This conclusion demands zero TCs. Given that assumption, gains equalling the two hatched areas are obtained. The quota price P_q will be equal to an optimal tax. If TCs are positive, the gains from trade are reduced and the amount of trade will be lower. If total TCs are larger than the gains – the two hatched areas of Figure 13.12 – there will be no trade. Our example assumes grandfathering the quotas – that is, they are given to the firms for free. They could also be sold directly from the principal to the agents. The main difference between these two quota schemes is the distributional effect – see Section 13.4.3. Basically, this reflects a difference in the (implied) rights distribution.

In principle, taxes and tradable quotas give the same incentives. Either the principal determines the efficient price (the tax) and the agents find the

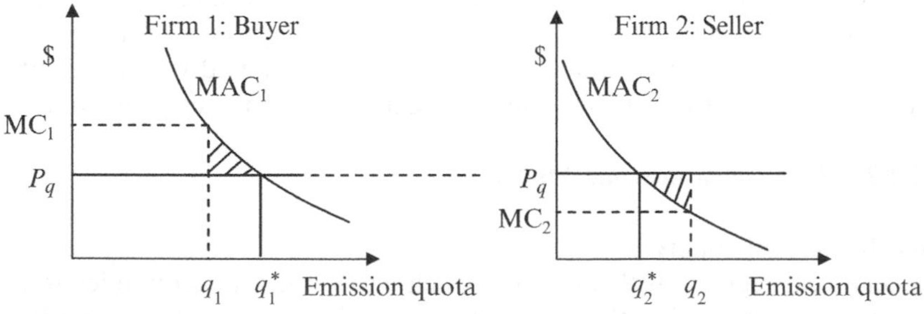

Figure 13.12 Tradable emission quotas

efficient abatement volumes or the principal determines the efficient volume (the total quota) and the agents determine the efficient price. Transaction costs will normally vary, though, since the quota system demands setting up and running a trading arrangement.

In the neoclassical regulation literature, environmental standards (legal regulations) are also studied (for example, Baumol and Oates 1975; Hanley et al. 1997). Such types of regulation are deemed inefficient since they tend to result in different marginal abatement cost levels across agents. The issue of what is 'right' behaviour is not focused on in this literature. To secure efficiency, one should be free to pollute as long as it is profitable to buy emission permits or pay taxes.

Furthermore, this literature clarifies which control and punishment schemes are optimal to use given that agents have information not available to the principal. We have the problems of *asymmetric information* and *moral hazard*. The first issue relates to the fact that the agent will have better knowledge about both own abatement costs and own emissions than the principal. The second issue points towards the possibility this gives for the agent to 'cheat' – to utilize this information for his or her benefit. Different penalty structures are discussed to control the agent in such situations. These are based on the assumption that increased control/punishment will result in increased compliance. Figure 13.13 illustrates the situation if this relationship is of the linear kind.

This literature discusses the possibility of reducing control costs by reducing the number of controls, but instead instituting higher fines, thus keeping the risk of not complying intact. It also discusses different strategies to reduce control costs by basing the control scheme on reputation – that is, that some firms have a history indicating that they are more prone to violating the regulation than others (Greenberg 1984; Russell 1990).

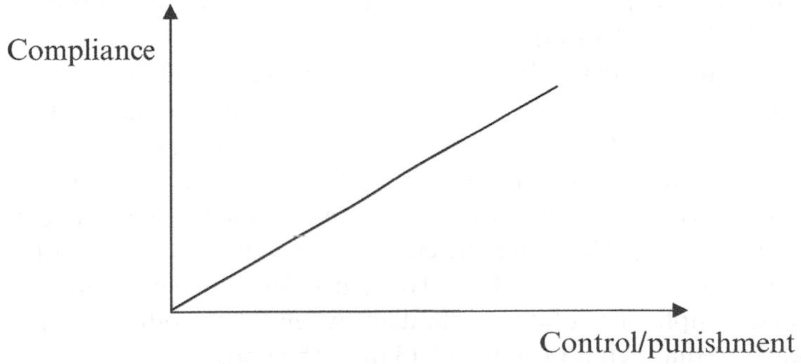

Figure 13.13 The effect of control/punishment on compliance: the neoclassical perspective

Observed behaviour does not always fit the standard model

In Chapter 6 we presented and discussed several situations where individual behaviour did not fit well to that of maximizing individual utility. Instead, reciprocity and norm following seemed to be important for determining choices in many situations – for example, the results from ultimatum games, public goods games, crowding out and so on. In general it was observed that the institutional structures influenced the rationality or logic evoked. Consequently, the same individual could act more or less calculatively or more or less reciprocally depending on the institutional setting. The observed deviations from what is expected from the neoclassical model do not imply that this model is unable to explain behaviour in many settings satisfactorily. Instead it implies that settings are different, and that this fact must be accounted for when formulating operative policies.

The literature on these issues is developing. While further research is needed to draw more specific conclusions, a rather consistent pattern seems to be evolving. The existing data suggest that the response to policy measures depends as much on their legitimacy as on the involved punishment structures. A striking illustration of this is found in the so-called 'Chicago study' by Tyler (1990). He found that compliance with the law bore little relationship to the level of punishment. Instead he observed situations with high compliance when punishments were low and low compliance when they were high. He found that the 'willingness to follow the law' was strongly dependent on the legitimacy of that law in society. He concluded that normative issues are important for explaining behaviour. The law is more than an external punishment structure.

In a study of sanctioning systems including fines for environmentally bad conduct, Tenbrunsel and Messick (1999) observed that control may actually reduce compliance up to a certain level, from where it then starts increasing. Similar findings are documented by others, for example, Fehr et al. (1997). Figure 13.14 illustrates this.

The literature indicates that control may reduce compliance because it signals that the principal does not trust the agent. The (internal) motivation for compliance is actually reduced. Beyond a certain level, the negative effect of the heavier control becomes so high that the agent must take account of it. The act of instituting control/punishment appears to destroy internal motivation. Therefore the curve may not be smooth as in Figure 13.14, but rather showing a high level of compliance when no control is executed, and compliance falling substantially when it is introduced. The logic is switched to that behind Figure 13.13 from then on.

Certainly, control of those (notoriously) breaking the law or agreements does not have this effect since there is no internal motivation to be destroyed.

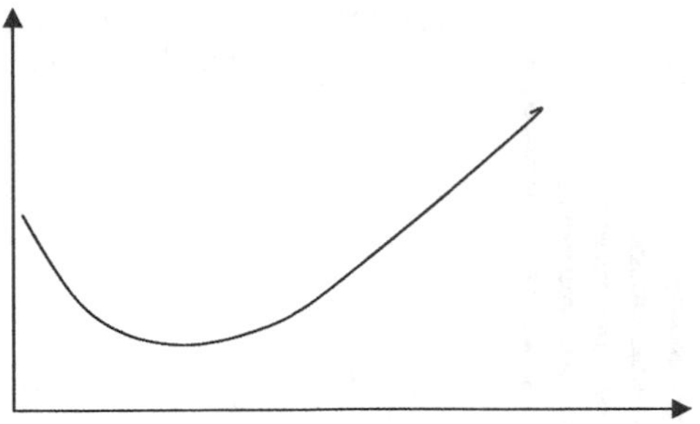

Figure 13.14 The effect of control/punishment on compliance: the institutional perspective

The fact that the principal punishes those with behaviour in line with that described in Figure 13.13, seems moreover to reinforce the positive self-image of those complying freely – that is, those who have internalized the logic of the law or the contract. Hence, the principal needs to be careful when developing the control scheme if there are some agents acting strategically and following the logic behind Figure 13.13 and others who act cooperatively/act on the basis of internal motivation. It may warrant a different control regime for the two groups. If defining who belongs to which group cannot be done independently of making controls, this is indeed no simple task.

Results that are similar to the above are also observed in other kinds of studies of incentive structures. Fehr and Gächter (2002) analysed the effect of 'incentive wages' – that is, wages that depend on monitored effort. In their research they have consistently observed the existence of what they term 'reciprocity-driven voluntary cooperation'. On the basis of this they ask the following question: do explicit incentives (such as 'incentive wages') leave the willingness to cooperate voluntarily intact, or do they increase or decrease it?

They compared a situation where the employees involved in the experiment were faced with three different wage structures and expected efforts in each case as determined by the employers. In the first case the employees were offered fixed wages at different levels and an expected effort for each of these levels. Actual effort was then observed. In the second situation the employees faced a situation with a maximum wage from which deductions were made if shirking was observed. The chance of being caught was 1/3. Those of the third group had a similar type of incentive, but formulated as a premium for greater effort and not as a punishment for lower effort than expected. Again the chance of being observed was 1/3.

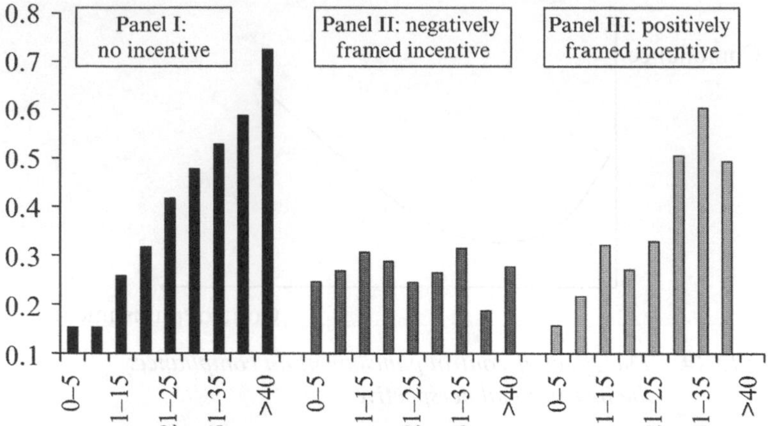

Sources: Fehr and Gächter (2000); Fehr and Falk (2002).

Figure 13.15 The effect of different wage structures on effort

The effort–payment structure of the last two was thus identical, just framed differently as punishment versus reward.

Fehr and Gächter's findings can be summarized as follows (see Figure 13.15). The first group – panel I – increased effort linearly with the increased payment. The effort was, however, somewhat lower than expected effort as set by the employers. In the figure, expected effort is standardized so that expected effort at maximum wage equals 1. Since there were no punishments involved, minimal effort would be expected from a reasoning based on individual calculative behaviour. However, the difference between the expected and the actual effort levels shows that some 'shirking' was involved. Their paper does not say anything about how shirking was distributed across the involved employees.

In the case of the negatively framed incentive – panel II – no clear relationship between effort and the final wage was observed. The effort was generally low. In the case of the positively framed incentive – panel III – effort again increased with final payment, but even in this case it was significantly less than that observed under the first kind of incentive structure. In another paper the authors conclude that 'in the presence of non-pecuniary motives, there are important and, relative to the predictions of the economic model, unexpected interactions between material incentives and non-pecuniary motives' (Fehr and Falk 2002: 695). They refer to several studies (such as Bohnet et al. 2001 and Schulze and Frank 2003) with similar results. Bowles (1998) and Gneezy and Rustichini (2000b) offer other observations supporting these kinds of findings. The latter paper, 'Pay enough or don't pay at all', describes a much observed situation where people are willing to perform a task for free. When payment is offered, this willingness immediately evaporates, while it increases gradually with increased payment. They conclude that 'The result has been that the usual prediction of higher performance with higher compensation, *when one is offered*, has been confirmed: but that the performance may be lower because of the introduction of the compensation' (ibid.: 807; original emphasis).[20] Therefore the picture is parallel to that depicted in Figure 13.14 if the vertical axis is shifted to 'performance' or 'effort' and the horizontal to payment or compensation.[21]

The work of Fehr and Falk (2002) gives further insights into the effects of control as observed by Tenbrunsel and Messick (1999). Fehr and Falk make a distinction between declaring a general threat of punishment to everybody and having a more diffuse opportunity to punish those acting badly. They therefore differentiate between an *ex ante* commitment given by employers to control and punish and an *ex post* opportunity to do so. In the first case, everybody is told that the employer (principal) does not have trust in the employees (agents) by declaring that all will be controlled. This tends to destroy the willingness to act reciprocally among those so willing. In the second case, those cooperating observe that those shirking are punished. This may even increase the willingness to cooperate among those acting reciprocally since they observe that the shirkers are caught. Thus, there seems to be an 'intricate' relationship between the methods used to signal trust and distrust.

In the case of environmental issues more specifically, studies show that people indicate willingness to cooperate, be it in the form of reciprocal behaviour or more normatively founded reasoning. Bruvoll et al. (2002: 348) show that an important motive for recycling was that 'I should do myself what I want others to do'. In a series of papers, Frey has shown how paying for doing environmentally friendly acts may reduce willingness to undertake such acts (Frey 1992, 1997a; Frey and Oberholzer-Gee, 1997). This is called 'crowding-out' (see also Chapter 6) and is related to the effect of using monetary rewards in situations where the motivation is more normative, or 'intrinsic' as Frey denotes it. Paying 'crowds out' the internal moral motivation.[22]

While 'crowding out' may describe situations where normative motivations become 'distorted' as monetary rewards are involved, 'crowding in' is used to describe the opposite mechanism: that external incentives result in the establishment of a cooperative norm. Nyborg and Rege (2003) document a study of smoking behaviour in Norway. In 1988, smoking was forbidden in certain public spaces (transport, meeting rooms and so on). This then influenced attitudes to smoking in other arenas such as private homes, where a greater disapproval of smoking was observed or smokers started to ask if they could smoke, went outdoors without asking for permission to smoke indoors and so on. Nyborg and Rege propose an explanation that builds on the assumption that non-smokers experienced a more negative effect of smoking given that it was now experienced less often. Hence, a stronger motive to say 'no' to smokers developed. This may be part of the explanation. However, an alternative explanation would suggest that a new norm structure was established in the wake of the law. Banning smoking in some spaces resulted in smoking becoming less acceptable generally. Those smoking may also acknowledge that smoking is negative for non-smokers. The latter feel similarly that there is a supportive environment for saying no. This reaction has – due to the new law – become socially acceptable. A new norm is internalized. The 'right' is shifted to the non-smokers.

One would expect some differences between situations with different types of agents involved. The behaviour of nations that are parties to an international treaty might be expected to be quite different from that of individuals involved in polluting a local recipient. Firms might be supposed to act more in line with the idea of revenue maximization, while households or individuals might be more inclined to act reciprocally or morally. However, documentation already shows that the mechanism of reciprocity is also active when firms are involved (for example, Tenbrunsel and Messick 1999). As emphasized in Chapter 10, Young (2002) makes the comment that the follow-up of many international treaties cannot be explained by only referring to the gains each country makes. Certainly, a lot of strategic action is observed, not least at this level. However, the story is far more complex. Indeed, given the lack of explicit punishment structures at the level of international environmental agreements, no or very low compliance should be expected. This does not seem to be the case.

An institutional interpretation

How can we explain the observations described above and in Chapter 6? While the story is too complex to offer simple explanations that make sense in all situations, Figure 13.16 captures much of what is going on when again restricting ourselves to the principal–agent framework.[23]

The figure distinguishes between 'I' and 'We' motivated behaviour – between on the one hand strategic behaviour and on the other reciprocal or norm-based behaviour. In quadrants A and D, the principal and agents perceive the rationality or logic of the situation similarly. In A they see it as governed by strategic action. In D they see it as ruled by reciprocity or norms. In the first case (A) economic incentives will function well. In the last, reciprocity or voluntary compliance is the rule.

Motivational structures (rationality) as perceived by agents

Motivational structures (rationality) implied by the policy instrument

	I	We
I	A: Strategic or instrumental behaviour	B: Crowding out/ retaliation
We	C: Crowding in/ internalization of norms	D: Reciprocity/ moral commitments

Figure 13.16 Motivational structures, policy measures and behaviour

Quadrant B describes a situation covering many of the examples given here and in Chapter 6 that are 'counterintuitive' from an economic perspective. The principal believes that the agent is acting strategically (economically rational), while the agents look at the situation as being one of reciprocity or of norm following. In such a situation the response is likely to be 'perverted' as in the 'crowding-out' examples or the examples presented concerning control and economic incentive structures. People may be offered money to do something that has been considered a duty. Subsequently offering money ruins the basis for the duty.

When interpreting the various examples, one should recognize that there is a crucial difference between 'We' motivation as reciprocity and as norm following. In the first case, economic incentives may already be involved as in the case of wage structures. Thus, it is not the introduction of economic incentives *per se* that may cause reactions, but the issue of whether the *change* in economic motivation structures is considered *fair* or friendly. In the cases presented under the heading of 'crowding out' this is different. Here the act is initially motivated by normative reasoning and not by monetary reward. Then it is the *introduction* of monetary motivations that causes the effect. Paying destroys the moral or 'intrinsic' motivation.

A comparison of a tax on fertilizers and pesticides, respectively, may serve as an illustration. Since these are commodities, one would believe that no crowding out would be observed. Reactions related to fairness considerations may still be important. Vatn et al. (2002) study the process concerning the introduction of a fertilizer tax in Norway. Their study indicates that – even though the tax was on a commodity – farmers interpreted it more as a punishment than a change in economic incentives, and they protested fiercely about the proposal. A similar tax on pesticides did not result in such reactions. While the fertilizers were considered good and necessary in production by making soils more fertile, the use of pesticides was perceived negatively at the outset by farmers themselves.

Quadrant C describes a situation where the principal may support a process of norm building where no norm previously existed. The case of smoke regulation serves as a good example. Certainly, if all individuals perceive the issue as an 'I' issue, this policy is unlikely to be successful. If, however, some view it as a 'We' issue, a 'We' orientated public policy may turn the issue into a social one.

An interesting issue is the dynamics of each quadrant. A and D depict what seems to be stable situations. In the case of B, deterioration of the social capital embedded in reciprocity or the norms involved is likely. A situation characterized by cooperation will deteriorate into a situation of strategic behaviour where a need for control and direct punishment becomes necessary. We end up in quadrant A. Etzioni (1988) emphasizes this as a general problem in modern societies. He goes one step further by underlining the danger that this process in the end will undermine the markets themselves (the 'I'–'I' relations) since these also depend on some level of trust – on some level of cooperative will.

These observations raise the question of the stability of quadrant A, which was taken as given above. If TCs are positive and therefore some trust is necessary in any contracting situation, conditions characterized by pure strategic behaviour are in danger of becoming unstable. Increased levels of TCs – more detailed contracts, more control and so on – are necessary for maintaining the same level of compliance.[24]

While reciprocity and commitments on behalf of the agents may deteriorate into strategic behaviour by setting up incentive schemes assuming such behaviour, building trust, reciprocity and normative commitment seems to be a much more complex and difficult process. Thus, there is no symmetry between 'crowding out' and 'crowding in' – between tearing down and establishing social behaviour/norms. Existing commitments may be undermined by using incentives where those playing strategically will systematically win. This accentuates the potentially high costs of using the wrong instruments. It also tells us that building trust is hardly possible by just choosing 'the right instrument'. It rather depends on a communicative process parallel to that of a deliberative institution and an active civil society as discussed in Chapter 12.

13.4.3 Policy Measures and Distributional Effects

As already indicated, the choice of policy measures will in practice also depend on their (implicit) distributional effects. This is not only

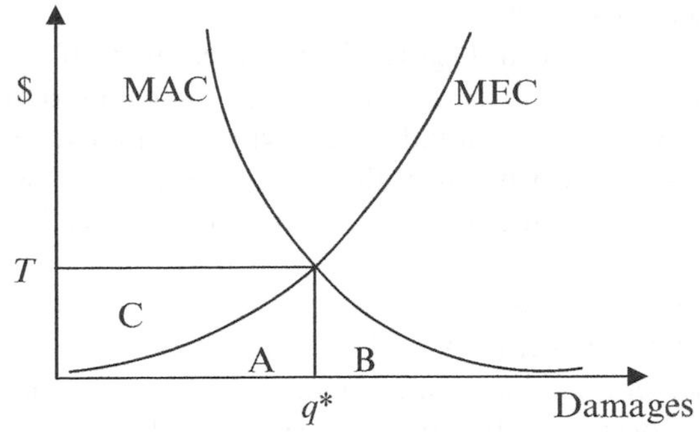

Figure 13.17 The costs for the agents of an ambient tax

because distribution is in itself important, but also because fairness influences agents' responses. We shall illustrate the potential distributional effects of different policy measures by comparing taxes with legal regulations/standards, taxes with tradable quotas, and finally input taxes with ambient/emission-based taxes. Again the picture will not in any sense be complete. However, the main mechanisms will be covered.

Taxes versus legal regulations

Figure 13.17 illustrates the costs facing agents in a situation with an ambient tax – that is, a tax based on the consequences of some damaging activity – for example, polluting emissions. In the standard optimum q^*, the damages are represented by the area A and the abatement costs amount to B. Given the tax T – rights are with the victim – the agents pay area A + C to the principal, and the total costs for polluting agents are A + B + C.

In the case of a legal regulation – that is, in the form of an emission standard with which the agents must comply – agents only face the abatements costs B. It is therefore not difficult to understand why industries tend to favour standards over taxes. The argument from an economic point of view is that standards rarely lead to the least-cost solution as one is unable to distribute the responsibilities for abatement between agents so that the rule of equal marginal abatement costs is obtained. Furthermore, while the tax gives a rather strong incentive to reduce abatement costs, the standard gives a motive that is weaker. In the case of a standard, a reduction in abatement costs is only rewarded by reductions in the area B. In the case of a tax, there will also be a motive to reduce the amount of taxes paid. A full evaluation of this must also include the effect of positive TCs.

Taxes versus tradable quotas

If an emission quota q^* as in Figure 13.17 is auctioned off and distributed to the involved agents by the principal, the price per quota unit will equal T. Hence, the distributional effects will be equal to that of a tax. Certainly, if TCs are positive, not all emitters may find it worthwhile to buy emission rights or some will buy less than otherwise. Thus the quota price will fall.

Quotas may be grandfathered. This implies that emission rights are given for free to the involved agents. As in Figure 13.12, trading will then appear between the different agents until all hold a quota where individual marginal abatement costs equal the quota price. Again the conclusion depends on zero TCs. With positive TCs deviations will occur.

A grandfathered tradable quota scheme has much of the same dynamics as that of a tax, while it has distributional effects more like the legal standard. This may explain why the business sector favour grandfathered emission quotas before taxes (Stavins 1998). Compared to the legal regulation, there is one difference concerning distributional effects. Due to the trade involved, sellers will have a pecuniary gain and the buyers an outlay of similar size. Because it combines lower cost for the business sector with increased cost-efficiency, grandfathered tradable quotas may be a favoured 'compromise' between principals and agents. However, the rule of initial distribution becomes an important political issue. It will give a direct gain to some – the net sellers of quotas – and put cost on other agents – the net buyers.

There have been some negative reactions to establishing private rights in environmental assets such as air. They are common goods, and should not be owned by individuals/firms. We see this reflected in many regulations where it is emphasized that the right is not a property right, but an 'allowance' (the American Clean Air Act) or a 'licence' as is the wording in many fisheries regulations. This practice is criticized by some economists since it may create uncertainty for the firms holding such allowances or licences, for example, Árnason and Gissurarson (1999) and Leal (2000). Other economists, like Weitzman (2002: 326), were 'shocked' by the extreme property rights interpretation of quotas since it pre-empts a serious discussion of letting society capture the rents of such a system.

We also observe different reactions to tradable quota systems dependent on which problem or sector we look at. Individually tradable quotas (ITQs) have, as an example, created much more conflict in the fishing industry than in the field of air pollution. This may be due both to differences in distributional consequences and the dynamics of the environmental problem. Box 13.2 discusses the experiences with quota systems within these two fields.

BOX 13.2 QUOTA TRADING: EFFICIENCY AND ETHICS

Throughout the 1980s there was increased interest in tradable emission quotas (often called individual tradable quotas) and in the 1990s such programmes were initiated, especially in the field of air pollution (for example, Klaassen and Nentjes 1997; Schmalensee et al. 1998; Tietenberg 1998). Regulations on acid rain (especially SO_2) and ozone-depleting chemicals are core examples. The basic idea behind this development was to move away from standard legal regulations to market-based instruments to secure more cost-effective solutions. Economic evaluations of these policies conclude that the effects have been clearly positive in that sense (for example, Conrad and Kohn 1996; Klaassen and Nentjes 1997; Schmalensee et al. 1998). While there are debates over which form the initial distribution of quotas and the system for trading should take and the ability of firms to handle uncertainties and so on, the conclusion is that the system has generally worked as expected.

However, it is also interesting to see how distributional and wider ethical concerns have influenced these programmes. A basic argument against establishing a system of private rights in emissions to air is that air is a common good. Thus, there has been a serious debate about the legitimacy of the system. Tietenberg (1998) shows how this argument has been reflected in policy documents – for example, the American Clean Air Act where the emission quota is defined as an 'allowance' with limited authorization to emit. It is explicitly stated that it is not a property right. Tietenberg shows how similar concerns influence the rules concerning the rights to sell emission quotas if shutting down. He also offers examples of cases where trading of emission rights is not accepted if the firm thereby reaches very high levels of emissions. This occurs despite the fact that it would be cost-effective to allow such trade. Finally, he emphasizes that ethical concerns are very important when the rules governing initial distribution of emission allowances are being formulated. All this suggests that while the market mechanism is used, it is structured to facilitate wider concerns than just trade.

The idea of ITQs has also been introduced in other sectors. The fishing industry is a prominent example following the collapse or threat of collapse in many fisheries. In this case, the conflict between 'efficiency' and 'equity' is pronounced. The basic idea is the same as in the case of air pollution. Sustainable catches will be secured via defining a total allowable catch (TAC), while an efficient way of catching the total quota is attained via establishing the right to trade permits. These positive effects are emphasized in the economics literature – for example, Árnason and Gissurarson (1999)– and are generally similar to the evaluations of air pollution programmes.

The reasons for larger conflicts here seem to be several. First, the right to trade has resulted in rapid concentration of the right to fish. This has had consequences both for the viability of many local communities and for the possibility of entering the fishing industry. Pálsson and Helgason (1997) document the high speed of concentration in Icelandic fisheries, and the negative reactions among many fishermen and communities to the whole process. The rapid concentration can partly be explained by the fact that the TAC has been reduced substantially over the years, and many fishermen with smaller quotas had in the end little alternative but to sell. Thus, instituting an ITQ system in a situation with a resource crisis – which is typically the case – seems to accelerate the concentration processes (Helgason and Pálsson 1998). However, not all fishermen are against the system. It certainly creates winners, too. As Brox (1997) emphasizes, those surviving tend to control the fishermen's organizations, thereby establishing pressure to maintain the system.

It has also been suggested that the ITQ system may be as much a cause as a solution to the management problems (for example, Macinko and Bromley 2002). ITQs are normally grandfathered on the basis of historical catches. Thus, when the debate began on instituting such a solution, increasing catches became important to get a large quota. Furthermore, as the number of companies involved is reduced, their capacity to influence the TAC may actually increase and therefore the long-run effect on total catches is uncertain. The internal control systems and the interest in long-run management of the fish stocks may erode as fishing is transformed from being society based to being firm and capital based (Jentoft 2004).

Some have argued that the problem is rather that ITQs are not a fully-fledged property rights system (for example, Leal 2000). It is more a type of licensing where the licence holder does not have a long-run security for his/her catches. Macinko and Bromley (2002) counter by arguing that it is not a property rights problem. The state owns the fish within the EEZ. The resource problem is instead a *management problem*; that of setting appropriate TACs and being able to control final harvests. The dynamics of the fish stocks makes it impossible to establish long-run individual rights in these stocks except for the case where there is only one owner. But that is already the case for the EEZs. Makinco and Bromley argue that such a 'monopoly right' should continue to rest with the state.

In the debate over these issues, it has been strongly emphasized by many that it is wrong to sell quotas. Helgason and Pálsson (1998) document strong sentiments among fishermen in this regard. A common resource should not give individuals who may not fish themselves the opportunity to acquire rent-earnings. Certainly, resource rents could be captured by the owner – the state – just by auctioning the permits, thus enabling society to obtain the rents. Whether they then should be general state income or redistributed to fishing communities would be a core topic. An alternative would be to establish co-management systems based on local participation by fishermen themselves. There are some interesting tendencies in this direction, mainly as a way to reform the ITQ system (for example, Hanna 1995; Jentoft and McCay 1995; McCay et al. 1995). The issue of how to organize co-management for fishermen outside of the ITQs and the communities dependent on them is still to be addressed.

The various literatures on the fishing case show a wide variety of foci. Economists tend to detach their analysis from the community aspects of fishing. This follows from the model, but may raise doubts about the relevance of the conclusions drawn. I do not want to argue that ITQs cannot be a useful tool. Rather, a wider analysis is necessary for establishing when they are and when they are not. When applicable, the ITQ scheme must be sensitive to the wider issues involved, not just cost-effectiveness.

Looking at our two cases, there are some important reasons why ITQs in air emissions seem to have a greater capacity to solve air pollution problems in an acceptable way than to solve the fisheries crisis. First, the distributional consequences and the consequences for local communities seem to be much greater in the fisheries case. Second, there are different environmental dynamics involved. In the case of air pollution it is much easier to set the total quota and it can be kept constant over time, while in the fisheries case the basic problem is to deal with the large variations almost from season to season. Both the social issues and the ecological characteristics must be evaluated.

Input versus ambient taxes

Input taxes usually have distributional effects that are different from ambient or emission-based taxes, even in the special case when the instruments are equally precise. Stevens (1988) shows that more fees will have to be collected in the case of an input tax as compared with a charge on emissions to get the same effect on the environment if emissions are convex in inputs. When emissions are concave in inputs, the opposite conclusion can be drawn.

Emissions will normally be convex in inputs – that is, increasing marginal emissions as a function of inputs. Figure 13.18 illustrates this. We observe that with low input levels – up to q – no emissions occur. This seems, for example, to be the situation for nitrogen fertilizers in agriculture.[25] Taxing inputs will therefore imply that taxes will also have to be paid for environmentally harmless uses. Moreover, if environmental damages are convex in emissions, the likelihood that the relation between inputs and damages is convex increases.[26] Taking nature's cleaning capacity into consideration, most relationships between inputs and environmental damages will be convex.

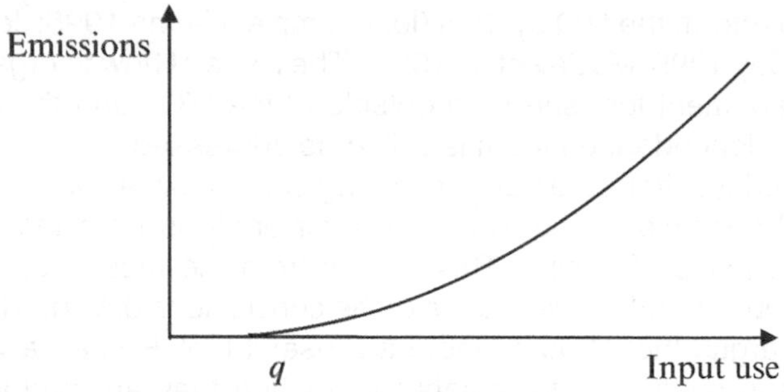

Figure 13.18 The relationship between inputs and emissions: a principal characterization

While input regulations therefore may be strongly favourable due to reductions in TCs, this may come about at the cost of increased distributional effects. Data given in Vatn et al. (1997) on nitrogen fertilizers indicate that an input tax will result in a collected tax volume 2–3 times higher than a tax issued directly on the damages. This illustrates that there are some really difficult questions involved when making trade-offs between efficiency and fairness. One may ask whether it is justifiable to add the extra 'burden' arising from using an input tax because the principal is unable to tax emissions or the change in recipient qualities. Again, the problem is invisible in a model with zero TCs. The above example is another illustration of the problem of making a strict distinction between efficiency and distribution.

13.5 SUMMARY

The basic policy measure or policy choice is that of choosing and forming resource regimes. The regime structure sets the stage concerning both which values and interests will flourish, and which values are articulated. Furthermore, it influences the costs of cooperating, communicating and transacting.

The market has some strong capacities concerning allocating easily demarcatable goods when ethical issues are not important. It is a great simplifier. The problem is that it is not good at handling interconnections – either physical or social. Therefore, making the market work well, securing a competitive environment, must result in an increasing number of physical interferences across economic units in a world characterized by physical interrelations. This causes a series of contradictions in modern societies, and concerns the creation of selfishness, the motive of cost shifting and the level of transaction costs.

In this chapter, we have focused mainly on two regime structures – the bargaining environment of markets and state regulation of market agents. The latter is also called a principal–agent framework. This narrows down the number of regimes studied, but nevertheless focuses on core structures and core debates of today.

One issue has been the interrelated effects of transaction costs and rights on what becomes an optimal allocation of resources – more specifically what become Pareto-relevant externalities. In a world of positive TCs, who has the right – the polluter or the victim – significantly influences optimal emission levels. We have also showed that the rights structure influences the environmental costs of various actions when measured in economic terms. If the victim has the right to a clean environment, WTA becomes the consistent monetary value measure, while if the right is with the polluter, WTP is the measure. Since these diverge substantially, the 'optimal' level of nuisance shifts even more between the two rights structures/institutional settings. Finally, since most environmental effects appear long after the (first) emissions have taken place, defining which costs should legitimately enter the calculation of the MAC curve becomes important. This issue is related to who has the burden of proof – whose actions are protected by the privilege of not having to prove harm.

Based on this, we can specify and expand the conclusions reached in Chapter 8:

1. optimal (Pareto-relevant) emissions $= f_1$ (TC, MEC, MAC). Then since:
2. TC $= f_2$ (institutional system),
3. MEC (WTP/WTA) $= f_3$ (rights/institutional system), and

4. MAC $= f_4$ (rights/institutional system), simple substitution gives
5. optimal emissions $= F$ (rights/institutional system).

What becomes optimal is a function of the institutional system. The relevance and consequences of the above varies between regimes. Thus, we have emphasized that TCs are generally much larger in a bargaining (Coasean) than in a principal–agent (Pigovian) framework. The core argument for the principal–agent model is its ability to reduce TCs. The failure to examine such costs in the literature based on the latter model is curious. Moreover, we have observed that the rights issue is inconsistently treated by the way the polluter pays principle is normally applied via the Pigovian model.

Given the principal–agent framework, the question of choosing policy measures can be divided into three: (a) how can we secure solutions with the lowest TCs?; (b) what are the distributional effects of the various policy measures? and (c) which interests should be protected and what incentives fit the motivational structures of the agents?

Concerning TCs, it has been shown that this issue is basically about making a trade-off between obtaining high precision and low TCs. We have looked at the environmental problem mainly as one of matter and energy conversion in a chain stretching from extraction via transformations and use in the economy to loss to the environment as waste. Then it becomes clear that emissions can be regulated by attaching instruments to various points in this chain. Moreover, it has been shown that the core technical issue is how the characteristics of this chain of transformations influence the trade-off between precision and TCs. In some cases it is best to institute regulations on inputs into the economy, in other situations the best point of instrument application is on the technology used in firms or households. Finally, there are cases where the standard Pigovian solution – regulating on emissions or the effect of emissions – is preferable.

In the principal–agent structure, defining who has the right to the environment also influences the choice of solution. If the right is with the polluter, increased environmental quality should be obtained by rewarding the polluter – for example, using subsidies for reduced emissions. If the right is with the victim, the polluter should be punished for emitting damaging substances, destroying habitats and so on. In many situations, formulating who has to carry the burden of a regulation is the core issue. It is often not explicitly treated, but is rather implicit in the choice of policy instrument. While the polluter pays principle is dominantly adhered to, this solution can be supported by many different types of policy instruments, again with very different distributional effects. Legal instruments in the form of prescriptions or prohibitions have very different distributional effects compared with the use of taxes. Taxes paid on inputs will have different – normally higher – distributional effects compared to taxes on emissions to obtain the same environmental effects. Tradable emission quotas have different distributional effects depending on, whether they are grandfathered or sold on, for example, auctions.

How the instrument used influences the motivational structures of the agents is the final issue. If profit/utility maximization is assumed to govern independently of which issues are at stake and which regime is involved, using instruments based on calculative rationality becomes the only sensible option. Taxes, subsidies or tradable emission quotas are highly preferable. If bounded rationality is assumed, these instruments are still favourable, but informational measures may be an important supplement.

The alternative perspective is to accept that (a) rationalities vary between problem areas and regime structures and (b) the policy measure itself influences the perceptions applied by the agents. If so, the principal must be sensitive to the existence of norms/internal motivations. This has been empirically supported by 'crowding-out' and 'crowding-in' effects, and by reactions to various incentive structures and control schemes. If people think in 'I' terms, using economic instruments seems safe. If they consider an issue to be morally important – that is, their behaviour is directed by concern for others – using economic incentives may result in effects that are opposite to what is expected. Similarly, using normative instruments may 'crowd in' a 'We' perspective on the issues involved. In the same way, controls may destroy internal motivations when these are important, while controls on those acting in a purely calculated manner may reinforce the positive self-image of those who voluntarily and unselfishly restrict themselves.

While this wider perspective is very helpful in explaining observations that are 'counterintuitive' from a neoclassical perception, it does not make it any simpler to formulate environmental policies. One has to abandon the pure technical or very distanced 'social engineering' perspective of modern regulation theory. Instead, one has to engage in dialogue and try to understand what motivates people in different situations. This also makes us realize that the fundamental problem is not to find the technically right instrument, but to create a social environment where trust and engagement in the preservation and development of the common good becomes the core issue. Existing neoclassical regulation theory is based on a contradiction in the sense that it uses instruments that foster individual calculation in a situation where engagement in solving collective problems is the core challenge. The institutionalist perspective advocated here is engaged in developing ways that makes it possible to cut this knot.

NOTES

1. It should be emphasized that Coase (1960) discusses both zero and positive TCs. While he was rather hostile concerning state intervention due to various 'policy failures', he was nevertheless very clear that looking for efficient policies implied discovering solutions with the lowest TCs possible. It was more the so-called 'hyper-Coaseans' who continued to stay in the fictitious world of zero TCs.
2. To simplify we also assume then that the involved agents can transform all costs involved into a single dimension.
3. If a liability rule is used and the right is with the victim, the polluter must compensate the victim if damage is observed. In such a case the court system would normally have to be used to set the compensation since the solution will not be the result of an *ex ante* bargain between the parties. If the right is with the polluter, a liability rule makes no sense.
4. The approximate number of people involved is six billion. Added to that are all entities in the form of firms, public administrative bodies and so on, which are also emitters and victims.
5. The reader might reply that if the climate problem has the capacity to cause an environmental disaster, losses would be enormous – virtually infinite. TCs could not be greater. However, this observation does not increase our capacity to negotiate. What is indicated in the text is that the TCs are so large that even if each of us used all our resources to do the individual negotiating, it would not help. We would not have the resources available to strike the necessary deals. The level of consequences does not change that fact. Put the other way: in economic terms – that is, willingness to pay terms – a catastrophe is not infinite in costs. It would be bound by our total income. If the TCs are larger than the gains thus measured, it would be optimal to let the catastrophe happen.
6. From a TC point of view, a state is no more than a 'merger' of all firms and households that it represents.
7. The regulator will not know the MAC and MEC curves before it has already done a lot of information gathering – that is, TCs will occur. It is possible that the state representatives may later observe that it did not pay to gather this information. The net gain does not pay for the engagement. This takes us back to the self-reference problem discussed in Chapter 5. This observation is of importance for both the bargained and the regulated cases. It may be argued that the problem is greater for the bargained one since information costs are generally higher in this situation.
8. While the gain of regulation in this case is also reduced by the level of the marginal TCs, it would be wrong to talk about net offer curves in the current setting – see Figure 13.2. This follows from the fact that the state does not bargain. It just sets the tax. Instead one could call the two new curves corrected marginal gains and corrected marginal costs from regulation.
9. Compensation could also be given under the Pigovian solution. It would, however, increase TCs.
10. It is easy to see that in the bargained situation – that is, where victims will be compensated – victims' rights would give a motive to become a victim. This is the logical consequence of standard rationality assumptions. Gains (compensations) to be collected will by definition be (equal to or) larger than the costs. In the situation with state regulation, the effect on victims, is a bit trickier since compensation is not given. However, moving into a polluted area may still be profitable. Vatn and Bromley (1997) discuss the case of a firm that plans to move into a polluted town. The firm is vulnerable to this pollution – that is, environmental costs will increase if the firm moves. But, it will also experience, for example, reduced transportation costs. Depending on the relative size of these costs and gains there will be situations where the firm finds it profitable to move if victims' rights are assumed, but not if the rights are with the polluter. This situation gives rise to a rather intricate decision problem for the regulator since which rule (right) to apply actually shifts from situation to situation. More fundamentally, the case illustrates the difficulty in drawing a clear distinction between rights and efficiency.

11. If R_v in a bargaining situation was protected by a liability rule and not a property rule, a somewhat similar solution would occur. The compensation would be given after damage is observed. Also, in this case the burden of proof would have to be with the victim.
12. The polluter is protected by a property rule until harm is proved.
13. More on this can be found in Vatn (2002).
14. Certainly, instituting a regulation based on retrospective or *ex post* regulation as described, would most probably influence the strategy of the firms. They would be less inclined to develop production processes where there is a likelihood that environmental costs will arise simply to avoid the higher taxes. Therefore by instituting that rule, the $MAC_{R_p}^{PPP}$ would most probably fall and the tax ($T_{R_{v2}}$) would become lower.
15. The equi-marginal principle is normally said to ensure cost-efficiency – that is, a certain level of abatement is reached with least cost. I avoid making this connection since TCs are omitted from the cost-efficiency analyses. In the case of positive TCs, it will normally be the case that equal marginal abatement costs across firms is not the solution with least total costs.
16. Williamson's (1985) focus on asset specificity is similar to this reasoning – that is, asset specificity becomes important when TCs are positive.
17. However, this increase in precision is most probably not the direct case for national regulations. Instead we observe the effect of the absence of international 'government'.
18. I have established a collection of citations found in the literature over the years indicating that a failure to understand the effect of TCs on what is efficient may cause confusion. The following may be instructive: 'While optimal instruments [that is, emission taxes] will achieve a specified pollution target at least cost, they may not always be easy to implement'; '[t]he challenge is therefore seen as designing indirect incentives that achieve environmental goals at reasonable rather than least cost, since efficiency is a utopian goal given today's technology'; '[i]t is recognized that implementation of efficient policy instruments for controlling agricultural pollution will generally be impractical'. If the efficient solution cannot be achieved, is it then efficient? The quotes stand as 'open wounds' that can only be healed by including positive TCs.
19. Certainly, there is a relationship between the economic and the legal spheres in that legally defined rights are a prerequisite for market transactions. In our case we focus on legal instruments as alternatives to economic ones given the necessary legal framework for any instrument to be politically legitimate is in place.
20. Gneezy and Rustichini (2000b) set up an experiment where high school students collecting donations for a charity participated. One group got no pay, a second 1 per cent of their collection, and a third group 15 per cent. The second group collected less than the first. The third group collected more than the second group, but less than the first. It should be mentioned that the payment was not deducted from the money collected, but paid by the researchers.
21. The findings of Gneezy and Rustichini (2000b) indicate a relationship more in line with the alternative discussed in the comments to Figure 13.14. Zero compensation gives a certain level of effort, while shifting to payment reduces effort to almost zero immediately. From there on, effort increases with payment.
22. While Frey seems to base his analysis on more psychological or individual explanations for 'intrinsic' motivation, I think the explanation is institutional. We are observing norms in action.
23. By this, I do not imply that the dynamics would be principally different in, for example, an agent–agent situation. It is just that we would have had to present it differently.
24. I tend too look at increasing rates of crime as an example of this. Keeping crime down is more dependent on the way society is able to integrate people and make a strong 'We', than on a strong police force. As soon as the 'We' element is eroded, the only way is to manage the situation is by instituting increasing levels of control and punishment. While this in itself destroys trust and commitment, one is forced onto a slippery slope of ever-increasing levels of control.

25. See Vatn et al. (1997). The situation will vary across pollutants. CO_2 emissions seem, as an example, to be linear in inputs.
26. Stevens's (1988) conclusions assume environmental damages to be proportional to emissions. If they are convex, there will be a difference in distributional effects between an ambient tax and a tax on inputs, despite the fact that the emissions might be proportional to the inputs.

14. Policies for a sustainable future

> We are the first generation to influence the climate and the last generation not to pay the price thereof. (Jostein Gaarder 2004; my translation)[1]

The aim of this book has been to develop a consistent understanding of the role of institutions in the economic process in general, for the protection and use of the environment in particular. In closing, I shall try to look into the future and raise a set of questions concerning what institutional reforms are needed to increase our ability to solve urgent environmental problems. In doing so, I start with a short summary of what we have observed so far (Section 14.1). Based on this insight I shall then present a model developed to support the evaluation of institutional reforms (Section 14.2). Finally, I shall use that structure to present some ideas concerning necessary improvements in economic and environmental policy making (Section 14.3). Certainly, this last issue is a very large one and warrants a book in itself. However, I shall present some ideas to kindle an awareness of the serious problems we are facing, and in which direction we should look if we think that something should be done about them.

This book has been based on the idea that economics must (again) become a science that puts *the issues of values and interests up front*. The focus on efficiency as Pareto efficiency is far too narrow. We have seen that reducing it to this issue has created inconsistencies. More importantly, it offers advice that is given power far beyond its bounds. Hence, problems appear both in theory and practice. Efficiency does not offer the supposed value-neutral haven for economics. Instead, the danger is that significant value issues are treated in inappropriate ways. As economists we should not leave value issues to philosophers. We should rather engage in a discussion with them.

14.1 WHAT WE HAVE OBSERVED

At the beginning of this book we raised a series of questions concerning what characterizes social science theories. We have emphasized three classes of issues: *the process of choosing*, the *understanding* of the *world* in which people act and what kind of *interactions* there are between individual choices. While the answers given to these questions form the theories, they also play a core role when we formulate ideas about a better society and what advice we give concerning the construction of institutions to obtain this.

In relation to the *process of choosing*, we have suggested that rationality is a plural concept. Rationality means different things in different institutional contexts and can be influenced by changes in these contexts. This is a fundamental insight for policy. On the one hand we have rationality as maximizing individual utility. On the other we have social rationality implying doing what is expected or appropriate – either through following social norms and rules that characterizes the setting or through acting reciprocally. While markets, accounting devices and firm structures advance the first kind of rationality, the forum with its various communicative elements fosters the latter.

While maximization is the dynamic aspect of individual rationality, communication and evaluation of arguments have this core role in the case of social rationality. Communication is crucial in both developing and changing the norms or rules defined for a certain situation. So while social rationality implies following the norms or rules of a setting, it has been equally important to show how learning and communicating can foster changes in these norms and rules. While the ideal is communication without domination, power relations inherent in existing institutional structures will always influence these processes. It is in itself an important institutional issue to develop arenas that function as checks against this.

Institutions are both produced and reproduced. While reproduction is in a way automatic and often unreasoned, it is suggested that the production of institutions is foremost intentional. Institutions are developed to solve difficult coordination problems, to establish order and to support specific interests or values.

Institutions rarely define the logic of a situation 100 per cent. People must interpret the contexts. There are always some individual variations in this, as we also observe variation concerning the willingness to act as expected. However, as the institutional context defines the meaning of a situation, it also defines what is or is not proper behaviour. It strongly influences which acts can be expected.

Certainly, while human capacities are great, we do not have the ability to be fully informed and to do all the necessary calculations in contexts where that is relevant. We are boundedly rational. That is a characteristic of our abilities and influences behaviour in all situations. We may not judge the information offered or do the trade-off calculation in a marketplace correctly. We may not know the norm or we may misunderstand the situation. Nevertheless, there is a distinction to be made here. Institutions like norms and conventions may themselves be seen as solutions to complicated value questions and coordination issues. They support the boundedly rational individual in a complex world, whether in individualized or social contexts.

From the plural understanding of rationality follows the social understanding of preferences. They are influenced by the socialization and the enculturation of the individual. They shift between social settings. While some desires have their basis in our physiological and psychological characteristics, the way in which we choose to satisfy even these wants is strongly influenced by the institutions and culture we live in.

Turning to the *external world*, we have emphasized its complexity. This applies to both the natural world and the social sphere. As a consequence of this, information gathering and transacting are costly activities that demand time and resources – similarly for communication and coordination. Institutions influence the levels of these costs. They can be instruments in reducing costs of information, transaction and communication. Institutions not only offer meaning and support values, they also influence the costs of acting together.

Natural resources offer the material and the energy necessary to sustain individuals and societies. Furthermore, natural resources are interlinked processes. This implies that uses are directly competing and resource regimes become crucial in defining both who has access to natural resources and how the effect of one agent's use of these resources is allowed to influence other agents' opportunities. In a world of interconnected physical processes, a large set of questions concerning ethics, values and conflicting interests is forced upon humanity. Institutions are constructed to deal with these issues. They define the distribution of rights, which interests get protection and which values are fostered. They influence the ways people can interact. This is about what power various people have to protect themselves in daily life situations, what possibilities there are to solve ongoing coordination problems, and how individuals and groups are able to influence necessary institutional change as problems accumulate.

We have seen that there are a wide variety of ways in which rights and capacities to coordinate are instituted. A main theme of this book is that when building resource regimes, one must take explicit account of the characteristics of the resources and the values involved. There is more to the story than commodities, exchange and prices. It is a challenge to see when one institutional system serves our interests and when it does not. The idea that there is only one institutional structure that is really the best solution to any problem, and that other solutions are used only because we fall short of realizing that ideal solution, is a great obstacle against both realism and creativity when the development of regimes is at stake. This is the case for those saying that private ownership and market allocation is always the solution, as it is for those believing that the state can allocate all goods or that more community is the solution to any problem.

From the above we see that the *way we interact* is in many respects *the* core aspect of institutions. The capacity to find cooperative solutions is to a very large extent institutionally dependent. The issue, however, goes deeper. We also interact normatively by defining, for example, which norms should exist in a certain domain. We both coordinate and communicate. We both organize our activities and develop our thoughts about what is right or preferable in an interactive way.

Institutions may foster instrumental or strategic interaction. Then other people appear as 'things'. More precisely, the behaviour of other people is something that is important to us only in the sense that it enters our calculations about what is best to do for ourselves. Institutions may also foster cooperation and communication. They transform 'the other' from being 'a thing' to becoming something we engage in, with which we reason together and develop common solutions.

Concerning the issue of interaction, we have observed several paradoxes. While the model of the perfect market is thought to create equilibrium states between isolated, non-communicating individuals with stable preferences, the most characteristic feature of modern markets is how it involves us in continuous changes in products, technologies, tastes, norms and other social relations. These dynamics offer opportunities, but also problems. While markets have the capacity to coordinate individual acts, they are unable to treat interlinkages well and are blind to reason. They may maximize commodity production, but following the same logic they may also maximize 'externalities'. They may lead to coordination in exchange, but likewise establish obstacles concerning coordinated action directed at creating or protecting the common good.

While different positions within economics and institutional theory have been contrasted throughout the book, we have also showed how the different positions link together. While there is a great difference between individual, strategic rationality and social rationality, it has also been suggested that these rationalities should all be understood as institutionally influenced. The market, as envisaged by the core neoclassical model, is a good model for issues where only individual desires, zero (low) TCs and demarcatable goods are involved. While there are several gains in simplifying issues to fit this 'special case', the main challenge we face is to define when markets are good and proper institutions, and when decisions must follow other logics and forms of interaction.

14.2 THE BASIC ISSUE IS CHOOSING THE REGIME

14.2.1 Core Dimensions to Consider

The fundamental choices we make concerning resource use are those about which regime should be in place to secure the best uses. If one regime cannot be used to solve all allocations, we need a meta theory to support the choice between regimes. This implies that we need a theoretical structure that can help us decide which dimensions are important and how to treat these dimensions consistently. This is crucial if we want to establish a basis that fosters sustainable futures. While the aim of this book has not been to suggest such a theory, nevertheless we have produced insights that can help us move in that direction (see Figure 14.1).

The dimensions in the figure are the three familiar ones. First we have the type of rationality involved, distinguishing between individual and social rationality. Next we have the type of human interaction involved, ranging from pure instrumental or strategic behaviour to communicative action and dialogue. Third, we have the character of the goods/the external world involved. In this last case I have distinguished between two interlinked characteristics. At the one end we have simple systems and individual items. Related to these dimensions are, for example, certainty/ordinary risk and low TCs. On the other end are complex systems and common goods. Related to these are, for example, radical uncertainty and high TCs.

The fundamental policy problem for sustainable development is the second-order problem – that of choosing and forming regimes. This issue is about finding a solution that fits the character of the good or resource at

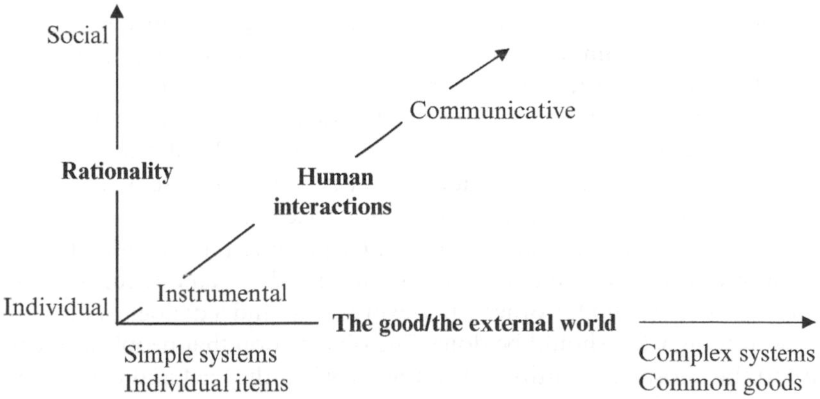

Figure 14.1 Three core dimensions when choosing resources regimes

hand. More precisely, it is about choosing which values and interests related to the use and protection of these resources should get protection. It implies choosing an institutional structure that supports a rationality modus and a type of interaction that is consistent with the problems we face. It is at the institutional level that these links can be made. Certainly, nobody can offer a fixed point from where to make such an evaluation. Specifically, people may disagree about what characterizes the good and which values are involved. Sorting this out is an issue that can only be treated in open dialogue – through testing and re-testing of arguments. What Figure 14.1 offers is a structure which can frame that discussion.

Following the neoclassical model and thinking in terms of only one type of rationality (individual) and only one type of interaction (instrumental), the problem of choosing an institutional framework tends to vanish. The market is the only relevant element to consider. Given this perspective, there is, furthermore, only one way the good can be consistently viewed – as a single item. Moreover, it must be possible to handle the risks the individual faces in calculative terms. Then it can all be captured by the origin of Figure 14.1. Disregarding bounded rationality, the market solves all problems in a way that produces sustainable equilibria.

While we have documented repeatedly throughout this book how the perspective on rationality and interaction influences the way goods are perceived by various theories, we have also seen that neoclassical economics both accepts the existence of public goods and offers ways of handling their allocation. It is only the dedicated property rights economists and the Hayekians who seem to deny that there is anything the market cannot treat. Neoclassical theory still seems to handle public goods as if they could be managed by the market. It is not a true *common* good. It takes on a special, abstracted form which fits the tenets of the model. It is a good that is common to a group of people only in the meaning that no one can be excluded from consuming it. It is not about forming the common good under the perspective that it should serve a community of people. It is instead about what the individual *qua* individual thinks the good might be worth to him/her given that others cannot be excluded. Its allocation is not treated through communicating about what is a reasoned use of the good we have in common, but through individual price bids.

This is the only way it can be consistently perceived if the individual is taken as given and individual rationality is assumed. When we accept social rationality we open up for other value expressions and a dialogical process of determining what should be done. The observation that people are sensitive to the chosen institutional framework when they act and communicate becomes of great importance. The market and the forum support different rationalities.

A basic argument of this book has been that to handle the common goods aspect, social rationality and some form of communicative process must be taken into account. It is the only institutional structure that can be true to the choice problem at hand. While social rationality and dialogue may also be important for the allocation of single items, individual rationality and instrumental action tend to pervert the care for the common good. The strength of the market is that it is a great simplifier. It brings it all down to one value – the exchange value. Problems occur when this solution is expanded to cover situations where this induces non-trivial information loss and/or where there are arguments for treating the good as a common one.[2]

There are problems related to accepting exchange value – a single value dimension – *both* at the individual and the group levels. First, not all aspects of a good may be treated unidimensionally by the individual. Complex goods may carry value dimensions that are difficult to transform into one scale. There may be ethical issues involved, making it impossible for the individual to treat all values using the logic of a trade-off. Second, we face the problem of 'aggregating' individual values. If goods fit the commodity concept, aggregating individual demands make sense, even though one cannot make a waterproof distinction between efficiency and distribution even in that case. The distribution of the gain is of another character than the measurement of the gain, so accepting the rule of one dimension and simple summation implies loss of information even in this case.

If the good is common, the issue of plural dimensions confronts us even more fundamentally. Then the priorities made by one directly influences the opportunities left for others. Then the gain for some will to a large extent be a loss for someone else. A wetland cannot be both a habitat and a location for a road. A dam cannot both store water and sustain the old pattern of water flow. The air cannot be both clean and a deposit for smoke. A river cannot be both a sewer and an unspoiled environment. Since some gain and others lose, reducing the issue to just summing monetized losses and gains provides little insight into the social realities lying behind the appearing figures.

From this we see that while individual rationality, instrumental behaviour and goods as demarcatable items are elements that consistently fit together, allocation of complex, common goods are equally linked to social rationality and communication. While the market can be used to allocate individual items, institutions of the forum type are the only ones that can treat complex, common goods. Their allocation must be based on fostering social rationality and reasoning over which principles and values should apply. We must move from *aggregating* individual measures or bids to reasoning over, and potentially *agreeing* on a common set of priorities. The observation that rationality is institutionally dependent – is plural – offers the key for thinking along these lines. Moreover, the story is a positive one.

It implies that people respond to being brought into an institutional setting where a communicative process is facilitated. It implies that people are able to change their views as a result of trying out various arguments. Building proper institutions becomes an important and powerful task.

14.2.2 State, Market and the Civil Society

The analyses presented in this book are based on the idea that a society needs three main institutional arenas. First, we need arenas for making the day-to-day allocation of resources. Second, we need an arena that can offer the necessary authority structures to make these allocations indisputable, otherwise we will be engaged in continuous fights over them. Finally, we need an arena that can offer the necessary agreements about the values that are fundamental to establishing this authority structure; that can offer legitimacy.

Concerning the issue of day-to-day allocations, we have looked at a varied set of solutions ranging from markets, firms, via state agency allocations to common property regimes. While ideal markets are a pure exchange mechanism under the direct authority of the state, all other structures are based on various internal authority and communicative structures. Nevertheless, the fundamental authority problem in a society concerns who owns what and who can do what to fellow citizens. This issue goes beyond the internal command of any property regime. It concerns first of all which issue should be treated under which regime type. It concerns also which specific rules should govern the interaction of the parties involved within each structure, be it the firm, the state agency or the owners of a commons. This is the ultimate responsibility for the state in today's societies, while international agreements have been delegated some responsibility, too.

The legitimacy of all these choices must still come from the third arena – the civil society. Neither the allocative instruments (the regime) nor the authority structure (the state) can offer an evaluation of which value issues are involved. The combination of open dialogue and the citizen reflecting over what are important values and interests to protect is the core element of the civil society, as observed in the form of, for example, mass media communication, a wide variety of voluntary organizations and political parties.

An imbalance between the three levels of society causes serious problems. We saw this in the Soviet Union where the prominence of state authority gradually undermined the vitality of the society. Not only were there fundamental problems concerning the control of centralized powers and open access to information, but engagement and creativity withered in many respects.

However, the balance is also becoming a challenge for market economies. The development we observe at present is that markets are expanding into areas of life where their presence seems problematic in a fundamental sense. Their expansion makes it increasingly difficult to treat community issues in a reasonable way. Two core issues are at the heart of today's developments.

First, there is a dynamic inherent in the market system leading towards expansion – towards transforming issues into a form such that markets can treat them. Such a development demands individual property rights. Today the issue of expansion concerns not least the privatization of community services, of knowledge, and the rights to the diversity of life – the common gene pool. It is envisaged that the future expansion for business lies not least in the knowledge sector and in biotechnology. This demands that knowledge and genes become commodities. There are vast ethical issues involved in this, but we are gradually sliding into a situation where individual property rights are more and more accepted. Knowledge as private property is problematic in two senses. One is that its potential cannot be fully utilized. While it is a true public good – actually its use tends to increase rather than reduce its value – use will be restricted to those who own it. The other issue is that it becomes increasingly difficult to check the quality of the knowledge in use the more it becomes exempted from the public sphere. This effect is amplified when the distribution of knowledge to the citizen – the mass media – is itself controlled by market forces.

Second, the expansion of markets depends on decisions by state authorities. The rights concerning, for example, who owns knowledge or who owns the genes have to be decided by political authorities. At the same time we observe that markets increasingly extend beyond state jurisdictions. Thus, states often have no choice but to accept the demands of business. This, however, reduces the societal legitimacy of policy making. The arena for solving political problems – the state – is demonstrating its lack of power as its representatives are continually forced to accept compromises that go against the general political will. Thus, the state becomes more responsible to business than to the citizen. Such a state–market structure runs a great danger of destroying its fundamental legitimacy. Without an engaged and vibrant community – a civil society – the basis from which to make authority decisions erodes. The long-run durability of such a civilization is at stake.

Is it, however, so simple? Firms may take social responsibility. As suggested earlier, firms are or can be more than profit maximizers. Corporate responsibility has been a core concept since the 1990s. Certainly, firms need social legitimacy to stay in business. Consumers may boycott them – despite the fact that it is not individually rational to do so. States may control them if they do not act responsibly. Nevertheless, there are many paradoxes here. It is very difficult for consumers to control large corporations. It demands great organizational efforts. If access to knowledge is restricted, people may not even be aware of the issues. Thus, the power relations go much more the other way around – the corporation controls the consumer. Moreover, the issue of social responsibility is counter to the basic rationale of the corporation, which was constructed not least to simplify the focus, and to obtain a rational evaluation of just one thing – the 'bottom line'.

This created a specific and very strong dynamic. In his recent book *The Corporation: The Pathological Pursuit of Profit and Power* Joel Bakan of the University of British Columbia analyses the character of the corporation (Bakan 2004). This institution is mainly a creation of the nineteenth century where in the beginning it was occasionally used to finance capital-demanding tasks such as railway and bridge constructions. A 'corporate charter' gave the participants the right to act as 'one person'. Today the corporation has reached the position of the dominant institution in economic life, into which even education, research and care is now being transformed.

On the basis of an analysis made by the psychologist Dr Robert Hare, Bakan actually concludes that the corporation fits well to the diagnosis of a psychopath. It thinks only about itself. It is one-sidedly focused on the bottom line, and does not refrain from bringing costs upon others in its pursuit of owners' profits. It lacks empathy. It is characterized by irresponsibility and manipulates everything to its own gain. It moves around the globe in a restless search for more profits, takes no voluntary responsibility for its own actions and shows no regret . . .

Certainly, the message emerges very strongly when formulated this way and may especially provoke reactions from the leaders of corporations: we are not like this! The analysis, however, says nothing about the people leading the corporations. It says something about the institutions – the norms and rules leaders must follow to stay in business.

Bakan emphasizes that within the family arena, business leaders may be very caring people. He thus concludes that '[b]usiness people should therefore take some comfort from their ability to compartmentalize the contradictory moral demands of their corporate and non-corporate lives' (ibid.: 56). We could add, it is the capacities inherent in institutions that make this possible. The corporation was invented to focus on the bottom line and to create investment with restricted responsibility. This created a very dynamic institutional structure, while its one-sidedness has also become problematic over the years. These problems concern not only the narrowness in what counts. They are certainly also an effect of the ever-increasing power that corporations have obtained over the last 100 years. While democracy implies a direct link between the civil society and the state, there is no such link between civil society and the corporate sector. Things are being turned on their head as societies risks ending up as a big consumer serving the needs of the corporations – not the other way around.

This story illustrates the importance of institutions as motivational structures. More specifically it illustrates the strengths and the weaknesses of creating a system based on a one-dimensional goal. It made it possible to support a specific narrowness in running the business in a way that could create the highest possible profits. Focus was forced on the costs and revenues appearing in the balance sheet. Administrators and executives should not be tempted to care for other issues. Pure calculative rationality was fostered. This is, on the other hand, problematic if society pursues goals other than those congruent with or fostered by profit maximization. While corporations are constructed to act one-sidedly, societies are multidimensional. This is the core problem of today's economies. The cost of simplifying has become too great. This applies to the social as well as for the environmental spheres.

Certainly, we have already seen examples where firms internalize social goals such as reduced pollution. As mentioned, the development of concepts like corporate social responsibility shows that the business sector sees the need for restrictions on anti-social and anti-environmental behaviour. Nevertheless, this does not eliminate the importance of the bottom line. It rather drives firms that also emphasize social responsibility to pursue market opportunities where being responsible and creating profits are not competing ends. While important, this is at best a partial solution to the problem.

The whole environment we live in has over the last centuries been transformed from being governed by the inherent dynamics of a life-creating system (Chapter 9) into being fit to the demands of the bottom line. A vast transformation is taking place in the image of this calculative procedure. Almost no square foot of the globe's surface has escaped its influence. It has even recently become an issue for the gene pool itself. While creative for the construction of immediate wealth, it seems ill-suited to ensure the long-run sustainability of the natural environment, which is also a necessary base for creating wealth in the long run.

14.3 THE FUNDAMENTAL CHALLENGES FOR ENVIRONMENTAL POLICY

The aim of this book has been to better understand the role of institutions in managing the environment. A conclusion has been that the present institutional structures and the theories built around them are poorly suited to solving either urgent environmental problems or the issue of long-run sustainability. The underlying rationale of that system is continuous growth. It is to take place in a world of limited environmental capacities.

Environmental policies of today are trying to *correct* instead of *direct* this development. This may have been acceptable in a situation where economic activity only had a marginal influence on the physical environment, but it is far from acceptable now. While I do not possess a fully-fledged cure for the problem, I think the diagnosis is an important step in itself. It gives direction to future institutional changes.

Growth seems to be the engine of the present economic system. To motivate investments in a system based on bottom-line revenues, growth seems a necessity. It creates the required conditions for investors: that the investment will pay off in the form of acceptable returns. Growth is also good for the state. It increases the tax basis. It makes maintaining social security systems easier and it offers employment in a world of rapid technological change. These issues seem fundamental to sustain political support or legitimacy of the system.

Growth has served important social goals including the elimination of poverty and it will be important in the future as far as poverty continues to be a dominant problem for vast numbers of people. However, there are at least three problems involved in what otherwise is a good thing. First, growth is not only a goal. As emphasized above, it is rather what fuels the system. If it stops, the system is thrown into a crisis so often illustrated by the recurring recessions over the last 150 years. Second, while it has the potential of eradicating poverty, it is instituted in a way that it must first serve those who are already rich – that is, those who have the necessary capital to invest in future growth and because of this command the bottom line. The solution it offers to poverty eradication is at best that of 'trickling down'. In a world of restricted environmental capacities this is not a sustainable solution. Third, while growth increases material consumption, it is doubtful whether it creates better lives or more satisfied people when consumption is well beyond the fulfilment of basic needs. If it is relative income more than absolute consumption that determines how well we feel we live, then the goal of growth is misguided.

14.3.1 A Simple Scenario

To illustrate the possibilities and problems involved in continuing on the present track, let us look at a simple scenario. The premise for the scenario is that to keep the present system going, with its fundamental motivation structures, a certain level of growth is necessary. Let us just assume that a yearly growth of 2 per cent is what is necessary for the continuation of the system both economically and politically. While perhaps somewhat moderate, what does this imply? Well, it implies that production will be doubled in approximately 35 years.[3] It is roughly eight times as high approximately

100 years from now and 64 times as high if we expand our horizon to 100 more years. While the figures are only illustrative, they show the tremendous dynamics of the system. Such a development will bring vast opportunities for consumption, but can it be sustained? The immediate response by most of us is probably that this must be impossible. The globe cannot sustain this. None the less, let me briefly discuss the counterarguments.

First, it may be argued that technological development and the creativity of the market may solve the environmental challenges following from such tremendous aggregated growth. Certainly, research may find new processes and so on that will reduce the need for environmental resources *per unit of output* substantially. Competition itself will have such an effect for *resources that are priced.* Then it becomes competitively advantageous to reduce resource use. However, is it reasonable to believe that it can be reduced to less than 2 per cent of today's level in 200 years? If not, then environmental stress will increase. Following the warnings of, for example, the IPCC, this stress should instead be reduced substantially. Moreover, expansion is not only about quantity, it is also about quality. To reduce matter and energy per unit of production by such a magnitude, new qualities – new compounds and new production processes – are very important. This involves us in a continuous production of solutions to past problems, but also in creating potential future risks so well documented over the last 100 years. The faster this development proceeds, the greater the chance of a mistake. Consequently, the solution offered here may turn into a source of the problem. The discussion about genetic engineering is again a typical case. The burgeoning discussions about potential environmental consequences of the most important new technological frontier – that of nano technology – offer another illustration.

Second, one may argue that the state, by establishing environmental policies, will be able to direct resource uses towards solutions that do not threaten the functioning of the system. It just has to monitor the development and correct the prices to internalize the future externalities. Certainly, again there are possibilities. However, this solution can easily be caught in a fundamental conflict as long as the corporate bottom line is the basic motivation driving the system. The economic agents will demand strong evidence before they are willing to accept restrictions. As long as effects are generally delayed and often hard to prove with certainty before a long time has elapsed, this strategy risks being bogged down in continuously chasing an ever-faster moving target. Furthermore, if some countries or regions restrict companies operating from their territory, while others do not, the regulating countries will be losers in a game that still does not protect the global environment. We see this tension over and over again in international markets and the attempts to create international regimes for trade and the environment. The strategy of restricting an economy where expansion is the basic motive is a very demanding one. It tends to be trapped in the role of correcting afterwards, not directing up front.

Third, it may be thought that consumers will solve the problem. They may not want to increase demand at a speed indicated in the growth scenario, and as production increases they will also buy more services and fewer goods. There is something also to this argument. One may have some difficulty envisaging a consumption level of, for example, 64 times that of today's average. Nevertheless, some do consume that amount already. While I tend to believe that a good life is one with a much more modest consumption than this, I am thus not sure that the 'restricted ability to consume' will put much restriction on development. Moreover, if we stopped consuming, the engine of the economy would also stop. Who has not heard politicians on the brink of a recession urging their citizens to consume more. Hence, the satisfied consumer could not be allowed to materialize, not because it would be a bad thing in itself, but because it would create a crisis for the system. Advertising agencies would have to be offered ever more resources to stop us thinking like this. This is actually the negative side of the social shaping of preferences, which can as easily be changed by instrumental manipulation as by open communication and reason. This is exactly what we observe today. The information sector – the mass media – is dominated by marketing and advertisement. Certainly, the tendency to increased demand for services and relatively fewer material goods as we grow richer, may reduce the potential environmental pressures of growth. However, services are not dematerialized goods. Rather it is uncertain whether the 'service economy' is less bound to material inputs and waste production.

Fourth and finally, another possible counter effect rests in the environmental damages themselves. A big future market for the bottom line to exploit will be the repair sector – repair of damaged environmental services and supplying various services to cure health problems following environmental degradation, direct exposure to toxins and so on. This is exactly what should be expected, but also feared. It is not that growth cannot continue. It is instead about the price thereof. So if the economy is, for example, 64 times larger, it may be engaged largely in repairs and still not offer a better world to live in. I do not primarily fear a collapse of the environment. While gigantic attractor shifts with unforeseeable consequences is a possibility within my 200-year scenario (some would say that it is not only probable but inevitable), I think it is as important to think about gradual degradation where the poor or weak lose out systematically as conditions become more difficult. The scenario here is that the rich parts of the world, as environmental problems increase, detach themselves from the rest and become more and more engaged in sustaining their privileges. In doing this they use increasing amounts of resources to repair or even construct their immediate environment. If enough resources are directed to it, I do not think it is impossible to create a totally artificial world for the few. Nevertheless, according to the laws of thermodynamics, creating this kind of local order and functioning will have to imply increasing disorder for the rest of the world.

Perhaps a mix of the above four 'solutions' will suffice? Maybe the negative effect of each can be balanced by the positive potentials of the others? I shall not deny this possibility outright. Is it, however, reasonable? Is it the most likely scenario? We should start reflecting seriously about formulating some more fundamental institutional reforms, and do so before it is too difficult to change course. As the economy grows, commodity production becomes relatively less important and maintenance of the environment increases its significance for satisfying our needs. It then seems reasonable to develop institutional structures that can directly support this. I shall close by making a few inputs into the discussion about what such a reform might involve. I shall focus on changes at two levels which cross-cut the dimensions of Figure 14.1: the institutionalization of changed motivational structures, and changes in the information system.

14.3.2 Institutional Reform: Some Alternatives

Changes in the motivational structures

Concerning the motivation system, I see two main options that should be thoroughly considered. The first concerns institutionalizing *ex ante* limitations on economic activity. The second concerns institutionalizing social responsibility as part of the motivation structure of economic agents. *Ex ante* limitations could take a variety of forms. We have touched upon two restrictions of this kind – the *safe minimum standard* and the *precautionary principle*. The SMS implies that the state puts strict and direct restrictions on which resources to use or on extraction levels for specified resources. It could also take more indirect forms, demanding a high diversity in resource-use patterns. As this tends to diminish risks, less strict limitations on aggregate natural resource use could be accepted compared to a case where rather similar strategies are pursued by all economic agents or all states.

The precautionary principle will function somewhat differently. It can be institutionalized in various ways. Defining the rights with the potential victims and furthermore putting the burden of proof for no negative environmental consequences on the business sector is the most consistent and also the most far-reaching. Firms are given the responsibility to prove to society that new production processes and/or new products are environmentally safe in both production and consumption before permission is given. It also implies that they are made responsible for the waste generated as an effect of their products being consumed. With strict demands on what constitutes a proof, this solution might have very substantial effects on resource-use patterns.

These systems of *ex ante* limitations all operate as external restrictions. They do not alter the basic motivational structure of the firms. The one-dimensional bottom line is kept intact, but it is given reduced importance in the overall resource allocation process. In this way environmental issues are shifted to the fore in production planning and give society a much stronger say in the use of environmental resources. It does not correct, rather it directs. Both the above institutional structures are already in some use. However, as is apparent not least in the debates concerning the liberalizing of international trade, we are still very far from giving them the power necessary to make any importance to the dynamics of the game.

The other option is to change the institutional set-up to foster other or wider motivation structures directly underlying the production and consumption decisions. This could again be instituted in various ways. It could take the form of changes of the legal underpinning of existing ownership structures, restricting or abandoning certain forms of ownership, demanding community representation on boards, and so on. These changes would imply that the business sector is not only responsible to the direct owners for its results, but also to the larger society. More fundamentally, it could imply changes in ownership structures, giving an increased role to community-based ownership. As an aspect of all this, markets and corporations would have to play a role in the allocation of goods and services that is much smaller than today. Issues concerning core natural resources should be decided through public decision making.

All the above changes in motivational structures have consequences for international regimes, for environmental issues and for trade. To have any power, precaution must be given a fundamental role. Typically, environmental agreements should then be given priority above the right to free trade. Some issues related to this warrant special attention. One concerns the rules governing capital movements between countries. Another concerns the fact that producing the same amount of goods may cause less environmental stress in one place compared with another. Hence, trade and protecting environmental goods are not fundamentally at odds with each other. This is the case even if we take into account the environmental costs of transport. Instead a certain level of trade, varying from sector to sector, would be favourable. The point is that free trade is unable to secure this kind of selective gain. It is a general, not specific use of its forces that should be instituted.

Formulating a good trade regime implies establishing institutions that make combined evaluations of opportunities concerning reduced production costs, reduced environmental stress and defending community goals related to security and jobs. Due to the great variations in local ecosystems and social conditions, this would constitute resource-use patterns that are very different from that of free trade. Moreover, if we establish free trade first and then try to regulate *ex post* the problems that appear afterwards, we are in an even weaker position than that of a state undertaking *ex post* controls of its national industries. International competition would make it impossible for states to regulate with any force due to the continuous threat of firms moving to countries with less strict regulations.

Changed information system

Regarding the issue of information and information systems, we move from defining the logic of the motivational arrangements to looking at the materials that are inputs to these systems. Three questions come to the fore. First we have the production of information or knowledge. Second, we are confronted with the issue of evaluating the quality and consequences of the information. Finally, we have the question of information dissemination. Again I shall only offer a few observations related to the crucial need of securing public or civil engagement in this field.

Knowledge is a true public good in the sense that when produced, 'consumption' by some will not reduce its value for others. Therefore keeping research a public responsibility is a core issue. Turning knowledge into a commodity implies either secrecy or some restriction on general use – as in the case of patents. In this case it is necessary simply for financial reasons. No private business would support research that afterwards becomes free for all to use. As already mentioned, commoditizing information also increases the uncertainty concerning its quality, since manipulating unwanted results becomes a source of potential 'success' in the meaning that there is a strong motive to suppress information that is negative for a certain product or production process. The costs of quality control guard the swindler. The scientific quality control must rest with the research community itself – the system of peer review.

Especially in situations with radical uncertainties involved, knowledge is not value neutral. In the 'risk society', evaluating the normative aspects related to the production and use of new knowledge is a core question which cannot be solved only by public finance and quality controls within the research community itself. Here the civil society and the citizen must also be involved. This calls for the development of public or 'extended' peer review in the form of, for example, citizens' juries. The role of this institution is to support the policy makers in their decisions about what precaution implies in the various situations. So, while I have argued for a shift in the burden of proof from the public to the business sector as a way to strengthen precautions, producing the data for such an evaluation should be made by independent, public researchers, and the results should be evaluated by the community.

Finally, dissemination of information should be a public responsibility much beyond that of today. The argument is similar to the above. The channel through which information is spread must not be guided by any other motive than producing as true a picture as possible of the status of various fields of knowledge. This implies documenting what uncertainties are involved and what major scientific conflicts there are. It should further be given a critical distance both from the research community and from the political institutions of a country. Establishing a system with competing public information channels could be a necessary part of this.

The above points are all counter to the present trends. Private funding of research is rapidly increasing in importance. Public research institutes are privatized. Patenting is becoming a positive merit on the CV even of public scientists. Public information channels are losing out in the media sector. Why is this so? I have three suggestions to offer.

First, due to the deliberate construction of institutions that favour international competition and the fact that modern industries are very much knowledge based, competing well in the arena of knowledge production has become a necessity. In relation to this, open, public research has two disadvantages. As emphasized, its results are free for all to use. Its broader motivation structure – its responsibility to the research community and society more at large – makes it somewhat 'slow-moving'. While this is a good thing for securing sustainability, it is not so for supporting a business sector that needs to always be some months ahead of the competitors. This even influences public research programmes themselves where supporting the business sector has become a core objective – maybe *the* core objective – of these programmes. I am not implying that public research should not be involved in the development of new production processes and so on. From the above it is clear that I find it extremely important that public research is thus engaged. The question is on whose premises this takes place.

Second, we live in a time when the ideology of markets and private enterprise is dominant. Instead of developing its own qualities, the public sector tends to copy the institutional structures of the private business sector. This may be an effect of the increased status of private enterprise. Public authorities are required to act more like the private sector. Public servants may find such copying a necessity to maintain some legitimacy of their activity.

Finally, the public sector has a problem with financing its activities and it is also at a disadvantage compared to private enterprises when it comes to restructuring. A way out of both these problems is to privatize. This also applies to research. Then the market governs and the role of social considerations when restructuring will be reduced. The problem with this strategy is that to solve a minor, but immediate problem – that of restructuring public research as new needs develop – one creates a larger problem in the long run. One throws away the possibility of securing public research as the dominant producer of new knowledge.

In relation to the latter, certain reforms in the public research sector are important. I believe that 'slowing down', taking the time necessary to secure the quality of the research and to understand its consequences before putting it into practice is important. However, even more important is changing its focus, especially in the phase we are in now. We need to move onto a more sustainable path, which among other things demands changed research foci. Then increased flexibility in the public research community itself is needed.

14.3.3 Necessary, but Unrealistic?

Necessary, but unrealistic – the answer may be 'yes' to both. Great challenges are indeed facing us. One obstacle is that the present institutional systems tend to marginalize the problems we are bound to meet. They demand priority for growth and hence for a continuous interruption into natural systems dynamics. They demand certainty about the negative effects of this practice before accepting restrictions. The resilience of ecosystems makes such a strategy look reasonable in the short run. It seems to go well. While this resilience is a great bonus for us, it is therefore also a problem. Troubles may become visible only when it is too late or too difficult to do anything.

One could certainly 'hope' for 'minor catastrophes', events that act as strong enough early warnings to raise our consciousness. Maybe we are lucky? Maybe an element of the resilience of ecosystems is to produce such events to attract our attention? This would be resilience of really high quality. I believe that the chance of this happening is small, indeed. Resilience is an effect of long-run trial and error processes. Human influence at today's level has never been observed before. The system has not been tested against these kinds of pressures. Therefore the idea that nature should have a built-in counter-reaction – and built-in resilience – to the human capacity of creating inappropriate or deficient institutions is certainly very, very unlikely.

So, humanity itself must take responsibility. While modern societies have developed great freedoms for the individual, this has created even greater responsibilities. This is the fundamental paradox of modernity. The trust lies in our ability to communicate and in reasoning together. We have seen that humankind has great social abilities. It is not just an egoistic calculator. Our hope lies here. However, to develop these abilities further, institutional support is needed. At this point we need to go beyond our historic abilities – that of communicating in small groups, that of caring only about the narrower community. The globalization we need is not foremost that of markets. It is that of open communication and community creation. This calls for agency. This calls for the creation of institutions not seen before that match the problems we face. We were able to create the large corporation. Why shouldn't we be able to create the grand cooperation? The choice is ours.

* * *

If it was possible to buy 'nothing', I thought. If I could buy 50 square feet of 'no thing', I could secure some space that was free of all this useless material they call things. Certainly, if nothing was on a wall you could say: 'Something should be put there. We cannot just leave it.' Then I could answer: 'Well, it isn't just left. "No thing" is already there.' I could even go on and claim the right to this 'no thing' since I had bought it. What a marvellous idea, I thought! Then business and environment could thrive together. Business could make profits from making 'no thing'. It would be really cheap, too. It could come in different colours and it could be sold under various brands. Competition could be maintained. What a fortune for creativity! What a challenge for marketing! And I would be happy since at present producing more of this thing called 'no thing' would be about the best the system could offer to the future.

* * *

NOTES

1. In a programme on Norwegian television – Dok1: 'Grønn strøm', 12 January 2004. Among other things, Gaarder is the author of *Sophie's World*, a novel about the history of philosophy.
2. This is actually not only dependent on the complexity of the good. Several health-care operations – for example, mending a broken leg – can be considered simple as the concept is used here. However, public treatment may be chosen in many societies as opposed to private, since health itself is considered a basic right.
3. As an example, the Norwegian government forecast a doubling of the Norwegian economy between 2004 and 2030, that is, assuming a growth beyond 2 per cent. Parallel, the Norwegian economy has grown about eight times during the twentieth century.

References

Ackerman, F. and L. Heinzerling (2004), *Priceless. On Knowing the Price of Everything and the Value of Nothing*. New York: The New Press.

Agrawal, A. (2002), 'Common resources and institutional sustainability'. In E. Ostrom (ed.): *The Drama of the Commons*. Washington, DC: National Academic Press, pp. 41–85.

Agrawal, A. and C.C. Gibson (eds) (2001), *Communities and the Environment. Ethnicity, Gender, and the State in Community-Based Conservation*. New Brunswick, NJ: Rutgers University Press.

Aguillera-Klinck, F., E. Péres-Moriana and J. Sánches-Garcia (2000), 'The social construction of scarcity. The case of water in Tenerife (Canary Islands)'. *Ecological Economics*, **34**(2): 233–45.

Ahn, T.-K., E. Ostrom, D. Schmidt and J. Walker (1999), 'Dilemma games: game parameters and matching protocols'. Working paper. Bloomington: Indiana University, Workshop in Political Theory and Political Analysis.

Ajzen, I., T.C. Brown and L.H. Rosenthal (1996), 'Information bias in contingent valuation: effects of personal relevance, quality of information, and motivational orientation'. *Journal of Environmental Economics and Management*, **30**: 43–57.

Alchian, A. (1961), *Some Economics of Property*. RAND D-2316. Santa Monica, CA.

Allen, R. (1992), *Enclosure and the Yeomen*. Oxford: Oxford University Press.

Althusser, L. (1965), *For Marx*. Harmondsworth: Penguin.

Aoki, M. (2001), *Toward a Comparative Institutional Analysis*. Cambridge, MA: MIT Press.

Arendt, H. (1958), *The Human Condition*. Chicago: University of Chicago Press.

Árnason, R. and H. Gissurarson (1999), *Individual Transferable Quotas in Theory and Practice*. Reykjavik: University of Iceland Press.

Arrow, K. (1963), *Social Choice and Individual Values*. New Haven, CT: Yale University Press.

Arrow, K. (1969), 'The organization of economic activity. Issues pertinent to the choice of market versus nonmarket allocation'. In *The Analysis and Evaluation of Public Expenditure: The PPB System*. Vol. 1. US Joint Economic Committee, 91st Congress, 1st Session. Washington, DC: US Government Printing Office, pp. 59–73.

Arrow, K. and F.H. Hahn (1979), *General Competitive Analysis*. Amsterdam: North-Holland.

Arrow, K., R. Solow, P.R. Portney, E.E. Leamer, R. Radner and H. Schuman (1993), 'Report of the NOAA panel on contingent valuation'. *Federal Register*, **58**: 4601–14.

Asheim, G.B. (1994), 'New national product as an indicator of sustainability'. *Scandinavian Journal of Economics*, **96**: 257–65.

Attfield, R. and K. Dell (eds) (1989), *Values, Conflict and the Environment: Report of the Environmental Ethics Working Party*. Oxford: Ian Ramsey Centre.

Aubert, V. (1979), *Sosiologi. 1. Sosialt samspill* (*Sociology I. Social Interaction*). Oslo: Universitetsforlaget.

Axelrod, R. (1984), *The Evolution of Cooperation*. New York: Basic Books.

Ayres, R.U. (1993), 'Cowboys, cornucopians and long-run sustainability'. *Ecological Economics*, **8**(3): 189–207.

Bakan, J. (2004), *The Corporation: The Pathological Pursuit of Profit and Power*, New York: Free Press.

Baland, J.M. and J.P. Platteau (1996), *Halting Degradation of Natural Resources: Is there a Role for Rural Communities?* New York: FAO/UN and Oxford University Press.

Bana e Costa, C.A. (ed.) (1990), *Readings in Multiple Criteria Decision Aid*. Berlin: Springer Verlag.

Bana e Costa, C.A. and P. Vincke (1990), 'Multiple criteria decision aid: an overview'. In Bana e Costa (ed.), pp. 3–16.

Banville, C., M. Laundry, J.-M. Martel and C. Baulaire (1998), 'A stakeholder approach to MCDA'. *Systems Research*, **15**: 15–32.

Baran, P.A. and P.M. Sweezy (1968), *Monopoly Capital: An Essay on the American Economic and Social Order*. Harmondsworth: Penguin.

Barbier, E. (1989), *Economics, Natural Resource Scarcity and Development. Conventional and Alternative Views*. London: Earthscan.

Barkow, J.H., L. Cosmides and J. Tooby (eds) (1992), *The Adapted Mind: Evolutionary Psychology and the Generation of Culture*. Oxford: Oxford University Press.

Barrett, M. and D.M. Hausman (1990), 'Making interpersonal comparison coherently'. *Economics and Philosophy*, **6**(2): 293–300.

Barrow, C.J. (1995), *Developing the Environment. Problems and Management*. London: Longman.

Bator, F. (1958), 'The anatomy of market failure'. *Quarterly Journal of Economics*, **87**: 495–502.

Baumol, W.J. and W.E. Oates (1975), *The Theory of Environmental Policy*. Englewood Cliffs, NJ: Prentice-Hall.

Beck, U. (1992), *Risk Society: Towards a New Modernity*. London: Sage.

Becker, G. (1976), *The Economic Approach to Human Behavior*. Chicago: University of Chicago Press.

Becker, G. (1993), 'Nobel lecture. The economic way of looking at behavior'. *Journal of Political Economy*, **101**(3): 385–409.

Becker, G. (1996), *Accounting for Tastes*. Cambridge, MA: Cambridge University Press.

Bentham, J. ([1789] 1970), *Introduction to the Principles of Morals and Legislation*. London: Methuen.

Berger, P. and T. Luckmann ([1967] 1991), *The Social Construction of Reality. A Treatise in the Sociology of Knowledge*. London: Penguin.

Bhaskar, R. (1989), *Reclaiming Reality: A Critical Introduction to Modern Philosophy*. London: Verso.

Bhaskar, R. (1991), *Philosophy and the Idea of Freedom*. Oxford: Blackwell.

Binmore, K. (1994), *Game Theory and the Social Contract, vol. 1: Playing Fair*. Cambridge, MA: MIT Press.

Blackhouse, R. (1985), *A History of Modern Economic Analysis*. Oxford: Blackwell.

Blaug, M. (1992), *The Methodology of Economics. Or How Economists Explain*. 2nd edn, Cambridge: Cambridge University Press.

Blount, S. (1995), 'When social outcomes aren't fair: the effect of causal attributions on preferences'. *Organizational Behavior and Human Decision Process*, **63**(2): 131–44.

Boadway, R.W. (1974), 'The welfare foundations of cost–benefit analysis'. *Economic Journal*, **84**: 926–39.

Boadway, R.W. and N. Bruce (1984), *Welfare Economics*. Oxford: Blackwell.

Boardman, A.E., D.H. Greenberg, A.R. Vining and D.L. Weimer (2001), *Cost–Benefit Analysis. Concepts and Practice*. 2nd edn, Englewood Cliffs, NJ: Prentice-Hall.

Bogetoft, P. and P. Pruzan (1997), *Planning with Multiple Criteria. Investigation, Communication and Choice*. Copenhagen: Copenhagen Business School Press.

Bohman, J. and W. Rehg (1997), *Deliberative Democracy: Essays on Reason and Politics*. Cambridge, MA: MIT Press.

Bohnet, I., B. Frey and S. Huck (2001), 'More order with less law: on contract enforcement: trust and crowding'. *American Political Science Review*, **95**(1): 131–44.

Boland, L.A. (1981), 'On the futility of criticising the neoclassical maximization hypothesis'. *American Economic Review*, **71**: 1031–6.

Boulding, K. (1966), 'The economics of the coming Spaceship Earth'. In H. Jarret (ed.): *Environmental Quality in a Growing World.* Baltimore, MD: Johns Hopkins University Press, pp. 3–14.

Boulding, K. (1970), 'The network of interdependence'. Paper presented at the Public Choice Society Meeting, 19 February.

Bowles, S. (1998), 'Endogenous preferences: the cultural consequences of markets and other economic institutions'. *Journal of Economic Literature*, **36** (March): 75–111.

Boyle, A. (2001), 'Problems of compulsory jurisdiction and the settlement of disputes related to straddling fish stocks'. In Stokke (ed.), pp. 91–112.

Boyle, K.J., W.H. Desvousges, F.R. Johnson, R.W. Dunford and S.P. Hudson (1994), 'An investigation of part–whole biases in contingent-valuation studies'. *Journal of Environmental Economics and Management*, **27**: 64–83.

Boyle, K.J., F.R. Johnson and D.W. McCollum (1997), 'Anchoring and adjustment in single-bounded contingent-valuation questions'. *American Journal of Agricultural Economics*, **79**(5): 1495–500.

Brack, D. (2002), 'Environmental treaties and trade: multilateral environmental agreements and the multilateral trading system'. In G.P. Sampson and W. Bradnee Chambers (eds): *Trade, Environment and the Millennium.* New York: United Nations University Press, pp. 321–52.

Brekke, K.A., S. Kverndokk and K. Nyborg (2003), 'An economic model of moral motivation'. *Journal of Public Economics*, **87**: 1967–83.

Bromley, D.W. (1989), *Economic Interests and Institutions. The Conceptual Foundations of Public Policy*. Oxford: Basil Blackwell.

Bromley, D.W. (1990), 'The ideology of efficiency: searching for a theory of policy analysis'. *Journal of Environmental Economics and Management*, **19**: 86–107.

Bromley, D.W. (1991), *Environment and Economy: Property Rights and Public Policy*. Oxford: Basil Blackwell.

Bromley, D.W. (2006), *Sufficient Reason: Volitional Pragmatism and the Meaning of Economic Institutions*. Princeton, NJ: Princeton University Press.

Broome, J. (1991a), 'Utility'. *Economics and Philosophy*, **7**(1): 1–12.

Broome, J. (1991b), 'A reply to Sen'. *Economics and Philosophy*, **7**(2): 285–7.

Brouwer, R., N. Powe, R.K. Turner, I.J. Bateman and I.H. Langford (1999), 'Public attitudes to contingent valuation and public consultation'. *Environmental Values*, **8**: 325–47.

Brown, T.C., G.I. Peterson and B.E. Tonn (1995), 'The values jury to aid natural resource decisions'. *Land Economics*, **71**: 250–60.

Brox, O. (1997), 'How the ITQ revolution is being implemented in the Norwegian fishing industry'. In G. Pálsson and G. Pétursdóttir (eds): *Social Implications of Quota Systems in Fisheries*. TemaNord report 1997:593. Copenhagen: Nordic Council of Ministers, pp. 51–9.

Bruvoll, A., B. Halvorsen and K. Nyborg (2002), 'Households' recycling efforts'. *Resources, Conservation and Recycling*, **36**: 337–54.

Buchanan, J. (1978), 'From private preferences to the development of public choice'. In Buchanan (ed.), *The Economics of Politics*. London: Institute of Economic Affairs, pp. 1–20.

Burgess, J., J. Clark and C. Harrison (1995), 'Valuing nature: what lies behind responses to contingent valuation surveys?' Working Paper, Department of Geography, University College London.

Burgess, J., M. Limb and C.M. Harrison (1988a), 'Exploring environmental values through the medium of small groups: 1. Theory and practice'. *Environment and Planning A*, **20**: 309–26.

Burgess, J., M. Limb and C.M. Harrison (1988b), 'Exploring environmental values through the medium of small groups: 2. Illustration of a group at work'. *Environment and Planning A*, **20**: 457–76.

Carson, R.T. (1997), 'Contingent valuation surveys and tests of insensitivity to scope'. In R.J. Kopp, W.W. Pommerehne and N. Schwarz (eds): *Determining the Value of Non-marketed Goods. Economic, Psychological, and Policy Relevant Aspects of Contingent Valuation Methods*. Boston/Dordrecht/London: Kluwer Academic, pp. 127–64.

Carson, R.T., N.E. Flores and W.M. Hanemann (1998), 'Sequencing and valuing public goods'. *Journal of Environmental Economics and Management*, **36**: 314–23.

Chang, R. (ed.) (1997), *Incommensurability, Incomparability and Practical Reason*. Cambridge, MA: Harvard University Press.

Cheung, S.N.S. (1983), 'The contractual nature of the firm'. *Journal of Law and Economics*, **26** (April): 1–21.

Chipman, J. and J. Moore (1978), 'Why an increase in GNP need not imply an improvement in potential welfare'. *Kyklos*, **29**(3): 391–418.

Cicero (2003), 'Internasjonale klimaforhandlinger' ('International negotiations on climate'), www.cicero.uio.no/background/negotiations.html, 10 March.

Ciriacy-Wantrup, S.V. (1968), *Resource Conservation: Economics and Policies*. Berkeley, CA: University of California Press.

Clark, J. (2002), 'Structuring deliberation in environmental decision making using a multi-criteria approach'. Paper presented at the European Science Foundation/Standing Committee for the Social Sciences exploratory workshop on new strategies for solving environmental conflicts, Leipzig, 26–28 June.

Clark, J., J. Burgess and C.M. Harrisson (2000), '"I struggled with this money business": respondents' perspectives on contingent valuation'. *Ecological Economics*, **33**: 45–62.

Coase, R.H. (1937), 'The nature of the firm'. *Economica*, **4**: 386–405.

Coase, R.H. (1960), 'The problem of social cost'. *Journal of Law and Economics*, **3**: 1–44.

Coase, R.H. (1984), 'The new institutional economics'. *Journal of Theoretical and Institutional Economics*, **140**(1): 229–31.

Cole, H.S. and R.C. Curnow (1973), 'An evaluation of the world models'. In H.S. Cole, C. Freeman, M. Johado and K.L.R. Pavit (eds): *Thinking About the Future: A Critique of the Limits to Growth*. London: Chatto & Windus, pp. 108–34.

Common, M. and C. Perrings (1992), 'Towards an ecological economics of sustainability'. *Ecological Economics*, **6**(1): 7–34.

Commons, J.R. ([1924] 1974), *Legal Foundations of Capitalism*. Clifton, NJ: Augustus M. Kelley.

Commons, J.R. (1931), 'Institutional economics'. *American Economic Review*, **21**: 648–57.

Commons, J.R. ([1934] 1990), *Institutional Economics. Its Place in Political Economics*. New Brunswick, NJ: Transaction.

Conrad, K. and R.E. Kohn (1996), 'The US market for SO_2 permits. Policy implications of the low price and trading volume'. *Energy Policy*, **24**(12): 1051–9.

Cornes, R. and T. Sandler (1996), *The Theory of Externalities, Public Goods and Club Goods*. Cambridge: Cambridge University Press.

Costanza, R. (1991), *Ecological Economics and the Science and Practice of Sustainability*. New York: Columbia University Press.

Crawford, S.E.S. and E. Ostrom (1995), 'The grammar of institutions'. *American Political Science Review*, **89**(3): 582–600.

Crosby, N. (1995), 'Citizen juries: one solution for difficult environmental questions'. In O. Renn, T. Webler and P. Wiedermann (eds): *Fairness and Competence in Citizen Participation*. Dordrecht: Kluwer, pp. 157–74.

Crowards, T. (1997), 'Nonuse values and the environment: economic and ethical motivations'. *Environmental Values*, **6**: 143–67.

Cummings, R.G., D.S. Brookshire and W.D. Schulze (1986), *Valuing Environmental Goods: An Assessment of the Contingent Valuation Method*. Totowa, NJ: Howman & Allanheld.

Cummings, R.G. and G.W. Harrison (1992), 'Identifying and measuring nonuse values for natural and environmental resources: a critical review of the state of the art'. Mimeo, Department of Economics, University of New Mexico.

Cyert, R.M. and J.G. March (1963), *A Behavioral Theory of the Firm*. Englewood Cliffs, NJ: Prentice-Hall.

Dahlman, C.J. (1979), 'The problem of externality'. *Journal of Law and Economics*, **22**: 141–62.

Daly, H.E. (1973), *Steady-State Economics*. San Francisco, CA: W.H. Freeman.

Dasgupta, P.S. and G.M. Heal (1974), 'The optimal depletion of exhaustible resources'. *Review of Economic Studies. Symposium on the Economics of Exhaustible Resources*, **41**: 3–28.

Dasgupta, P.S. and G.M. Heal (1979), *Economic Theory and Exhaustible Resources*. Cambridge: Cambridge University Press.

Dasgupta, P., K.-G. Mäler and S. Barrett (1999), 'Intergenerational equity, social discount rates, and global warming'. In Portney and Weyant (eds), pp. 51–77.

Dasgupta, P.S. and D.W. Pearce (1972), *Cost–Benefit Analysis: Theory and Practice*. London: Macmillan.

Davis, R. (1963), 'Recreation planning as an economic problem'. *Natural Resources Journal*, **3**(2): 239–49.

Dearlove, J. (1989), 'Neoclassical politics: public choice and political understanding'. *Review of Political Economy*, **1**: 208–37.

Deaton, A. and J. Muellbauer (1980), *Economics and Consumer Behavior*. Cambridge: Cambridge University Press.

Delap, C. (2001), 'Citizens' juries: reflections on the UK experience'. *PLA Notes*, **40** (February): 39–42, www.iied.org/sarl/planotes/sample.html, accessed January 2004.

Demsetz, H. (1967), 'Toward a theory of property rights'. *American Economic Review*, **57**: 347–59.

Dewey, J. (1927), *The Public and its Problems*. New York: Holt.

Diamond, P.A., J.A. Hausman, G.K. Leonard and M.A. Denning (1993), 'Does contingent valuation measure preferences? Experimental evidence'. In J.A. Hausman (ed.): pp. 41–85.

Dienel, P. and O. Renn (1995), 'Planning cells: a gate to "fractal" mediation'. In O. Renn, T. Webler and P. Wiedermann (eds): *Fairness and Competence in Citizen Participation. New Models for Environmental Discourse*. Dordrecht: Kluwer, pp. 117–40.

Douglas, M. (1966), *Purity and Danger: An Analysis of Concepts of Pollution and Taboo*. New York: Praeger.

Douglas, M. (1986), *How Institutions Think*. Syracuse: Syracuse University Press.

Douglas, M. and A. Wildavsky (1982), *Risk and Culture. An Essay on the Selection of Technical and Environmental Dangers*. Berkeley, CA: University of California Press.

Dryzek, J.S. (1990), *Discursive Democracy: Politics, Policy and Political Science*. New York: Cambridge Unversity Press.

Dryzek J.S. (2002), *Deliberative Democracy and Beyond. Liberals, Critics, Contestations*. Paperback edn, Oxford: Oxford University Press.

Dugger, W.M. (1989), 'Power: an institutional framework of analysis'. In M.R. Tool and W.J. Samuels (eds): *The Economy as a System of Power*. New Brunswick, NJ: Transaction, pp. 133–43.

Durkheim, É. ([1893] 1964), *Les Règles de la méthode sociologique*. English translation: *The Division of Labor in Society*. New York: Free Press.

Durkheim, É. ([1895] 1938), *Le Suicide: étude de sociologie*. English translation: *The Rules of Sociological Method*. Chicago: University of Chicago Press.

Eaton, B.C. and D.E. Eaton (1991), *Microeconomics*. 2nd edn, New York: Freeman.

Edgeworth, F.Y. ([1881] 1967), *Mathematical Physics: An Essay on the Application of Mathematics to the Moral Sciences*. New York: Augustus M. Kelley.

Edgeworth, F.Y. (1899), 'Utility'. In R.H.I. Palgrave (ed.): *Dictionary of Political Economy*, Vol. III. London: Macmillan, p. 602.

Edwards-Jones, G., B. Davies and S. Hussain (2000), *Ecological Economics. An Introduction*. Oxford: Blackwell.

Eek, D., A. Biel and T. Gärling (2001), 'Cooperation in asymmetric social dilemmas when equality is perceived as unfair'. *Journal of Applied Social Psychology*, **31**: 649–66.

Eggertsson, T. (1990), *Economic Behavior and Institutions*. Cambridge: Cambridge University Press.

Ehrlich, P.R., A.H. Ehrlich and J.P Holdren (1977), *Ecoscience: Population, Resources, Environment*. San Fransisco, CA: Freeman.

Elster, J. (1979), *Forklaring og dialektikk. Noen grunnbegreper i vitenskapsteorien* (*Explanation and Dialetics. Some Fundamental Concepts in the Theory of Science*). Oslo: Pax forlag.

Elster, J. (1983a), *Explaining Technical Change: A Case Study in the Philosophy of Science*. Cambridge: Cambridge University Press.

Elster, J. (1983b), *Sour Grapes: Studies in the Subversion of Rationality*. Cambridge: Cambridge University Press.

Elster, J. (1998), 'Introduction'. In Elster (ed.): *Deliberative Democracy*. New York: Cambridge University Press, pp. 1–18.

Engquist, L. and A. King (2003), 'Privat sjukvård kollapsar' ('Private nursing is collapsing'). Newspaper article, *Svenska Dagbladet*, 20 January, p. 5.

Etzioni, A. (1988), *The Moral Dimension: Toward a New Economics*. New York: Free Press.

European Environment Agency (2001), 'Late lessons from early warnings: the precautionary principle 1896–2000'. Environmental issue report no 22, Copenhagen.

Færøy, S.H. (2003), 'The ballast water problem: characteristics and negotiations'. Masters Thesis, Department of Economics and Social Sciences, Agricultural University of Norway, Ås.

Faustmann, M. (1849), 'Berechnung des Wertes welchen Waldboden, sowie noch nicht haubahre Holzbeständen für die Waldwirtschaft besitzen' ('Calculation of the value which forest land and immature stands possess for forestry'). *Allgemeine Forst- und Jagd-Zeitung*, **15**: 441–55.

Fehr, E. and A. Falk (2002), 'Psychological foundations of incentives. Joseph Schumpeter Lecture'. *European Economic Review*, **46**: 687–724.

Fehr, E. and S. Gächter (2000), 'Cooperation and punishment'. *American Economic Review*, **90**: 980–94.

Fehr, E. and S. Gächter (2002), 'Do incentive contracts undermine voluntary cooperation?' Working paper no. 34, Institute for Empirical Research in Economics, University of Zurich.

Fehr, E., S. Gächter and G. Kirchsteiger (1997), 'Reciprocity as a contract enforcement device: experimental evidence'. *Econometrica*, **65**(4): 833–60.

Field, A.J. (1994a), 'North, Douglass C.'. In G.M. Hodgson, W.J. Samuels and M.R. Tool (eds): *The Elgar Companion to Institutional and Evolutionary Economics*, vol. 2. Aldershot, UK and Brookfield, US: Edward Elgar, pp. 134–8.

Field, B. (1994b), *Environmental Economics. An Introduction*. New York: McGraw-Hill.

Finlayson, A.C. and B. McCay (2000), 'Crossing the threshold of ecosystem resilience: the commercial extinction of the northern cod'. In F. Berkes and C. Folke (eds): *Linking Social and Ecological Systems. Management Practices and Social Mechanisms for Building Resilience*. Paperback edn, Cambridge: Cambridge University Press, pp. 311–37.

Fischhoff, B., N. Welch and S. Frederick (1999), 'Construal processes in preference assessment'. *Journal of Risk and Uncertainty*, **19**(1): 139–64.

Fisher, A.C. (1990), *Resource and Environmental Economics*. Cambridge: Cambridge University Press.

Fishkin, J.S. (1997), *The Voice of the People*. New Haven, CT: Yale University Press.

Folke, C., F. Berkes and J. Colding (1998), 'Ecological practices and social mechanisms for building resilience and sustainability'. In Berkes and Folke (eds): *Linking Social and Ecological Systems. Management Practices and Social Mechanisms for Building Resilience*. Cambridge: Cambridge University Press, pp. 414–36.

Forsythe, R., J.L. Horowitz, N.E. Savin and M. Sefton (1994), 'Fairness in simple bargaining experiments'. *Games and Economic Behavior*, **6**: 347–69.

Frey, B.S. (1992), 'Pricing and regulating affect environmental ethics'. *Environmental and Resource Economics*, **2**: 399–414.

Frey, B.S. (1997a), *Not Just for the Money. An Economic Theory of Personal Motivation*. Cheltenham, UK and Lyme, USA: Edward Elgar.

Frey, B.S. (1997b), 'A constitution for knaves crowds out civic virtue'. *Economic Journal*, **107**(443): 1043–53.

Frey, B.S. and R. Jegen (2001), 'Motivation crowding theory'. *Journal of Economic Surveys*, **15**(5): 589–611.

Frey, B.S. and F. Oberholzer-Gee (1997), 'The cost of price incentives: an empirical analysis of motivation crowding-out'. *American Economic Review*, **87**(4): 746–55.

Friedman, M. (1962), *Capitalism and Freedom*. Chicago: University of Chicago Press.

Frohlich, N. and J.A. Oppenheimer (1996), 'Experiencing impartiality to invoke fairness in the N-PD: some experimental results'. *Public Choice*, **86**: 117–35.

Fudenberg, D. and J. Tirole (1991), *Game Theory*. Cambridge, MA: MIT Press.

Funtowicz, S. and J.R. Ravetz (1993), 'Science for the post-normal age'. *Futures*, **25**: 735–55.

Funtowicz, S. and J.R. Ravetz (1994), 'The worth of a songbird: ecological economics as a post-normal science'. *Ecological Economics*, **10**(3): 197–208.

Galbraith, J.K. (1971), *The New Industrial State*. Boston, MA: Houghton Mifflin.

Gandy, M. (1996), 'Crumbling land: the postmodernity debate and the analysis of environmental problems'. *Progress in Human Geography*, **20**(1): 23–40.

Georgescu-Roegen, N. (1971), *The Entropy Law and the Economic Process*. Cambridge, MA: Harvard University Press.

Gezelius, S.S. (1996), 'The northern cod crisis and the "turbot war"', Dissertation and Thesis No. 9, Department of Sociology, University of Oslo.

Giddens, A. (1984), *The Constitution of Society*. Cambridge: Polity.

Gintis, H. (2000), 'Beyond *Homo Economicus*: evidence from experimental economics'. *Ecological Economics*, **35**: 311–22.

Gneezy, U. and A. Rustichini (2000a), 'A fine is a price'. *Journal of Legal Studies*, **29**: 1–17.

Gneezy, U. and A. Rustichini (2000b), 'Pay enough or don't pay at all'. *Journal of Economic Behavior and Organization*, **39**: 341–69.

Goldman, M. (ed.) (1998), *Privatizing Nature. Political Struggles for the Global Commons*. New Brunswick, NJ: Rutgers University Press.

Goodin, R.E. (1992), *Motivating Political Morality*. Oxford: Basil Blackwell.

Gorman, W.M. (1955), 'The intransitivity of certain criteria used in welfare economics'. *Oxford Economic Papers, NS*, **7**(1): 25–35.

Goudie, A. (1993), *The Human Impact on the Natural Environment*. Oxford: Blackwell.

Gowdy, J. (2004), 'The revolution in welfare economics and its implications for environmental valuation and policy'. *Land Economics*, **80**(2): 239–57.

Gowdy, J., R. Iorgulescu and S. Onyeiwu (2003), 'Fairness and the retaliation in a rural Nigerian village'. *Journal of Economic Behavior and Organization*, **52**(4): 469–79.

Graves, J. and D. Reavy (1996), *Global Environmental Change*. London: Longman.

Gray, L.C. (1914), 'Rent under the assumption of exhaustibility'. *Quarterly Journal of Economics*, **28**: 466–89.

Greenberg, J. (1984), 'Tax avoidance: a (repeated) game theoretic approach'. *Journal of Economic Theory*, **32**(1): 1–13.

Gregory, R. (1986), 'Interpreting measures of economic loss: evidence from contingent valuation and experimental studies'. *Journal of Environmental Economics and Management*, **13**: 325–37.

Gregory, R., S. Lichtenstein and P. Slovic (1993), 'Valuing environmental resources: a constructive approach'. *Journal of Risk and Uncertainty*, **7**(2): 177–97.

Griffin, R.C. and D.W. Bromley (1982), 'Agricultural runoff as a nonpoint externality: a theoretical development'. *American Journal of Agricultural Economics*, **64**(3): 547–52.

Griliches, Z. (1971), *Price Indexes and Quality Change*. Cambridge, MA: Harvard University Press.

Groenewegen, J., C. Pitelis and S.-E. Sjöstrand (eds) (1995), *On Economic Institutions: Theory and Applications*. Aldershot, UK and Brookfield, US: Edward Elgar.

Grubb, M.C., D. Brack and C. Vroljik (1999), *The Kyoto Protocol. A Guide and Assessment*. London: Earthscan.

Guba, E.G. (1990), 'The alternative paradigm dialogue'. In E.G. Guba (ed.): *The Paradigm Dialogue*. London: Sage, pp. 17–27.

Gustafsson, B. (1991), 'Introduction'. In Gustafsson (ed.): *Power and Economic Institutions. Reinterpretations in Economic History*. Aldershot, UK and Brookfield, US: Edward Elgar, pp. 1–50.

Güth, W., R. Schmittberger and B. Schwarze (1982), 'An experimental analysis of ultimatum bargaining'. *Journal of Economic Behavior and Organization*, **3**: 367–88.

Habermas, J. (1984), *The Theory of Communicative Action. Vol I: Reason and the Rationalization of Society*. Boston, MA: Beacon Press.

Habermas, J. (1995), *Between Facts and Norms. Contribution to a Discourse Theory of Law and Democracy*. Cambridge, MA: MIT Press.

Hahn, T. (2000), 'Property rights, ethics and conflict resolution. Foundations of the Sami economy in Sweden'. *Agraria* 258, Uppsala: Swedish University of Agricultural Sciences.

Halvorsen, B. (1996), 'Ordering effects in contingent valuation surveys'. *Environmental and Resource Economics*, **8**: 485–99.

Hammond, P.J. (1985), 'Welfare economics'. In G.R. Feiwel (ed.): *Issues in Contemporary Microeconomics and Welfare*. London: Macmillan, pp. 404–34.

Hanemann, W.M. (1991), 'Willingness to pay and willingness to accept: how much can they differ?'. *American Economic Review*, **81**(3): 635–47.

Hanemann, W.M. (1994), 'Valuing the environment through contingent valuation'. *Journal of Economic Perspectives*, **8**(4): 19–43.

Hanley, N., J.F. Shogren and B. White (1997), *Environmental Economics: Theory and Practice*. London: Macmillan.

Hanley, N. and C.L. Spash (1993), *Cost–Benefit Analysis and the Environment*. Aldershot, UK and Brookfield, US: Edward Elgar.

Hanna, S. (1995), 'User participation and fishery management performance within the Pacific Fisheries Management Council'. *Ocean Coastal Management*, **28**(1–3): 23–44.

Harcourt, G.C. (1972), *Some Cambridge Controversies in the Theory of Capital*. Cambridge: Cambridge University Press.

Hardin, G. (1968), 'The tragedy of the commons'. *Science*, **162**: 1243–8.

Hartwick, J.M. (1977), 'Intergenerational equity and the investing of rents from exhaustible resources'. *American Economic Review*, **66**: 972–4.

Hartwick, J.M. and N.D. Olwiler (1998), *The Economics of Natural Resource Use*. 2nd edn, New York: Addison Wesley.

Hausman, D.M. (1992), *The Inexact and Separate Science of Economics*. Cambridge: Cambridge University Press.

Hausman, J.A. (ed.) (1993), *Contigent Valuation. A Critical Assessment*. Amsterdam: North-Holland.

Hayami, Y. and V. Ruttan (1985), *Agricultural Development*. Baltimore, MD: Johns Hopkins University Press.

Hayek, F.A. (1948), *Individualism and Economic Order*. Chicago: University of Chicago Press.

Hayek, F.A. (1960), *The Constitution of Liberty*. Chicago: University of Chicago Press.

Hayek, F.A. (1967), *Studies in Philosophy, Politics and Economics*. London: Routledge.

Hayek, F.A. (1973), *Law, Legislation and Liberty, vol. 1: Rules and Order*. Chicago: University of Chicago Press.

Hayek, F.A. (1976), *Law, Legislation and Liberty, vol. 2: The Mirage of Social Justice*. Chicago: University of Chicago Press.

Hayek, F.A. (1988), *The Fatal Conceit: The Errors of Socialism, vol. 1 of Collected works of F.A. Hayek*. London: Routledge.

Helgason, A. and G. Pálsson (1998), 'Cash for quotas: disputes over the legitimacy of an economic model of fishing in Iceland'. In J.G. Carrier and D. Miller (eds): *Virtualism. A New Political Economy*. Oxford: Berg, pp. 117–34.

Henrich, J.R., B.S. Bowles, C. Camerer, E. Fehr, H. Gintis and R. McElrath (2001), 'In search of *Homo Economicus*: behavioral experiments in 15 small-scale societies'. *American Economic Review Papers and Proceedings*, **91**(2): 73–8.

Herriges, J.A. and J.F. Shogren (1996), 'Starting point bias in dichotomous choice valuation with follow-up questioning'. *Journal of Environmental Economics and Management*, **30**: 112–31.

Hicks, J.R. (1939), 'The foundations of welfare economics'. *Economic* Journal, **49**: 696–712.

Hinloopen, E. and P. Nijkamp (1990), 'Quality multiple criteria choice analysis, the dominant regime method', *Quality and Quantity*, **24**: 37–56.

Hinloopen, E., P. Nijkamp and P. Rietveld (1983), 'Qualitative discrete multiple criteria choice models in regional planning'. *Regional Science and Urban Economics*, **13**(1): 77–102.

Hirschman, A.O. (1982), 'Rival interpretations of market society: civilizing, destructive or feeble?'. *Journal of Economic Literature*, **20**: 1463–84.

Hobbes, T. ([1651] 1985), *Leviathan*. Harmondsworth: Penguin Books/ Penguin Classics.

Hobsbawm, E. (1983), 'Introduction: inventing traditions'. In E. Hobsbawm and T. Ranger (eds): *The Invention of Tradition*. Cambridge: Cambridge University Press, pp. 1–14.

Hodgson, G.M. (1988), *Economics and Institutions: A Manifesto for a Modern Institutional Economics*. Cambridge: Polity.

Hodgson, G.M. (1996), *Economics and Evolution. Bringing Life Back into Economics*. Ann Arbor, MI: University of Michigan Press.

Hodgson, G.M. (1999), *Evolution and Institutions: On Evolutionary Economics and the Evolution of Economics*, Cheltenham, UK and Northampton, MA, USA: Edward Elgar.

Hodgson, G.M. (2000), 'What is the essence of institutional economics'. *Journal of Economic Issues*, **34**(2): 317–29.

Hoehn, J.P. (1991), 'Valuing the multidimensional impacts of environmental policy: theory and methods'. *American Journal of Agricultural Economics*, **73**(2): 289–99.

Hoehn, J.P. and A. Randall (1989), 'Too many proposals pass the benefit cost test'. *American Economic Review*, **79**(3): 544–51.

Hoffman, E., K. McCabe, K. Shachat and V. Smith (1994), 'Preferences, property rights, and anonymity in bargaining games'. *Games and Economic Behavior*, **7**: 346–80.

Hohfeld, W.N. (1913), 'Some fundamental legal conceptions as applied in judicial reasoning'. *Yale Law Journal*, **23**: 16–59.

Hohfeld, W.N. (1917), 'Fundamental legal conceptions as applied in judicial reasoning'. *Yale Law Journal*, **26**: 710–70.

Holland, A. (1997), 'Substitutability, or why strong sustainability is weak and absurdly strong sustainability is not absurd'. In J. Foster (ed.): *Valuing Nature? Economics, Ethics and Environment*. London: Routledge, pp. 119–34.

Holland, A. (2002), 'Are choices tradeoffs?'. In D.W. Bromley and J. Paavola (eds): *Economics, Ethics and Environmental Policy. Contested Choices*. Oxford: Blackwell, pp. 17–34.

Holling, C.S. (1973), 'Resilience and stability of ecological systems'. *Annual Review of Ecological Systems*, **4**: 1–24.

Holling, C.S. (1986), 'The resilience of terrestrial ecosystems: local surprise and global change'. In W.C. Clark and R.E. Munn (eds): *Sustainable Development of the Biosphere*. Cambridge: Cambridge University Press, pp. 292–317.

Holmes T.P. and R.A. Kramer (1995), 'An independent sample test of yea- saying and starting point bias in dichotomous-choice contingent valuation'. *Journal of Environmental Economics and Management*, **29**: 121–32.

Honoré, A.M. (1961), 'Ownership'. In A.G. Guest (ed.): *Oxford Essays in Jurisprudence*. Oxford: Clarendon, pp. 107–47.

Horowitz, J.K. and K.E. McConnell (2002), 'A review of WTA/WTP studies'. *Journal of Environmental Economics and Management*, **44**: 426–47.

Hotelling, H. (1931), 'The economics of exhaustible resources'. *Journal of Political Economy*, **39**: 137–75.

International Forum on Globalization (2002), *Alternatives to Economic Globalization: A Better World Is Possible*. San Francisco, CA: Berrett-Koehler.

Jacobs, M. (1997), 'Environmental valuation, deliberative democracy and public decision-making'. In J. Foster (ed.): *Valuing Nature? Economics, Ethics and Environment*. London: Routledge, pp. 211–31.

Jankulovska, A., P. Vedeld and J. Kabbogoza (2003), '"Coaching – poaching?" governance, local people and wildlife around Mount Elgon National Park, Uganda'. Noragric working paper no. 31, Ås: Noragric/NLH.

Janssen, R. (1994), *Multiobjective Decision Support for Environmental Management*. Dordrecht: Kluwer.

Jentoft, S. (2004), 'Institutions in fisheries: what they are, what they do, and how they change'. *Marine Policy*, **28**(2): 137–49.

Jentoft, S. and B. McCay (1995), 'User participation in fisheries management; lessons drawn from international experiences'. *Marine Policy*, **29**(3): 227–46.

Jevons, W.S. ([1871] 1957), *Theory of Political Economy*. 5th edn, New York: Augustus M. Kelly.

Jevons, W.S. (1909), *The Coal Question: An Inquiry Concerning the Progress of the Nation and the Probable Exhaustion of Our Coal Mines*. London: Macmillan.

Jodha, N. (1987), 'A case study of the degradation of common property resources in Rajasthan'. In P. Blaikie and H. Brookfield (eds): *Land Degradation and Society*. London and New York: Methuen, pp. 196–205.

Johansson, S. and N.-G. Lundgren (1998), *Vad kostar en ren? En ekonomisk och politisk analys* (*What does a reindeer cost? An economic and political analysis*). Ds 1998:8. Stockholm, Regjeringskansliet/Ministry of Finance.

Joss, S. (1998), 'Danish consensus conference as a model of participatory technology assessment: an impact study of consensus conferences on Danish parliament and Danish public debate'. *Science and Public Policy*, **25**(1): 2–22.

Joss, S. and J. Durant (1995), 'The UK National Consensus Conference on Plant Biotechnology'. *Public Understanding of Science*, **4**: 195–204.

Kagel, J. and A. Roth (eds) (1995), *The Handbook of Experimental Economics*. Princeton, NJ: Princeton University Press.

Kahneman, D. (1986), 'The review panel's assessment'. Comments by Professor Daniel Kahneman. In Cummings, Brookshire and Schulze (eds), pp. 185–94.

Kahneman, D. and J.L. Knetsch (1992), 'Valuing public goods: the purchase of moral satisfaction'. *Journal of Environmental Economics and Management*, **22**: 57–70.

Kahneman, D., I. Ritov and D. Schkade (1999), 'Economic preferences or attitude expressions? An analysis of dollar responses to public issues'. *Journal of Risk and Uncertainty*, **19**(1–3): 203–35.

Kaldor, N. (1939), 'Welfare propositions in economics'. *Economic Journal*, **49**: 549–52.

Kant, I. ([1785] 1981), *Grundlegung zur Metaphysik der Sitten*. English translation: *Grounding for the Metaphysics of Morals*. Indianapolis, IN: Hackett.

Kapp, K.W. (1971), *The Social Costs of Private Enterprise*. New York: Schoken Books.

Kealy, M.J. and R.W. Turner (1993), 'A test of the equality of closed-ended and open-ended contingent valuation'. *American Journal of Agricultural Economics*, **75**: 221–31.

Keeney, R.L. and H. Raiffa (1976), *Decisions with Multiple Objectives: Preferences and Value Trade-Offs*. New York: John Wiley & Sons.

Kenyon, W., N. Hanley and C. Nevin (2001), 'Citizens' juries: an aid to environmental valuation?' *Environment and Planning C: Government and Policy*, **19**: 557–66.

Kerr, A., S. Cunningham-Burley and A. Amos (1998), 'The new genetics and health: mobilising lay expertise'. *Public Understanding of Science*, **7**: 41–60.

Keynes, J.M. (1936), *The General Theory of Employment, Interest and Money*. London: Macmillan.

Klaassen, G. and A. Nentjes (1997), 'Creating markets for air pollution control in Europe and the USA'. *Environmental and Resource Economics*, **10**: 125–46.

Klappholz, K. (1964), 'Value judgements and economics'. *British Journal for the Philosophy of Science*, **15**: 97–114.

Klein, N. (2000), *No Logo*. Toronto: Random House.

Kneese, A.V. and W.D. Schulze (1985), 'Ethics and environmental economics'. In J.L. Sweeney (ed.): *Handbook of Natural Science*. New York: Elsevier Science, pp. 191–220.

Knetsch, J.L. (2000), 'Environmental valuation and standard theory: behavioural findings, context dependence and implications'. In T. Tietenberg and H. Folmer (eds): *The International Yearbook of Environment and Resource Economics 2000/2001: A Survey of Current Issues*. Cheltenham, UK and Northampton, MA, USA: Edward Elgar, pp. 267–99.

Knight, F.H. (1922), 'Ethics and the economic interpretation'. *Quarterly Journal of Economics*, **36**: 454–81.

Knudsen, C. (1993), 'Equilibrium, perfect rationality and the problem of self-reference in economics'. In U. Mäki, B. Gustafsson and C. Knudsen (eds): *Rationality, Institutions and 'Economic Methodology'*. London: Routledge, pp. 133–70.

Kolstad, C. (2000), *Environmental Economics*. Oxford: Oxford University Press.

Krutilla, J.V. and O. Eckstein (1958), *Multiple Purpose River Development: studies in applied economic analysis*. Baltimore, MD: Resources for the Future/Johns Hopkins University Press.

Kula, E. (1998), *History of Environmental Economic Thought*. London: Routledge.

Lakatos, I. (1974), 'Falsification and the methodology of scientific research programmes'. In I. Lakatos and A. Musgrave (eds): *Criticism and the Growth of Knowledge*. Cambridge: Cambridge University Press, pp. 91–196.

Lancaster, K.J. (1966), 'A new approach to consumer theory'. *Journal of Political Economy*, **74**: 132–57.

Lane, R.E. (1991), *The Market Experience*. Cambridge: Cambridge University Press.

Lane, C. and R. Moorehead (1994), 'New directions in rangeland resource tenure and policy'. In I. Scoones (ed.): *Living with Uncertainty: New Directions in Pastoral Development in Africa*. London: Intermediate Technology Publication, pp. 116–33.

Leal, D.R. (2000), 'Homesteading the oceans: the case for property rights in U.S. fisheries', PS-19 PERC Policy Series 1, www.perc.org/pdf/ps19.pdf, accessed July 2004.

Lemons, J. (1998), 'Burden of proof requirements and environmental sustainability: science, public policy, and ethics'. In J. Lemons, L. Westra and R. Goodland (eds): *Ecological Sustainability and Integrity: Concepts and Approaches*. Dordrecht/Boston/London: Kluwer, pp. 75–103.

Lenaghan, J. (2001), 'Participation and governance in the UK Department of Health'. *PLA Notes*, **40** (February), 29–31, www.iied.org/sarl/planotes/sample.html, accessed January 2004.

Lerner, A. (1972), 'The economics and politics of consumer sovereignty'. *American Economic Review*, **62**(2): 258–66.

Levi, M. (1990), 'A logic of institutional change'. In K. Schweers Cook and M. Levi (eds): *The Limits of Rationality*. Chicago: University of Chicago Press, pp. 402–18.

Lévi-Strauss, C. (1968), *Structural Anthropology*. London: Penguin.

Lewin, R. (1988), *In the Age of Mankind. A Smithsonian Book of Human Evolution*. Washington, DC: Smithsonian Institute.

Lichtenstein, S. and P. Slovic (1971), 'Reversals of preference between bids and choices in gambling decisions'. *Journal of Experimental Psychology*, **110**: 16–20.

Little, I.M.D. (1957), *A Critique of Welfare Economics*. Oxford: Clarendon.

Locke, J. ([1690] 1994), *Two Treatises of Government*. Cambridge: Cambridge University Press.

Macinko, S. and D.W. Bromley (2002), *Who Owns America's Fisheries?* Washington, DC: Island Press.

MacKenzie, D.A. (1981), *Statistics in Britain 1865–1930. The Social Construction of Scientific Knowledge*. Edinburgh: Edinburgh University Press.

Macpherson, C.B. (1973), *Democratic Theory: Essays in Retrieval*. Oxford: Clarendon.

Mäki, U., B. Gustafsson and C. Knudsen (eds) (1993), *Rationality, Institutions and 'Economic Methodology'*. London: Routledge.

Malinowski, B. ([1944] 1960), *A Scientific Theory of Culture*. New York: Oxford University Press.

Mallon, F. (1983), *The Defence of Community in Peru's Central Highlands: Peasant Struggle and Capitalist Transition 1860–1940*. Princeton, NJ: Princeton University Press.

Malthus, T.R. ([1803] 1992), *An Essay on the Principles of Population*. Edited by D. Winch on the basis of the 1803 version. Cambridge: Cambridge University Press.

Malthus, T.R. ([1836] 1968), *Principles of Political Economy: Considered with a View to Their Practical Application*. 2nd edn, New York: August M. Kelley.

March, J.G. (1994), *A Primer on Decision Making. How Decisions Happen*. New York: Free Press.

March, J.G. and J.P. Olsen (1989), *Rediscovering Institutions. The Organizational Basis of Politics*. New York: Free Press.

Marglin, S.A. (1963), 'The social rate of discount and the optimal rate of investment'. *Quarterly Journal of Economics*, 77: 95–111.

Marglin, S.A. (1991), 'Understanding capitalism: control versus efficiency'. In B. Gustafsson (ed.): *Power and Economic Institutions. Reinterpretations in Economic History*. Aldershot, UK and Brookfield, US: Edward Elgar, pp. 225–52.

Marshall, A. ([1890] 1949), *The Principles of Economics*. 8th edn, London: Macmillan.

Martin, H.-P. and H. Schumann (1996), *Die Globalisierungsfälle. Der Angriff auf Demokratie and Wholstand* (*The Global Trap. Globalization and the Assault on Democracy and Prosperity*). Reinbek by Hamburg: Rowholt Verlag.

Martinez-Alier, J. (1987), *Ecological Economics. Energy, Environment and Society*. Oxford: Basil Blackwell.

Martinez-Alier, J., G. Munda and J. O'Neill (1998), 'Weak comparability of values as a foundation of ecological economics'. *Ecological Economics*, **26**(3): 277–86.

Maslow, A. (1954), *Motivation and Personality*. New York: Harper & Row.

Mayhew, A. (1987), 'The beginnings of institutionalism'. *Journal of Economic Issues*, **21**: 971–98.

McCauley, C., P. Rozin and B. Schwartz (1994), 'On the origin and nature of preferences and values'. Unpublished paper, Pennsylvania: Bryn Mawr College.

McCay, B.J., C.F. Creed, A.C. Finlayson, R. Apostle and K. Mikalsen (1995), 'Individual transferable quotas (ITQs) in Canadian and US fisheries'. *Ocean Coastal Management*, **28**(1–3): 85–155.

McFadden, D. (1999), 'Rationality for economists?' *Journal of Risk and Uncertainty*, **19**(1–3): 73–105.

McFadden, D. and G.K. Leonard (1993), 'Issues in the contingent valuation of environmental goods: methodologies for data collection and analysis'. In J.A. Hausman (ed.): pp. 165–208.

McGoodwin, J.R. (1990), *Crisis in the World's Fisheries*. Stanford, CA: Stanford University Press.

McKean, M.A. (2000), 'Common property: what is it, what is it good for, and what makes it work?' In C.C. Gibson, M.A. McKean and E. Ostrom (eds): *People and Forests. Communities, Institutions and Governance*. Cambridge, MA: MIT Press, pp. 27–56.

Meade, W.J. (1993), 'Review and analysis of state-of-the-art contingent valuation studies'. In J.A. Hausman (ed.): *Contingent Valuation. A Critical Assessment*, Amsterdam: North-Holland, pp. 305–32.

Meadows, D., D.L. Meadows, J. Randers and W.W. Behrends III (1972), *Limits to Growth: A Report for the Club of Rome's Project on the Predicament of Mankind*. New York: Universe Books.

Melchior, A. (2001), *Nasjonalstater, globale markeder og ulikhet: Politiske spørsmål og institusjonelle utfordringer* (*National States, Global Markets and Inequality: Political Questions and Institutional Challenges*). Oslo: Norwegian Institute of Foreign Policy.

Menger, C. ([1871] 1981), *Grundsätze der Wirtscahftslehre*. English translation: *Principles of Economics*. New York: New York University Press. (Translated from the German edition by J. Dingwall and B.F. Hoselitz.)

Menger, C. ([1883] 1963), *Untersuchungen über die Methode der Sosialwisssenschaften und der Politischen Oeconomie insbesonderes*. English translation: *Problems of Economics and Sociology*. New York: New York University Press. (Translated from the German edition by F.J. Nock.)

Milgrom, P. (1993), 'Is sympathy an economic value? Philosophy, economics, and the contingent valuation method'. In J.A. Hausman (ed.): pp. 417–35.

Miliband, R. (1969), *The State in Capitalist Society*. London: Weidenfeld & Nicolson.

Mill, J.S. ([1848] 1965), *Principles of Political Economy: With Some of Their Applications to Social Philosophy*. Fairfield, NJ: A.M. Kelley.

Mill, J.S. ([1861] 1987), *Utilitarianism*. New York: Prometheus Books.

Mises, L. von (1949), *Human Action: A Treatise on Economics*. London: William Hodge.

Mishan, E.J. (1971), 'The postwar literature on externalities: an interpretive essay'. *Journal of Economic Literature*, **9**: 1–28.

Mitchell, R.C. and R.T. Carson (1989), *Using Surveys to Value Public Goods: The Contingent Valuation Method.* Washington, DC: Resources for the Future.

Morgenstern, O. ([1935] 1976), 'Vollkommene Voraussicht und wirtschaftliches gleichgewicht'. *Zeitschrift für Nationalökonomie*, **6**. English translation: 'Perfect foresight and economic equilibrium'. In A. Schotter (ed.): *Selected Economic Writings of Oscar Morgenstern*. New York: New York University Press, pp. 169–83.

Munda, G. (1995), *Multicriteria Evaluation and a Fuzzy Environment. Theory and Applications in Ecological Economics*. Heidelberg: Physica-Verlag.

Munda, G. (1996), 'Cost–benefit analysis in integrated environmental assessment: some methodological issues'. *Ecological Economics*, **19**(2): 157–68.

Navrud, S. (2004), 'Value transfer and environmental policy'. In T. Tietenberg and H. Folmer (eds): *The International Yearbook of Environmental and Resource Economics 2004/2005. A Survey of Current Issues*. Cheltenham, UK and Northampton, MA, USA: Edward Elgar, pp. 189–217.

Navrud, S. and R. Ready (eds) (2004), *Environmental Value Transfer: Issues and Methods*. Dordrecht: Kluwer.

Nawaz, M.K. (1980), 'On the advent of the exclusive economic zone: implications for a law of the sea'. In R.E. Anand (ed.): *Law of the Sea – Caracas and Beyond.* New Delhi: Radiant, pp. 180–203.

Nelson, R.R. and S.G. Winter (1982), *An Evolutionary Theory of Economic Change*. Cambridge, MA: Belknap Press of Harvard University Press.

Ng, Y.K. (1972), 'Value judgments and economists' role in policy recommendation'. *Economic Journal*, **82**: 1014–18.

Ng, Y.K. (1983), *Welfare Economics*. London: Macmillan.

Nicolis, G. and I. Prigogigne (1989), *Exploring Complexity. An Introduction*. New York: W.H. Freeman.

Niemeyer, S. and C.L. Spash (2001), 'Environmental valuation analysis, public deliberation, and their pragmatic synthesis: a critical appraisal'. *Environment and Planning C: Government and Policy*, **19**: 567–85.

Nijkamp, P., P. Rietveld and H. Voogd (1990), *Multicriteria Evaluation in Physical Planning*. Amsterdam: North-Holland.

Niskanen, W. (1971), *Bureaucracy and Represenative Government*. Chicago: Aldine-Atherton.

Nöel, J.F and M. O'Connor (1998), 'Strong sustainability and critical natural capital'. In S. Faucheux and M. O'Connor (eds): *Valuation for Sustainable Development*. Cheltenham, UK and Lyme, USA: Edward Elgar, pp. 75–98.

Norgaard, R.B. (1984), 'Coevolutionary development potential'. *Land Economics*, **60**(2): 160–73.

Norgaard, R.B. (1994), *Development Betrayed. The End of Progress and a Coevolutionary Revisioning of the Future*. New York: Routledge.

Nørretranders, T. (1991), *Mærk Verden. En beretning om bevisthed* (*Mind the World. A Story about Consciousness*). Copenhagen: Gyldendal.

North, D.C. (1981), *Structure and Change in Economic History*. New York: W.W. Norton.

North, D.C. (1990), *Institutions, Institutional Change and Economic Performance*. Cambridge: Cambridge University Press.

North, D.C. and R.P. Thomas (1973), *The Rise of the Western World: A New Economic History*. Cambridge: Cambridge University Press.

North, D.C. and R.P. Thomas (1977), 'The first economic revolution'. *Economic History Review*, **30**(2): 229–41.

Nyborg, K. and M. Rege (2003), 'On social norms: the evolution of considerate smoking behavior'. *Journal of Economic Behavior and Organization*, **52**: 323–40.

Oakersson, R.J. (1992), 'Analyzing the commons: a framework'. In D. Bromley (ed.): *Making the Commons Work. Theory, Practice and Policy*. San Francisco: ICS Press, pp. 41–59.

Oakeshott, M. (1962), *Nationalism in Politics and Other Essays*. London: Methuen.

O'Conor, R.M., M. Johannesson and P.-O. Johansson (1999), 'Stated preferences, real behavior and anchoring: some empirical evidence'. *Environmental and Resource Economics*, **13**: 235–348.

O'Neill, J. (1993), *Ecology, Policy and Politics. Human Well-Being and the Natural World*. London: Routledge.

O'Neill, J. (1998), *The Market. Ethics, Knowledge and Politics*. London: Routledge.

O'Neill, J. (2001), 'Representing people, representing nature, representing the world'. *Environment and Planning C: Government and Policy*, **19**: 483–500.

O'Neill, J., A. Light and A. Holland (2005), *Values and the Environment*. London: Routledge.

Ostrom, E. (1990), *Governing the Commons: The Evolution of Institutions for Collective Action*. Cambridge: Cambridge University Press.

Ostrom, E. (2000), 'Collective action and the evolution of social norms'. *Journal of Economic Perspectives*, **14**(3): 137–58.

Ostrom, E. (2001), 'Foreword'. In Agrawal and Gibson (eds), pp. ix–xii.

Ostrom, E., R. Gardner and J. Walker (1994), *Rules, Games, and Common-Pool Resources*. Ann Arbor, MI: University of Michigan Press.

Paavola, J. (2002), 'Rethinking the choice and performance of environmental policies'. In D.W. Bromley and J. Paavola (eds): *Economics, Ethics, and Environmental Policy. Contested Choices*. Oxford: Blackwell, pp. 87–102.

Page, T. (1997), 'On the problem of achieving efficiency and equity, Intergenerationally'. *Land Economics*, **73**(4): 580–96.

Pálsson G. and A. Helgason (1997), 'Figuring fish and measuring men: the ITQ system in the Icelandic cod fishery'. In Pálsson and Pétursdóttir (eds), pp. 189–218.

Pálsson, G. and G. Pétursdóttir (eds) (1997), *Social Implications of Quota Systems in Fisheries*. Tema Nord Report no. 593. Copenhagen: Nordic Council of Ministers.

Pareto, V. ([1906] 1971), *Manuale di economia politica; con una introduzione alla scienza sociale*. English translation: *The Manual of Political Economy*. New York: Augustus M. Kelley. (Translated by A.S. Schwier and edited by A.S. Schwier and A.N. Page.)

Parsons, T. (1937), *The Structure of Social Action*. New York: McGraw-Hill.

Parsons, T. (1951), *The Social System*. Glencoe, IL: Free Press.

Payne, J.W. and J.R. Bettman (1999), 'Measuring constructed preferences: towards a building code'. *Journal of Risk and Uncertainty*, **19**(1): 243–70.

Pearce, D.W. (1983), *Cost–Benefit Analysis*. 2nd edn, London: Macmillan.

Pearce D.W. and R.K. Turner (1990), *Economics of Natural Resources and the Environment*. Hemel Hempstead: Harvester Wheatsheaf.

Perman, R., Y. Ma, J. McGilvray and M. Common (1999), *Natural Resource and Environmental Economics*. 2nd edn, New York: Longman.

Perrings, C. (1997), 'Ecological resilience in the sustainability of economic development'. In Perrings (ed.): *Economics of Ecological Resources. Selected Essays*. Cheltenham, UK and Northampton, MA, USA: Edward Elgar, pp. 45–63.

Pigou, A.C. (1920), *The Economics of Welfare*. London: Macmillan.

Pinker, S. (1994), *The Language Instinct*. London: Penguin.

Pitelis, C. (ed.) (1993), *Transaction Costs, Markets and Hierarchies*. Oxford: Blackwell.

Platteau, J.-P. (2000), *Institutions, Social Norms, and Economic Development*. Chur: Harwood Academic Publishers.

Polanyi, K. ([1944] 1957), *The Great Transformation: The Political and Economic Origins of Our Time*. Boston, MA: Beacon Press.

Polanyi, M. (1967), *The Tacit Dimension*. London: Routledge.

Popper, K.R. (1945), *The Open Society and its Enemies*. London: Routledge.

Portney, P.R and J.P. Weyant (eds) (1999), *Discounting and Intergenerational Equity*. Washington, DC: Resources for the Future.

Posner, R.A. (1977), *Economic Analysis of Law*. 2nd edn, Boston, MA: Little, Brown.

Postlewaite, A. (1998), 'The social basis for interdependent preferences'. *European Economic Review*, **42**(3–5): 779–800.

Poulantzas, N., R. Miliband and E. Laclau (1976), *Kontroverse über den kapitalistischen Staat* (*Disputes Concerning the Capitalist State*). Berlin: Merve.

Pratt, A.C. (1995), 'Putting critical realism to work: the practical implications for geographical research'. *Progress in Human Geography*, **19**(1): 61–74.

Radcliffe-Brown, A.R. (1952), *Structure and Function in Primitive Society*. London: Cohen & West.

Randall, A. (1974), 'Coasian externality theory in a policy context'. *Natural Resources Journal*, **14**: 35–54.

Randall, A. (1983), 'The problem of market failure'. *Natural Resources Journal*, **23**(1): 131–48.

Rawls, J.A. (1971), *A Theory of Justice*. Cambridge, MA: Belknap Press of Harvard University Press.

Rawls, J.A. (1993), *Political Liberalism*. New York: Columbia University Press.

Renn, O. (1999), 'A model for an analytical deliberative process in risk management'. *Environmental Science and Technology*, **33**(18): 3049–55.

Renn, O., T. Webler, H. Rakel, P. Dienel and B. Johnson (1993), 'Public participation in decision-making: a three-step procedure'. *Policy Science*, **26**: 189–214.

Ricardo, D. ([1817] 1973), *The Principles of Political Economy and Taxation*. Foreword by D. Winch. London: Dent.

Riseth, J.Å. (2000), 'Sámi reindeer management under technological change 1960–1990. Implications for common-pool resource use under various natural and institutional conditions'. Dr Sci. Thesis 2000:1. Ås: Agricultural University of Norway.

Robbins, L. (1935), *An Essay on the Nature and Significance of Economic Science*. London: Macmillan.

Robinson, J. (1962), *Economic Philosophy*. London: Watts & Co Ltd.

Romanow, R.J. (2002), 'Building on values. The future of health care in Canada'. Final report on the Commission on the Future of Health Care in Canada, webhome.indirect.com/~janet 313/janetsjo/romanow/, accessed August 2003.

Romer, P. (1996), 'Preferences, promises and the politics of entitlement'. In V. Fuchs (ed.): *Individual and Social Responsibility: Child Care, Education, Medical Care and Long-term Care in America*. Chicago: University of Chicago Press, pp. 195–228.

Romp, G. (1997), *Game Theory. Introduction and Application*. Oxford: Oxford University Press.

Röpke, I. (2004), 'The early history of modern ecological economics'. *Ecological Economics*, **50**(3–4): 293–314.

Rosen, S. (1974), 'Hedonic prices and implicit markets: product differentiation in pure competition'. *Journal of Political Economy*, **82**: 34–55.

Ross, L. and A. Ward (1996), 'Naïve realism: implications for social conflict and misunderstanding'. In E.S. Reed, E. Turiel and T. Brown (eds): *Values and Knowledge*. Mahwah, NJ: Lawrence Erlbaum.

Rousseau, J.-J. ([1762] 1968), *The Social Contract*. Harmondsworth: Penguin.

Roy, B. (1990), 'Decision aid and decision making'. In Bana e Costa (ed.), pp. 17–35.

Runge, C.F. (1986), 'Common property and collective action in economic development'. *World Development*, **14**(5): 623–35.

Russell, C.S. (1990), 'Game models for structuring monitoring and enforcement systems'. *Natural Resources Modeling*, **4**(2): 143–73.

Ruth, M. (1993), *Integrating Economics, Ecology and Thermodynamics*. Dordrecht: Kluwer.

Rutherford, M. (1994), 'Commons, J.R.'. In G.M. Hodgson, W.J. Samuels and M.R. Tool (eds): *The Elgar Companion to Institutional and Evolutionary Economics*. Aldershot, UK and Brookfield, US: Edward Elgar, pp. 63–9.

Sagoff, M. (1988), *The Economy of the Earth: Philosophy, Law and Environment*. Cambridge: Cambridge University Press.

Sagoff, M. (1994), 'Should preferences count?' *Land Economics*, **70**: 127–44.

Sagoff, M. (1998), 'Aggregation and deliberation in valuing environmental public goods: a look beyond contingent pricing'. *Ecological Economics*, **24** (2,3): 213–30.

Samuels, W.J., S.G. Medema and A.A. Schmid (1997), *The Economy as a Process of Valuation*. Cheltenham, UK and Lyme, USA: Edward Elgar.

Samuelson, P.A. (1938), 'A note on the pure theory of consumers' behaviour'. *Economica*, **5**: 61–71.

Samuelson, P.A. (1947), *Foundations of Economic Analysis*. Cambridge, MA: Harvard University Press.

Samuelson, P.A. (1948), 'Consumption theory in terms of revealed preferences'. *Economica*, **15**: 243–53.

Samuelson, P.A. (1950), 'Evaluation of real national income'. *Oxford Economic Papers*, **2**: 1–29.

Schkade, D.A. and J.W. Payne (1993), 'Where do the numbers come from? How people respond to contingent valuation questions'. In J.A. Hausman (ed.): pp. 271–93.

Schlesinger, W.H. (1991), *Biogeochemistry. An Analysis of Global Change*. London: Academic Press.

Schmalensee, R., P.L. Joskow, A.D. Ellerman, J.P. Montero and E.M. Bailey (1998), 'An interim evaluation of sulfur dioxide emissons trading'. *Journal of Economic Perspectives*, **12**(3): 53–68.

Schmid, A.A. (1987), *Property, Power, and Public Choice. An Inquiry into Law and Economics*. New York: Praeger.

Schotter, A. (1981), *The Economic Theory of Social Institutions*. Cambridge: Cambridge University Press.

Schulze, G.G. and B. Frank (2003), 'Deterrence versus intrinsic motivation: experimental evidence on the determinants of corruptibility'. *Economics of Governance*, **4**: 143–60.

Schulze, W., G. McClelland and J. Lazo (1994), 'Methodological issues in using contingent valuation to measure non-use values'. Paper US Environmental Protection Agency/US Department of Energy workshop, Herndon.

Scitovsky, T. (1941), 'A note on welfare propositions in economics'. *Review of Economic Studies*, **9** (November): 77–88.

Scott, J. (1995b), *Sociological Theory: Contemporary Debates*. Aldershot, UK and Brookfield, US: Edward Elgar.

Scott, W.R. (1987), 'The adolescence of institutional theory'. *Administrative Science Quarterly*, **32**: 493–511.

Scott, W.R. (1995a), *Institutions and Organizations*. Newbury Park, CA: Sage.

Screpanti, E. (1995), 'Relative rationality, institutions and precautionary behaviour'. In: Groenewegen, Pitelis and Sjöstrand (eds), pp. 63–84.
Searing, D.D. (1991), 'Roles, rules, and rationality in the new institutionalism'. *American Political Science Review*, **85** :1239–60.
Sen, A. (1977), 'Rational fools: a critique of the behavioral foundations of economic theory'. *Philosophy and Public Affairs*, **6**(4): 317–44.
Sen, A. (1981), *Poverty and Famines: An Essay on Entitlements and Deprivation*. Oxford: Clarendon.
Sen, A. (1988), *On Ethics and Economics*. Paperback edn, Oxford: Blackwell.
Sen, A. (1991), 'Utility, ideas and terminology'. *Economics and Philosophy*, **7**(2): 277–83.
Sened, I. (1997), *The Political Institution of Private Property*. Cambridge: Cambridge University Press.
Shabman, L. and K. Stephenson (1996), 'Searching for the correct benefit estimate: empirical evidence for an alternative perspective'. *Land Economics*, **72**(4): 433–49.
Shaffer, G.C. (2001), 'The World Trade Organization under challenge: democracy and the law and politics of the WTO's treatment of trade and environmental matters'. *Harvard Environmental Law Review*, **25**(1): 1–93.
Simmons, I.G. (1989), *Changing the Face of the Earth: Culture, Environment, History*. Oxford: Blackwell.
Simon, H.A. (1947), *Administrative Behavior: A Study of Decision-making Processes in Administrative Organization*. New York: Macmillan.
Simon, H.A. (1957), *Models of Man*. New York: John Wiley & Sons.
Simon, H.A. (1959), 'Theories of decision making in economics and behavioral science'. *American Economic Review*, **49**: 253–8.
Simon, H.A. (1979), 'Rational decision making in business organizations'. *American Economic Review*, **69**(4): 493–513.
Sjöstrand, S.-E. (1995), 'Towards a theory of institutional change'. In Groenewegen, Pitelis and Sjöstrand (eds), pp. 19–44.
Slovic, P. (1995), 'The construction of preferences'. *American Psychologist*, **50**: 364–71.
Slovic, P., D. Griffin and A. Tversky (1990), 'Compatibility effects in judgment and choice'. In R.M. Hogarth (ed.): *Insights in Decision Making: A Tribute to Hillel J. Einhorn*. Chicago: University of Chicago Press, pp. 5–27.
Slovic, P. and S. Lichtenstein (1983), 'Preference reversals: a broader perspective'. *American Economic Review*, **73**(4): 596–605.
Smith, A. ([1759] 1976), *The Theory of Moral Sentiments*. (Edited by E. Cannan). London: Methuen.
Smith, A. ([1776] 1976), *An Inquiry into the Nature and Causes of the Wealth of Nations*. Chicago: University of Chicago Press.

Smith, G. (2003), *Deliberative Democracy and the Environment*. London: Routledge.
Smith, G. and C. Wales (1999), 'The theory and practice of citizens' juries'. *Policy and Politics*, **27**(3): 295–308.
Smith, V.K. and L.L. Osborne (1996), 'Do contingent valuation estimates pass a "Scope" test? A meta-analysis'. *Journal of Environmental Economics and Management*, **31**: 287–301.
Smith, V.L. (1991), *Papers in Experimental Economics*. Cambridge: Cambridge University Press.
Smith, V.L. (2000), *Bargaining and Market Behavior: Essays in Experimental Economics*. Cambridge: Cambridge University Press.
Söderbaum, P. (2000), *Ecological Economics: A Political Economics Approach to Environment and Development*. London: Earthscan.
Solow, R.M. (1974), 'Intergenerational equity and exhaustible resources'. *Review of Economic Studies. Symposium on the Economics of Exhaustible Resources*, **41**: 29–45.
Solow, R.M. (1993), 'Sustainability: an economist's perspective'. In R. Dorfman and N. Dorfman (eds): *Economics of the Environment*. New York: Norton, pp. 179–87.
Spash, C.L. (1999), 'The development of environmental thinking in economics'. *Environmental Values*, **8**: 413–35.
Spash, C.L. (2000), 'Multiple value expression in contingent valuation: economics and ethics'. *Environmental Science Technology*, **34**: 1433–8.
Spash, C.L. (2002), *Greenhouse Economics: Values and Ethics*. New York: Routledge.
Spash, C.L. and N. Hanley (1994), 'Preferences, information and biodiversity preservation'. Discussion Papers in Ecological Economics No. 94/1, University of Stirling.
Spash, C.L. and N. Hanley (1995), 'Preferences, information and biodiversity preservation'. *Ecological Economics*, **12**: 191–208.
Spulber, D.F. (1989), *Regulation and Markets*. Cambridge, MA: MIT Press.
Srole, L. (1975), 'Measurement and classification in socio-psychiatric epidemiology: Midtown Manhattan Study (1954) and Midtown Manhattan Restudy (1974)'. *Journal of Health and Social Behavior*, **16**: 347–63.
Stagl, S. (2003), 'Multicriteria evaluation and public participation: in search for theoretical foundations'. Paper presented at Frontiers 2: European Application in Ecological Economics. Tenerife, Spain, 12–15 February.
Stavins, R.N. (1998), 'What can we learn from the grand policy experiment? Lessons from SO_2 allowance trading'. *Journal of Economic Perspectives*, **12**(3): 69–88.

Stevens, B.K. (1988), 'Fiscal implications of effluent charges and input taxes'. *Journal of Environmental Economics and Management*, **15**: 285–96.

Stevens, T.H., J. Echeverria, R.J. Glass, T. Hager and T.A. Moore (1991), 'Measuring the existence value of wildlife: what do CVM estimates really show'. *Land Economics*, **67**: 390–400.

Stewart, J., E. Kendall and A. Coote (1994), *Citizens' Juries*. London: Institute of Public Policy Research.

Stewart, T.J. and L. Scott (1995), 'A scenario-based framework for multicriteria decision analysis in water resource planning'. *Water Resources Research*, **31**(11): 2835–43.

Stigler, G.J. and G.S. Becker (1977), 'De Gustibus non est disputandum'. *American Economic Review*, **76**(1): 76–90.

Stiglitz, J.E. (1974), 'The optimal depletion of exhaustible resources'. *Review of Economic Studies. Symposium on the Economics of Exhaustible Resources*, **41**: 123–37.

Stiglitz, J.E. (2003), *Globalization and its Discontents*. New York: W.W. Norton.

Stirling, A. and S. Mayer (2001), 'A novel approach to appraisal of technological risk: a multicriteria mapping study of genetically modified crops'. *Environment and Planning C: Government and Policy*, **19**: 529–55.

Stokke, O.S. (ed.) (2001), *Governing High Seas Fisheries: The Interplay of Global and Regional Regimes*. Oxford: Oxford University Press.

Strand, R. (2001), 'The role of risk assessment in the governance of genetically modified organisms of agriculture'. *Journal of Hazardous Materials*, **86**(1–3): 197–204.

Streit, M.-E. (1997), 'Constitutional ignorance: spontaneous order and rule-orientation: Hayekian paradigms from a policy perspective'. In S. Frowen (ed.): *Hayek. Economist and Social Philosopher. A Critical Retrospect*. London: Macmillan, pp. 37–58.

Sugden, R. (1986), *The Economics of Rights, Co-operation and Welfare*. Oxford: Basil Blackwell.

Sunstein, C.R. (1993), 'Endogenous preferences, environmental law'. *Journal of Legal Studies*, **22**(2): 217–54.

Tacconi, L. (1997), 'An ecological economic approach to forest and biodiversity conservation: the case of Vanuatu'. *World Development*, **25**(12): 1995–2008.

Tenbrunsel, A.E. and D.M. Messick (1999), 'Sanctioning systems, decision frames and cooperation'. *Administrative Science Quarterly*, **44**: 684–707.

Tester, K. (1991), *Animals and Society: The Humanity of Animal Rights*. London: Routledge.

Tietenberg, T.H. (1985), *Emission Trading: An Exercise in Reforming Pollution Policy*. Washington, DC: Resources for the Future.

Tietenberg, T.H. (1998), 'Ethical influences on the evolution of the US tradable permit approach to air pollution control'. *Ecological Economics*, **24**(2,3): 241–57.

Tjernshaugen, A. (2003), 'Synspunkt' ('Viewpoint'). *Cicerone*, **5**: 3.

Toman, M.A. (1994), 'Economics and "Sustainability": balancing trade-offs and imperatives'. *Land Economics*, **70**(4): 399–413.

Toman, M.A., J. Pezzey and J. Krautkraemer (1995), 'Neoclassical economic growth theory and "Sustainability"'. In D.W. Bromley (ed.): *The Handbook of Environmental Economics*. Oxford: Blackwell, pp. 139–65.

Tool, M.R. (1995), *Pricing, Valuation and Systems. Essays in Neoinstitutional Economics*. Aldershot, UK and Brookfield, US: Edward Elgar.

Torsvik, G. (1996), 'Why should governments redistribute income?' *Nordic Journal of Political Economy*, **23**: 105–20.

Torvanger, A. (1999), 'Byrdefordeling viktig i klimaforhandlingane' ('The distribution of burden is important for the negotiations on climate') *Cicerone*, **4**: 12–13.

Trice, A.H. and S.E. Wood (1958), 'Measurement of recreation benefits'. *Land Economics*, **34**: 195–207.

Tversky, A. (1969), 'Intransitivity of preferences'. *Psychological Review*, **76**: 31–48.

Tversky, A. and D. Kahneman (1986), 'Rational choice and the framing of decisions'. In R.M. Hogarth and M.W. Reder (eds): *Rational Choice: The Contrast Between Economics and Psychology*. Chicago: University of Chicago Press, pp. 67–94.

Tyler, T.R. (1990), *Why People Obey the Law*. New Haven, CT: Yale University Press.

United Nations (UN) (1982), 'Convention on the Law of the Sea'. www.un.org/Depts/los/convention_agreements/texts/unclos/unclos_e.pdf, accessed September 2003.

United Nations (UN) (1987), *Our Common Future*. Oxford: Oxford University Press.

United Nations (UN) (1992a), 'Convention on Biological Diversity'. www.biodiv.org/doc/legal/cbd-en.pdf, accessed September 2003.

United Nations (UN) (1992b), 'Rio Declaration on Environment and Development'. www.unep.org/Documents/Default.asp?DocumentID=78&ArticleID=1163, accessed September 2003.

United Nations (UN) (1994), 'International Tropical Timber Agreement'. www.itto.or.jp/live/Live_Server/144/ITTA1994e.doc, accessed September 2003.

United Nations (UN) (1997), 'The Kyoto Protocol'. www.unfccc.int/resource/docs/convkp/kpeng.pdf, accessed Semptember 2003.

United Nations (UN) (2000), 'The Cartagena Protocol on Biosafety to the Convention on Biological Diversity. Text and Annexes'. www.biodiv.org/doc/legal/cartagena-protocol-en.pdf, accessed September 2003.

United Nations (UN) (2003), 'Status of the United Nations Convention on the Law of the Sea'. www.un.org/Depts/los/reference_files/status2003.pdf, accessed January 2004.

Vadjnal, D. and M. O'Connor (1994), 'What is the value of Rangitoto island?' *Environmental Values*, **3**: 369–90.

van Vugt, M. (1997), 'Concerns about the privatization of public goods: a social dilemma analysis'. *Social Psychology Quarterly*, **60**: 355–68.

van Vugt, M., P.A.M. van Lange and R.M. Meertens (1996), 'Commuting by car or public transportation? A social dilemma analysis of travel mode judgements'. *European Journal of Social Psychology*, **26**: 373–95.

Varian, H.R. (1992), *Microeconomic Analysis*. 3rd edn, New York: W.W. Norton.

Vatn, A. (1984), *Teknologi og politikk. Om framveksten av viktige styringstiltak i norsk jordbruk 1920–1980* (*Technology and Politics. On the Developement of Important Measures in Norwegian Agricultural Policy 1920–1980*). Oslo: Landbruksforlaget.

Vatn, A. (1998), 'Input vs. emission taxes: environmental taxes in a mass balance and transactions cost perspective'. *Land Economics*, **74**(4): 514–25.

Vatn, A. (2000), 'The environment as a commodity'. *Environmental Values*, **9**: 493–509.

Vatn, A. (2001), 'Environmental resources, property regimes and efficiency'. *Environment and Planning C: Government and Policy*, **19**(5): 681–93.

Vatn, A. (2002), 'Efficient or fair: ethical paradoxes in environmental policy'. In D. Bromley and J. Paavola (eds): *Economics, Ethics and Environmental Policy: Contested Choices*. Oxford: Blackwell, pp. 148–63.

Vatn, A. (2004). 'Valuation and rationality'. *Land Economics*, **80** (February): 1–18.

Vatn, A. and D. Bromley (1994), 'Choices without prices without apologies'. *Journal of Environmental Economics and Management*, **26**: 129–48.

Vatn, A. and D. Bromley (1997), 'Externalities – a market model failure'. *Journal of Environmental and Resource Economics*, **9**(2): 135–51.

Vatn, A., L.R. Bakken, P. Botterweg, H. Lundeby, E. Romstad, P.K. Rørstad and A. Vold (1997), 'Regulating nonpoint-source pollution from agriculture. An integrated modelling analysis'. *European Review of Agricultural Economics*, **26**(2): 207–29.

Vatn, A., L. Bakken, P. Botterweg and R. Romstad (1999), 'ECECMOD: an interdisciplinary modelling system for analyzing nutrient and soil losses from agriculture'. *Ecological Economics*, **30**(2): 189–205.

Vatn, A., E. Krogh, F. Gundersen and P. Vedeld (2002), 'Environmental taxes and politics: the dispute over nitrogen taxes in agriculture'. *Environmental Policy Journal*, **12**: 224–40.

Veblen, T. (1898), 'Why is economics not an evolutionary science'. *Quarterly Journal of Economics*, **12**: 373–97.

Veblen, T. (1899), *The Theory of the Leisure Class: An Economic Study of Institutions*. New York: Macmillan.

Veblen, T. (1909), 'The limitation of marginal utility'. *Journal of Political Economy*, **17**: 620–36.

Veblen, T. (1919), *The Place of Science in Modern Civilisation and Other Essays*. New York: Huebsch.

Vedeld, P. (2002), 'The process of institution building to facilitate local biodiversity management'. Noragric working paper no 26. Ås: Noragric/NLH.

Vidal-Naquet, P. and J. Bertin (eds) (1987), *Atlas historique: histoire de l'humanité de la préhistoire à nos jours* (*Hisotric Atlas: The History of Humanity and the Prehistory of our Days*). Paris: Hachette.

Wakeford, T. (2001), 'A selection of methods used in deliberative and inclusionary processes'. *PLA Notes*, **40** (February), 29–31. www.iied.org/sarl/planotes/sample.html, accessed January 2004.

Wallis, J.J. and D.C. North (1986), 'Measuring the transaction sector in the American economy, 1970–1970'. In S.L. Engerman and R.E. Gallman (eds): *Long-Term Factors in American Economic Growth*. Vol. 51, Income and Wealth series. Chicago: University of Chicago Press, pp. 95–148.

Walras, L. ([1874] 1954), *Eléments d'économie politique pure*. English translation: *Elements of Pure Economics*. London: Allen & Unwin. (Translated from the French by W. Jaffe.)

Walzer, M. (1983), *Spheres of Justice. A Defence of Pluralism and Equality*. New York: Basic Books.

Ward, H. (1999), 'Citizens' juries and valuing the environment: a proposal'. *Environmental Politics*, **8**: 75–96.

Weibull, J. (1995), *Evolution Game Theory*. Cambridge, MA: MIT Press.

Weisbrod, B. (1968), 'Income redistribution effects and benefit–cost analysis'. In S.B. Chase (ed.): *Problems in Public Spenditure Analysis*. Washington, DC: Brookings Institute, pp. 177–209.

Weitzman, M.L. (1974), 'Prices vs. quantities'. *Review of Economic Studies*, **41**: 477–91.

Weitzman, M.L. (2002), 'Landing fees and harvest quotas with uncertain fish stocks'. *Journal of Environmental Economics and Management*, **43**: 325–38.

Williams, B. (1973), 'A critique of utilitarianism'. In J.J.C. Smart and B. Williams (eds): *Utilitarianism: For and Against*. Cambridge: Cambridge University Press, pp. 77–150.

Williamson, O.E. (1975), *Market and Hierachies: Analysis and Antitrust Implications*. New York: Free Press.

Williamson, O.E. (1985), *The Economic Institutions of Capitalism*. New York: Free Press.

Williamson, O.E. (2000), 'The new institutional economics: taking stock/looking ahead'. *Journal of Economic Literature*, **38**(3): 595–613.

Willig, R.D. (1976), 'Consumer surplus without apology'. *American Economic Review*, **66**(4): 589–97.

Wilson, E.O. (2001), *The Diversity of Life*. London: Penguin.

Woolgar, S. (1988), *Science: The Very Idea*. Chichester: Ellis Horwood and Tavistock.

World Trade Organization (WTO) (1994), 'Agreement on Trade-Related Aspects of Intellectual Property Rights'. www.wto.org/english/tratop_e/trips_e/t_agm0_e.html, accessed October 2003.

World Trade Organization (WTO) (1999), 'Review of the provisions of Article 27.3(b)', IP/C/W/163 (99-4812).

World Trade Organization (WTO) (2003), Taking forward the review of Article 27.3(b) of the TRIPS Agreement', IP/C/W/404 (03-3410).

Young, H.P. (1998), *Individual Strategy and Social Structure: An Evolutionary Theory of Institutions*. Princeton, NJ: Princeton University Press.

Young, O.R. (2002), *The Institutional Dimension of Environmental Change: Fit, Interplay, and Scale*. Cambridge, MA: MIT Press.

Name index

Subject index

Printed and bound by CPI Group (UK) Ltd, Croydon, CR0 4YY
27/10/2024
14580411-0005